Lecture Notes in Bioinformatics

Edited by S. Istrail, P. Pevzner, and M. Wate

T0238631

Editorial Board: A. Apostolico   S. Brunak   M. Gelfand

T. Lengauer   S. Miyano   G. Myers   M.-F. Sagot   D. Sankoff

R. Shamir   T. Speed   M. Vingron   W. Wong

Subseries of Lecture Notes in Computer Science

Zhipeng Cai   Oliver Eulenstein
Daniel Janies   Daniel Schwartz (Eds.)

Bioinformatics
Research and Applications

9th International Symposium, ISBRA 2013
Charlotte, NC, USA, May 20-22, 2013
Proceedings

 Springer

Series Editors

Sorin Istrail, Brown University, Providence, RI, USA
Pavel Pevzner, University of California, San Diego, CA, USA
Michael Waterman, University of Southern California, Los Angeles, CA, USA

Volume Editors

Zhipeng Cai
Georgia State University
Atlanta, GA 30303, USA
E-mail: zcai@gsu.edu

Oliver Eulenstein
Iowa State University
Ames, IA 50011, USA
E-mail: oeulenst@iastate.edu

Daniel Janies
University of North Carolina at Charlotte
Charlotte, NC 28223, USA
E-mail: djanies@uncc.edu

Daniel Schwartz
University of Connecticut
Storrs, CT 06269, USA
E-mail: daniel.schwartz@uconn.edu

ISSN 0302-9743   e-ISSN 1611-3349
ISBN 978-3-642-38035-8   e-ISBN 978-3-642-38036-5
DOI 10.1007/978-3-642-38036-5
Springer Heidelberg Dordrecht London New York

Library of Congress Control Number: 2013935987

CR Subject Classification (1998): J.3, H.2.8, H.3-4, F.1, F.2.2, I.5

LNCS Sublibrary: SL 8 – Bioinformatics

© Springer-Verlag Berlin Heidelberg 2013
This work is subject to copyright. All rights are reserved, whether the whole or part of the material is
concerned, specifically the rights of translation, reprinting, re-use of illustrations, recitation, broadcasting,
reproduction on microfilms or in any other way, and storage in data banks. Duplication of this publication
or parts thereof is permitted only under the provisions of the German Copyright Law of September 9, 1965,
in its current version, and permission for use must always be obtained from Springer. Violations are liable
to prosecution under the German Copyright Law.
The use of general descriptive names, registered names, trademarks, etc. in this publication does not imply,
even in the absence of a specific statement, that such names are exempt from the relevant protective laws
and regulations and therefore free for general use.

Typesetting: Camera-ready by author, data conversion by Scientific Publishing Services, Chennai, India

Printed on acid-free paper

Springer is part of Springer Science+Business Media (www.springer.com)

Preface

The 9th edition of the International Symposium on Bioinformatics Research and Applications (ISBRA 2013) was held during May 20–22, 2013, at Charlotte, North Carolina, USA. The symposium provides a forum for the exchange of ideas and results among researchers, developers, and practitioners working on all aspects of bioinformatics and computational biology and their applications.

The technical program of the symposium included 25 contributed papers, selected by the Program Committee from a number of 46 full submissions received in response to the call for papers. Additionally, the symposium included poster sessions and featured invited keynote talks by five distinguished speakers: Tanya Berger-Wolf from the University of Illinois at Chicago spoke on computational behavioral ecology; Martha L. Bulyk from Brigham & Women's Hospital and Harvard Medical School spoke on transcription factors and DNA regulatory elements; Luonan Chen from the Chinese Academy of Sciences spoke on identifying critical transitions of biological processes by dynamical network biomarkers; Stephen C. Harvey from the Georgia Institute of Technology spoke on unusual RNA structures and information content in RNAs from the "prebiotic ribosome" to modern viruses; and Bin Ma from the University of Waterloo spoke on peptide identification from mass spectrometry.

We would like to thank the Program Committee members and external reviewers for volunteering their time to review and discuss the symposium papers. We would like to extend special thanks to the Steering and General Chairs of the symposium for their leadership, and to the Finance, Publication, Publicity, and Local Organization Chairs for their hard work in making ISBRA 2013 a successful event. Last but not least, we would like to thank all authors for presenting their work at the symposium.

May 2013

Zhipeng Cai
Oliver Eulenstein
Daniel Janies
Daniel Schwartz

Organization

Steering Chairs

Dan Gusfield	University of California, Davis, USA
Yi Pan	Georgia State University, USA
Ion Mandoiu	University of Connecticut, USA
Marie-France Sagot	INRIA, France
Alexander Zelikovsky	Georgia State University, USA

General Chair

Cynthia Gibas	The University of North Carolina at Charlotte, USA

Program Chairs

Zhipeng Cai	Georgia State University, USA
Oliver Eulenstein	Iowa State University, USA
Daniel Janies	The University of North Carolina at Charlotte, USA
Daniel Schwartz	University of Connecticut, USA

Publication Chair

Zhipeng Cai	Georgia State University, USA

Finance Chair

Larry Mays	The University of North Carolina at Charlotte, USA

Local Organization Chair

Zhengchang Su	The University of North Carolina at Charlotte, USA

Program Committee

Srinivas Aluru	IIT Bombay/Iowa State University, India/USA
Danny Barash	Ben-Gurion University, Israel
Robert Beiko	Dalhousie University, Canada

Daniel Berrar   Tokyo Institute of Technology, Japan
Paola Bonizzoni   Università di Milano-Bicocca, Italy
Daniel Brown   University of Waterloo, Canada
Doina Caragea   Kansas State University, USA
Tien-Hao Chang   National Cheng Kung University, Taiwan
Matteo Comin   University of Padova, Italy
Ovidiu Daescu   University of Texas at Dallas, USA
Jorge Duitama   International Center for Tropical Agriculture
                             (CIAT), Colombia
Guillaume Fertin   University of Nantes, France
Vladimir Filkov   University of California Davis, USA
Katia Guimaraes   Universidade Federal de Pernambuco, Brasil
Jiong Guo   Universität des Saarlandes, Germany
Robert Harrison   Georgia State University, USA
Jieyue He   Southeast University, China
Steffen Heber   North Carolina State University, USA
Jinling Huang   East Carolina University, USA
Lars Kaderali   University of Technology Dresden, Germany
Iyad Kanj   DePaul University, USA
Yury Khudyakov   Centers for Disease Control and Prevention,
                             USA
Danny Krizanc   Wesleyan University, USA
Jing Li   Case Western Reserve University, USA
Min Li   Central South University, China
Guohui Lin   University of Alberta, Canada
Ion Mandoiu   University of Connecticut, USA
Fenglou Mao   University of Georgia, USA
Osamu Maruyama   Kyushu University, Japan
Ion Moraru   University of Connecticut Health Center, USA
Giri Narasimhan   Florida International University, USA
Yi Pan   Georgia State University, USA
Bogdan Pasaniuc   UCLA, USA
Andrei Paun   Louisiana Tech University, USA
Nadia Pisanti   Università di Pisa, Italy
Maria Poptsova   University of Connecticut, USA
Teresa Przytycka   NIH, USA
Sven Rahmann   University of Duisburg-Essen, Germany
David Sankoff   University of Ottawa, Canada
Russell Schwartz   Carnegie Mellon University, USA
Joao Setubal   Virginia Bioinformatics Institute, USA
Ileana Streinu   Smith College, Northampton MA, USA
Raj Sunderraman   Georgia State University, USA
Wing-Kin Sung   National University of Singapore, Singapore
Sing-Hoi Sze   Texas A&M University, USA

Ilias Tagkopoulos   University of Califronia, Davis, USA
Marcel Turcotte   University of Ottawa, Canada
Stéphane Vialette   Université Paris-Est, France
Panagiotis Vouzis   Carnegie Mellon University, USA
Xiang Wan   Hong Kong Baptist University, SAR China
Jianxin Wang   Central South University, China
Li-San Wang   University of Pennsylvania, USA
Lusheng Wang   City University of Hong Kong, SAR China
Fangxiang Wu   University of Saskatchewan, Canada
Yufeng Wu   University of Connecticut, USA
Dechang Xu   Harbin Institute of Technology, China
Jinbo Xu   Toyota Tech Inst at Chicago, USA
Zhenyu Xuan   University of Texas at Dallas, USA
Alex Zelikovsky   Georgia State University, USA
Fa Zhang   Institute of Computing Technology, China
Yanqing Zhang   Georgia State University, USA
Leming Zhou   University of Pittsburgh, USA

Additional Reviewers

Bernauer, Julie   Mohamed Babou, Hafedh
Campos, Jaime   Mohamed, Nabeel
Chirita, Claudia   Montangero, Manuela
Dondi, Riccardo   Roman, Theodore
Fang, Ming   Ryvkin, Paul
Jaric, Melita   Skums, Pavel
Knapp, Bettina   Tang, Xiwei
Leung, Fanny   Warren, Andrew
Liu, Li Zhi   Xie, Lu
Ma, Qin   Zhou, Chan
Marschall, Tobias

Table of Contents

Peptide Identification from Mass Spectrometry

Bin Ma

David R. Cheriton School of Computer Science
University of Waterloo
binma@uwaterloo.ca

Abstract. Mass spectrometry is nowadays the method of choice for protein characterization in proteomics. Computer algorithms and software have played an essential role in analyzing the large amount of mass spectrometry data produced in any proteomics experiment. The fundamental task of such analyses is to identify the peptide for each spectrum in the data. Such identification is called "database search" if it requires the assistance of a protein database, and called "de novo sequencing" if not. In the past 20 years, many database search software tools have been developed for peptide identification; and a particular one, Mascot, that was developed in 1999, became dominant in the market. While new tools were continuously published in the following decade, none has significantly improved Mascot. The situation was disrupted around 2010, when the field witnessed a flurry of new database search tools that significantly improved Mascot in terms of both accuracy and sensitivity. In the first part of the talk, the peptide identification problem will be introduced, and the history briefly reviewed. In the second part of the talk, some practical concerns for using the bioinformatics tools in a proteomics lab are discussed. Properly dealing with these concerns resulted into the significant improvement we witnessed in the past few years. The second part of the talk will be focused on the research conducted at the author's own group.

Z. Cai et al. (Eds.): ISBRA 2013, LNBI 7875, p. 1, 2013.
© Springer-Verlag Berlin Heidelberg 2013

Identifying Critical Transitions of Biological Processes by Dynamical Network Biomarkers

Luonan Chen

Key Laboratory of Systems Biology
Shanghai Institutes for Biological Sciences
Chinese Academy of Sciences
lnchen@sibs.ac.cn

Abstract. There are non-smooth or even abrupt state changes during many biological processes, e.g., cell differentiation process, proliferation process, or even disease deterioration process. Such changes generally signal the emergence of critical transition phenomena, which result in drastic transitions in system states or phenotypes [1-4]. Therefore, it is of great importance to identify such transitions and further reveal their molecular mechanism. Recently based on dynamical network biomarkers (DNBs), we developed a novel theory as well as the computational method to detect critical transitions even with a small number of samples. We show that DNBs can identify not only early-warning signals of the critical transitions but also their leading networks, which drive the whole system to initiate such transitions [1-4]. Examples for complex diseases are also provided to detect pre-disease stages (or detect early-signal of complex diseases) for which traditional methods failed, for demonstrating the effectiveness of this novel approach.

References

[1] Chen, L., Liu, R., Liu, Z., Li, M., Aihara, K.: Detecting early-warning signals for sudden deterioration of complex diseases by dynamical network biomarkers. Scientific Reports 2, 342 (2012), doi:10.1038/srep00342
[2] Liu, R., Li, M., Liu, Z.-P., Wu, J., Chen, L., Aihara, K.: Identifying critical transitions and their leading networks in complex diseases. Scientific Reports 2, 813 (2012), doi:10.1038/srep00813
[3] Wang, J., Sun, Y., Zheng, S., Zhang, X.-S., Zhou, H., Chen, L.: APG: an Active Protein-Gene Network Model to Quantify Regulatory Signals in Complex Biological Systems. Scientific Reports 3, 1097 (2013), doi:10.1038/srep01097
[4] Liu, R., Aihara, K., Chen, L.: Dynamical network biomarkers for identifying critical transitions of biological processes. Quantitative Biology (2013)

Z. Cai et al. (Eds.): ISBRA 2013, LNBI 7875, p. 2, 2013.
© Springer-Verlag Berlin Heidelberg 2013

Computational Behavioral Ecology

Tanya Berger-Wolf

Department of Computer Science
University of Illinois at Chicago
tanyabw@uic.edu

Abstract. Computation has fundamentally changed the way we study nature, from molecules to ecosystems. Recent advances in data collection technology, such as GPS and other mobile sensors, high definition cameras, satellite images, and genotyping, are giving biologists access to data about the natural world which are orders of magnitude richer than any previously collected. Such data offer the promise of answering some of the big questions about why animals do what they do, among other things. Unfortunately, in the domain of behavioral ecology and population dynamics, our ability to analyze data lags substantially behind our ability to collect it. In this talk I will show how computational approaches can be part of every stage of the scientific process of understanding animal sociality, from data collection (identifying individual animals from photographs by stripes and spots) to hypothesis formulation (by designing a novel computational framework for analysis of dynamic social networks).

Z. Cai et al. (Eds.): ISBRA 2013, LNBI 7875, p. 3, 2013.
© Springer-Verlag Berlin Heidelberg 2013

Unusual RNA Structures: Information Content in RNAs from the "Prebiotic Ribosome" to Modern Viruses

Stephen C. Harvey

School of Biology
Georgia Institute of Technology, Atlanta, GA
steve.harvey@biology.gatech.edu

Abstract. About a dozen years ago, Robin Gutell proposed that comparative sequence analysis of all ribosomal RNAs pointed to the existence of a "minimal ribosome". Together we mapped the structure of the minimal ribosome in three dimensions by computer-based modeling. The resulting model (Mears et al. (2002) J Mol Biol 321:215-234) was subsequently confirmed in direct structural studies (Mears et al. (2006) J Mol Biol 358:193-212). The minimal ribosome points back toward the "prebiotic ribosome", i.e., the RNA molecule that catalyzed peptide bond formation during the RNA World, before the appearance of the genetic code. We have been pursuing the prebiotic ribosome in collaboration with a group of scientists at the Georgia Tech Astrobiology Center, under the leadership of Loren Williams, using a variety of experimental and computational approaches. This work has led to recognition of the role of unusual RNA structures and interactions in a previously unexpected catalytic activity: RNA is capable of catalyzing coupled oxidation-reduction reactions, a key requirement for any system of prebiotic molecules with the capacity for metabolism and replication.

Many small icosahedral RNA viruses assemble spontaneously, without the need for a specialized apparatus or the input of energy. An understanding of assembly mechanisms would have obvious implications for the development of new antiviral treatments, and for the design of novel nanoparticles capable of self-assembly. We have developed the first complete model of any real virus in full atomic detail (Zeng et al. (2012) J Struct Biol 180:110-116), and we have also determined the structure of the viral RNA in vitro (Athavale et al. (2013) PLoS ONE 8:e54384). These studies contribute to the growing understanding that unusual RNA secondary structures play important functional roles in the viral life cycle.

In this talk, I will discuss the convergence of these lines of research, with particular emphasis on the search for novel signals in RNA sequences and secondary structures.

Z. Cai et al. (Eds.): ISBRA 2013, LNBI 7875, p. 4, 2013.
© Springer-Verlag Berlin Heidelberg 2013

The Radiation Hybrid Map Construction Problem Is FPT

Iyad Kanj[1], Ge Xia[2], and Binhai Zhu[3]

[1] School of Computing, DePaul University, Chicago, IL 60604-2301
ikanj@cs.depaul.edu
[2] Department of Computer Science, Lafayette College, Easton, PA 18042
gexia@cs.lafayette.edu
[3] Department of Computer Science, Montana State University, Bozeman,
MT, 59717-3880
bhz@cs.montana.edu

Abstract. The Radiation Hybrid Map Construction problem (RHMC) is of prime interest in the area of Bioinformatics, and is concerned with reconstructing a genome from a set of given gene clusters. The problem is $\mathcal{NP}$-complete, even for the special case when the cardinality of each cluster is 2. Recently, Zhang et al. considered the case when the cardinality of each cluster is at most three, and proved that RHMC in this case is fixed-parameter tractable. They asked whether RHMC is fixed-parameter tractable for any fixed upper bound on the cluster cardinality.

In this paper, we answer the question of Zhang et al. in the affirmative by showing that RHMC is fixed-parameter tractable when the cardinality of each cluster is at most d, for any nonnegative integer-constant d.

Keywords: Radiation Hybrid Mapping, fixed-parameter tractability, path decomposition.

1 Introduction

Sequencing a genome is a fundamental problem in modern genetics. Radiation hybrid (Rh) mapping is an early technique for mapping unique DNA sequences onto chromosomes and whole genome. The technique has been used since 1990 for constructing maps of small chromosomal regions for human and several other mammals [8, 23, 24]. In Rh mapping experiments, chromosomes of the target organism are randomly broken into small DNA fragments through gamma radiation. The underlying assumption is that, when two genes are physically close to each other on the chromosome, the probability that these two corresponding gene markers are broken down by the gamma radiation is low, and so with a high probability they are either co-present in, or co-absent from, a DNA fragment. The *Radiation Hybrid Map Construction* problem (RHMC) is to determine, based on the observed co-occurrences, whether or not a linear order of the markers exist that would interpret the observed co-occurrences in the DNA fragments, accounting for the possibility of a certain number of errors in the

Z. Cai et al. (Eds.): ISBRA 2013, LNBI 7875, pp. 5–16, 2013.
© Springer-Verlag Berlin Heidelberg 2013

observed co-occurrences/fragments. The problem is formally defined as follows, where a gene is represented by a "marker" (symbol), and a DNA fragment is represented by a "cluster":

RADIATION HYBRID MAP CONSTRUCTION (RHMC)
Given: A set $\mathcal{M}$ of symbols called *markers*, a set $\mathcal{C}$ of *clusters*, where each cluster is a subset of $\mathcal{M}$, and $k \in \mathbb{N}$
Parameter: k
Question: Decide if we can delete a set $\mathcal{C}'$ of at most k clusters from $\mathcal{C}$ so that there exists a linear order $\mathcal{L}$ on the markers in $\mathcal{M}$ satisfying that the markers in each (remaining) cluster in $\mathcal{C} \setminus \mathcal{C}'$ appear consecutively/adjacently with respect to $\mathcal{L}$

To clarify, when in the above problem definition we say that the markers in a cluster "appear consecutively" with respect to the linear order, we mean that there exists a permutation of the markers in the cluster such that the permuted markers appear consecutively in this linear order.

For $d \in \mathbb{N}$, we denote by $\mathsf{RHMC_d}$ the restriction of RHMC to the instances in which the cardinality/size of each cluster is at most d.

Most of the traditional Rh map construction methods use heuristics, and often lead to the construction of maps for a small subset of the set of markers (see, for example, [13]). Slonim et al. [24] proposed a hidden Markov model on the Rh mapping data and used a maximum-likelihood approach to construct the map. Givry et al. [9] employed known sequence information for target chromosomes to construct the map.

Recently, RHMC was shown to be $\mathcal{NP}$-complete, even when the size of each cluster is 2, that is, $\mathsf{RHMC_2}$ [7]. An approximation algorithm of ratio 2 was also given for the optimization version of $\mathsf{RHMC_2}$ [7]. More recently, this approximation ratio was improved to $10/7$ in [6].

In this paper, we consider the RHMC problem from the parameterized complexity perspective. Parameterized complexity was developed in the 1990's by Downey and Fellows [11] to cope with the NP-hardness of a problem. It was motivated by the core observation that in many practical instances of hard problems, certain parameters remain small even when the input instances are large. For example, one naturally expects the number of clusters that need to be deleted, that is the parameter k, in an instance of the RHMC problem (due to possible errors in the Rh process) to be rather small compared to the total number of clusters in the instance. Therefore, a natural question to ask is whether we can take advantage of this observation and design an algorithm for RHMC such that the exponential term in its running time is a function of the parameter (k) *only* rather than the instance size. Such algorithms would be computationally feasible when the parameter is moderately small, and they have the advantage (over approximation algorithms) of computing exact solutions to the input instances. More formally, a parameterized problem is said to be *fixed-parameter tractable* if there exists an algorithm that, given (x, k) where x is an input instance of the problem and k is a parameter, the algorithm decides the instance (x, k)

in time $f(k)|x|^{O(1)}$, where f is a function of k *only*. Parameterized complexity has witnessed rapid growth and development, and has become one of the main tools for coping with the computational intractability of NP-hard problems. It has found numerous applications in database systems, VLSI design, games, robotics, computational biology, linguistics, cryptography, and computational learning (see [2, 3, 5, 10, 14–16, 20], to name a few).

Zhang et al. [25] studied parameterized and kernelization algorithms for the RHMC problem. They gave a parameterized algorithm for RHMC$_3$ that runs in time $O(6^k k + n)$, thus proving that the problem is fixed-parameter tractable; they also proved that the problem has a kernel with at most $22k$ clusters. For RHMC$_2$, they gave a parameterized algorithm that runs in time $O^*(2.45^k)$.[1] Jiang and Zhu [19] proved that RHMC$_2$ has a weak kernel with at most $5k$ clusters. Zhang et al. [25] posed the question of whether RHMC$_d$ is fixed-parameter tractable for any integer constant $d > 3$.

In this paper, we answer Zhang et al.'s [25] question in the affirmative by showing that RHMC$_d$ is fixed-parameter tractable for any $d \in \mathbb{N}$. Our approach proceeds as follows. With any instance of RHMC$_d$, we associate an auxiliary graph defined based on the set of markers and the set of clusters in the instance. We first show that, for any yes-instance of RHMC$_d$, the pathwidth of the auxiliary graph associated with this instance must be upper bounded by a function of d and k only. We then use Bodlaender's [1] algorithm to compute a path decomposition of this auxiliary graph whose width meets the proven upper bound (if such a path decomposition does not exists, we reject); this algorithm is fixed-parameter tractable with the pathwidth as the parameter, and hence is fixed-parameter tractable in k. Finally, we apply a sophisticated dynamic programming algorithm based on the computed path decomposition whose running time is fixed-parameter tractable in the width of the computed path decomposition, and hence in k.

The paper is organized as follows. Section 2 presents the necessary background and notations. Section 3 proves that a yes-instance of the problem must have the pathwidth of its auxiliary graph upper bounded by a function of the parameter. Section 4 presents the dynamic programming algorithm that is based on the path decomposition of the auxiliary graph. Section 5 concludes the paper with some questions that remain open.

2   Preliminaries

We assume familiarity with basic graph theory and parameterized complexity notation and terminology. For more information, we refer the reader to [11, 18, 21, 26].

A *parameterized problem* is a set of instances of the form (x, k), where $x \in \Sigma^*$ for an alphabet Σ, and k is a non-negative integer called the *parameter*. A parameterized problem Q is *fixed-parameter tractable*, or simply FPT, if there exists an algorithm that on input (x, k) decides if (x, k) is a yes-instance of Q

---

[1] $O^*(t(n))$ denotes time complexity of the form $O(t(n) \cdot p(n))$, where $p(n)$ is a polynomial.

in time $f(k)|x|^{O(1)}$, where f is a computable function independent of $|x|$. By *fpt-time*, we denote time complexity of the form $f(k)|x|^{O(1)}$, where $|x|$ is the input length, k is the parameter, and f is a computable function of k.

For a graph G, $V(G)$ and $E(G)$ denote the vertex-set and the edge-set of G, respectively. A *tree* is a connected acyclic graph. The maximum degree of a graph G will be denoted by $\Delta(G)$. A graph G is said to be an *interval graph* if there exists a set of (real) intervals $\mathcal{I}$ whose intervals correspond to the vertices of G, and such that for any two vertices $u, v \in G$: $(u, v) \in E(G)$ if and only if the two intervals corresponding to u and v in $\mathcal{I}$ intersect.

A *path decomposition* of G is a sequence $\mathcal{P} = (B_1, \ldots, B_t)$, where $B_i \subseteq V(G)$, $i = 1, \ldots, t$, that satisfies the following conditions:

(i)   each vertex $v \in V(G)$ is contained in some B_i, $i \in \{1, \ldots, t\}$;
(ii) for each edge $uv \in E(G)$ there exists $i \in \{1, \ldots, t\}$ such that both $u, v \in B_i$; and
(iii) if a vertex v is contained in B_i and B_j, where $i, j \in \{1, \ldots, t\}$ and $i < j$, then $v \in B_r$ for every r satisfying $i \leq r \leq j$. (Put it differently, $B_i \cap B_j \subseteq B_r$.)

We call each B_i in $\mathcal{P}$, $i = 1, \ldots, t$, a *bag*. The *width* of a path decomposition $\mathcal{P} = (B_1, \ldots, B_t)$ of G, denoted $pw(\mathcal{P})$, is $\max\{|B_i| : i = 1, \ldots, r\} - 1$. The *pathwidth* of G, denoted $pw(G)$ is the minimum width over all path decompositions of G. A path decomposition of G is said to be *nice* [4] if: $|B_1| = 1$, and either $|B_i \setminus B_{i-1}| = 1$ or $|B_{i-1} \setminus B_i| = 1$, for $i = 2, \ldots, t$; in the case when $|B_i \setminus B_{i-1}| = 1$ we say that B_i is an *introduce bag*, otherwise ($|B_{i-1} \setminus B_i| = 1$), we say that B_i is a *forget bag* [4]. It is well known [4] that, given a path decomposition $\mathcal{P}$ of G with $O(V(G))$ bags and width w, there is an $O(|V(G)|)$-time algorithm that computes a nice path decomposition of G of width at most w and with at most $2|V(G)|$ bags. Therefore, without loss of generality, we will always consider nice path decompositions in this paper.

3   Bounding the Pathwidth

Let $(\mathcal{M}, \mathcal{C}, k)$ be an instance of the RHMC$_\mathsf{d}$ problem, where $\mathcal{M} = \{m_1, \ldots, m_\ell\}$ is a set of markers, $\mathcal{C} = \{C_1, \ldots, C_n\}$ is a set of clusters each of cardinality at most d, where $d \geq 1$ is an integer constant, and k is a nonnegative integer. Without loss of generality, we shall assume that every marker in $\mathcal{M}$ appears in some cluster in $\mathcal{C}$ (otherwise the marker can be discarded), and that every cluster in $\mathcal{C}$ has a nonempty intersection with at least one other cluster in $\mathcal{C}$ (otherwise, the markers in $\mathcal{C}$ can be prefixed/suffixed to any valid ordering).

We define two auxiliary graphs, one is defined based on the set of clusters $\mathcal{C}$ and is denoted $\mathcal{G}_\mathcal{C}$, and the other is defined based on the set of markers $\mathcal{M}$ and is denoted by $\mathcal{G}_\mathcal{M}$. The graph $\mathcal{G}_\mathcal{C}$ has a vertex for each cluster in $\mathcal{G}_\mathcal{C}$, and two vertices in $\mathcal{G}_\mathcal{C}$ are adjacent if and only if their corresponding clusters have a nonempty intersection (i.e., overlap). The graph $\mathcal{G}_\mathcal{M}$ has a vertex for each marker in $\mathcal{M}$, and two vertices in $\mathcal{G}_\mathcal{M}$ are adjacent if and only if their corresponding markers appear together in at least one cluster in $\mathcal{C}$. For simplicity, we will often refer to the vertices of $\mathcal{G}_C$ by their corresponding clusters, and to the vertices in $\mathcal{G}_{\mathcal{M}}$ by their corresponding markers. We have the following lemma:

Lemma 1. *If the instance $(\mathcal{M}, \mathcal{C}, k)$ is yes-instance of $\mathsf{RHMC_d}$ then there exists a path decomposition of $\mathcal{G}_C$ with $n^{O(1)}$ bags and whose width is at most $k + 2d^2$.*

Proof. Suppose that $(\mathcal{M}, \mathcal{C}, k)$ is yes-instance of the problem. Then there exists a subset $\mathcal{C}' \subseteq \mathcal{C}$ of cardinality at most k, and a linear ordering $\mathcal{L}$ on the markers in $\mathcal{M}$ for which the markers in every cluster in $\mathcal{C} \setminus \mathcal{C}'$ are adjacent with respect to $\mathcal{L}$. Let $\mathcal{C}^- = \mathcal{C} \setminus \mathcal{C}'$, $\mathcal{M}^-$ be the set of markers that appear in the clusters in $\mathcal{C}^-$, and let $\mathcal{G}_{C^-}$ be the subgraph of $\mathcal{G}_C$ induced by the set of vertices corresponding to the clusters in $\mathcal{C}^-$.

We claim that $\mathcal{G}_C^-$ is an interval graph whose maximum degree is at most $2d^2$. In effect, consider the linear ordering $\mathcal{L} : m_{i_1} \prec \ldots \prec m_{i_p}$ on the markers in $\mathcal{M}^-$. Label the markers $m_{i_1}, \ldots, m_{i_p}$ with the integers $1, \ldots, p$, respectively. For every cluster C in $\mathcal{C}^-$, the markers in C appear adjacently with respect to $\mathcal{L}$; let m_{i_r} and m_{i_s} be the first and last markers in C, respectively, with respect to the ordering $\mathcal{L}$. We associate with C the interval $[r, s]$ with integer endpoints. Because the markers in each cluster C appear adjacently in $\mathcal{L}$, it is easy to see that two clusters in $\mathcal{G}_{C^-}$ are adjacent if and only if their corresponding intervals intersect. It follows that $\mathcal{G}_{C^-}$ is the intersection graph of a set of intervals, and hence, is an interval graph. Moreover, since each cluster contains at most d markers, the length of the corresponding intervals is bounded by d. Since the intervals have integer endpoints, it is easy to verify that every interval can intersect with at most $2d^2$ other intervals. It follows from this observation that the maximum degree of $\mathcal{G}_{C^-}$, $\Delta(\mathcal{G}_{C^-})$, is at most $2d^2$.

Because $\mathcal{G}_C^-$ is an interval graph, there exists an ordering of its maximal cliques $M_1, \ldots, M_t$ such that if a cluster C appears in $M_j \cap M_q$, where $j < q$, then C must appear in every M_i where $j \leq i \leq q$ [17]. Noting that every edge in $\mathcal{G}_C^-$ appears in some maximal clique, it is easy to see that the above ordering on the maximal cliques of $\mathcal{G}_{C^-}$ gives a path decomposition of $\mathcal{G}_{C^-}$, $\mathcal{P}^- = (B_1, \ldots, B_t)$, where bag B_i, $i = 1, \ldots, t$, consists of the clusters contained in M_i. Since any two clusters in a bag B_i, $i \in \{1, \ldots, t\}$, of $\mathcal{P}^-$ are adjacent, and since $\Delta(\mathcal{G}_{C^-}) \leq 2d^2$, it follows that each bag B_i contains at most $2d^2$ vertices (clusters), and hence $pw(\mathcal{P}^-) \leq 2d^2$. Moreover, since each bag contains at most $2d^2$ clusters, the total number of bags in $\mathcal{P}^-$ is at most $\sum_{i=1}^{2d^2} \binom{n}{i} = n^{O(1)}$.

To show that the pathwidth of $\mathcal{G}_C$ is $O(k)$, consider the path decomposition $\mathcal{P}^-$ of $\mathcal{G}_C^-$, and add to every bag B_i, $i \in \{1, \ldots, t\}$, of $\mathcal{P}^-$ all the clusters in $\mathcal{C}'$. It is easy to verify that we obtain a path decomposition $\mathcal{P}$ of $\mathcal{G}_C$ with $n^{O(1)}$ bags and whose width is at most $k + 2d^2$. This completes the proof. ☐

Remark 1. The proof of Lemma 1 may suggest to the reader that we can reduce the $\mathsf{RHMC_d}$ problem to the problem of testing whether we can delete k vertices from $\mathcal{G}_C$ so that the resulting graph is an interval graph, which is known as the k-INTERVAL VERTEX DELETION problem, and was very recently proven to be FPT [22]. This intuition, however, is incorrect, because the statement that we can delete k vertices from $\mathcal{G}_C$ so that the resulting graph is an interval graph is not a sufficient condition for the corresponding instance of RHMC$_d$ to be a yes-instance, as can be easily verified by the reader.

Lemma 2. *If the instance $(\mathcal{M}, \mathcal{C}, k)$ is yes-instance of the problem then there exists a path decomposition of $\mathcal{G}_\mathcal{M}$ with $n^{O(1)}$ bags and whose width is at most $d \cdot (k + 2d^2)$.*

Proof. Suppose that the instance $(\mathcal{M}, \mathcal{C}, k)$ is yes-instance of the problem. By Lemma 1, there exists a path decomposition of $\mathcal{G}_\mathcal{C}$ with $n^{O(1)}$ bags and whose width is at most $k + 2d^2$; let $\mathcal{P} = (B_1, \ldots, B_t)$ be such a path decomposition of $\mathcal{G}_\mathcal{C}$. Consider the following decomposition $\mathcal{P}_\mathcal{M} = (B_1', \ldots, B_t')$, where B_i' is the set of markers contained in the clusters in bag B_i; that is, $B_i' = \{m \in \mathcal{M} \mid m \in C$ for some $C \in B_i\}$. We prove that $\mathcal{P}_\mathcal{M}$ is a path decomposition of $\mathcal{G}_\mathcal{M}$.

First, every marker in $\mathcal{M}$ appears in some cluster in $\mathcal{G}_\mathcal{C}$, and hence, must appear in some bag in $\mathcal{P}_\mathcal{M}$. An edge in $\mathcal{G}_\mathcal{M}$ corresponds to two markers that appear together in some cluster. Since $\mathcal{P}$ is a path decomposition of $\mathcal{G}_\mathcal{C}$, each cluster appears in a bag in $\mathcal{P}$, and hence, any two adjacent markers in $\mathcal{G}_\mathcal{M}$ appear together in some bag of $\mathcal{P}_\mathcal{M}$. Suppose now that a marker m appears in bags B_i' and B_j', $i < j$, in $\mathcal{P}_\mathcal{M}$. We will show that $m \in B_r'$ for every r satisfying $i \leq r \leq j$. Since $m \in B_i'$, there exists a cluster C_i containing m such that $C_i \in B_i$. Similarly, there exists a cluster C_j containing m such that $C_j \in B_j$. If $C_i = C_j$, then since $\mathcal{P}$ is a path decomposition of $\mathcal{G}_\mathcal{C}$, every bag B_r, $i \leq r \leq j$, contains $C_i = C_j$, and hence every bag B_r', $i \leq r \leq j$, contains m. Assume that $C_i \neq C_j$. Since $m \in C_i \cap C_j$, C_i and C_j are adjacent, and hence must appear together in some bag B_q in $\mathcal{P}$. If $q \leq i$, then every bag B_r, $i \leq r \leq j$, contains C_j, and hence every bag B_r', $i \leq r \leq j$, contains m. On the other hand, If $q \geq j$, then every bag B_r, $i \leq r \leq j$, contains C_i, and hence every bag B_r', $i \leq r \leq j$, contains m. If $i < q < j$, then every bag B_r, $i \leq r \leq q$ contains C_i and every bag B_r, $q \leq r \leq j$ contains C_j; therefore, every bag B_r', $i \leq r \leq j$, contains m. It follows from the above that $\mathcal{P}_\mathcal{M}$ is a path decomposition for $\mathcal{G}_\mathcal{M}$.

Finally, the number of bags in $\mathcal{P}_\mathcal{M}$ is the same as that in $\mathcal{P}$. Since each cluster has size at most d, the width of each bag in $\mathcal{P}_\mathcal{M}$ is at most d times the width of the corresponding bag in $\mathcal{P}$. The statement of the lemma follows. $\square$

By Lemma 2, if $(\mathcal{M}, \mathcal{C}, k)$ is yes-instance of the problem then there exists a path decomposition of $\mathcal{G}_\mathcal{M}$ with $n^{O(1)}$ bags whose width is at most $d \cdot (k + 2d^2)$. Given an instance $(\mathcal{M}, \mathcal{C}, k)$, we can call the algorithm in [1] to check if the pathwidth of $\mathcal{G}_\mathcal{M}$ is at most $d \cdot (k + 2d^2)$; the algorithm in [1] runs in fpt-time in the pathwidth parameter $d \cdot (k + 2d^2)$, and hence, in fpt-time in k. If the algorithm rejects (i.e., the pathwidth is larger than $d \cdot (k + 2d^2)$) then we reject; otherwise, the algorithm returns a path decomposition of $\mathcal{G}_\mathcal{M}$ of width at most $d \cdot (k + 2d^2)$; we can assume, without loss of generality, that this path decomposition is nice. Therefore, we have:

Theorem 1. *There is an algorithm running in fpt-time that takes an instance $(\mathcal{M}, \mathcal{C}, k)$ of RHMC$_d$ and either decides correctly that the instance is a no-instance of RHMC$_d$, or returns a nice path decomposition $\mathcal{P} = (B_1, \ldots, B_t)$ of $\mathcal{G}_\mathcal{M}$ of width at most $d \cdot (k + 2d^2)$, and $n^{O(1)}$ nodes.*

We will see in the next section how we can use this nice path decomposition to solve the instance $(\mathcal{M}, \mathcal{C}, k)$ in fpt-time.

4   An FPT Algorithm for RHMC$_d$

Given an instance $(\mathcal{M}, \mathcal{C}, k)$ of RHMC$_d$, by Theorem 1, we can assume that a nice path decomposition $\mathcal{P} = (B_1, \ldots, B_t)$ of $\mathcal{G}_\mathcal{M}$ with width $w \leq d \cdot (k + 2d^2)$ has been computed.

We proceed with a dynamic programming approach based on the path decomposition $\mathcal{P}$. During the dynamic programming, each bag B_i keeps a table T_i of strings of length at most $4w + 3$ (see Corollary 1), composed of markers that appear in B_i, and two special symbols: $\boxtimes$ and $\square$. The $\boxtimes$ symbol represents a nonempty string of markers that have been "forgotten" during the dynamic programming process. When a $\boxtimes$ is adjacent to a marker, as in $u\boxtimes$ or $\boxtimes u$, it indicates that the place next to u on the right-hand or the left-hand side, respectively, is already occupied by markers that have been forgotten (in the dynamic programming process). The $\square$ symbol represents an open space, which may be filled with any number of markers. When a $\square$ is adjacent to a marker, as in $u\square$ or $\square u$, it indicates that the space on the right-hand side or left-hand side of u, respectively, is open and hence can be filled with markers to fulfill the requirements imposed by the clusters.

A string S can be simplified as follows:

- Rule 1. Replace any two or more consecutive $\boxtimes$'s in S by a single $\boxtimes$ and replace any two or more consecutive $\square$'s in S by a single $\square$.
- Rule 2. Remove every $\square$ that appears between two $\boxtimes$'s.

The correctness of the simplification rules will be proven in Lemma 5.

A simplified string has the following property.

Corollary 1. *Let S be a simplified string whose markers are $\mathfrak{m}_1, \ldots, \mathfrak{m}_p$ appearing in S in this order. Then $S = \mathfrak{s}_0 \mathfrak{m}_1 \mathfrak{s}_1 \mathfrak{m}_2 \mathfrak{s}_2 \ldots \mathfrak{m}_t \mathfrak{s}_p$ where each $\mathfrak{s}_i$, for $0 \leq i \leq p$, is an element of $\{\epsilon$ (empty string)$, \square, \boxtimes, \square\boxtimes, \boxtimes\square, \square\boxtimes\square\}$. In particular, if $p \leq w$, then the length of S is at most $w + 3(w + 1) = 4w + 3$.*

Lemma 3. *The size of any table T_i is at most $6^{w+1} w!$.*

Proof. The number of markers in B_i is at most w. By Corollary 1, any string in the table T_i contains a permutation of the markers in B_i as a subsequence. There are at most $w!$ such permutations. For each permutation, there are at most $w + 1$ places to insert special symbols. By Corollary 1, for each place, there are 6 possible strings of special symbols: $\epsilon, \square, \boxtimes, \square\boxtimes, \boxtimes\square, \square\boxtimes\square$. Therefore, the total number of different strings in table T_i is at most $6^{w+1} w!$. $\square$

In the following, we define a table T_i for every bag B_i of the path decomposition. The table T_i contains a set of strings S. Each string S is associated with a set $\mathcal{U}_S$ and an integer v_S. Intuitively, $\mathcal{U}_S$ captures the set of clusters whose statuses are undecided in the computation leading to S, and v_S is the number of clusters that are violated in the computation leading to S. After enumerating all valid strings for the leaf bag B_1, at each step of the dynamic programming, we compute the table T_i for B_i based on the table T_{i-1} for B_{i-1}. In particular, we will compute strings $S \in T_i$ based on strings $S' \in T_{i-1}$. The set $\mathcal{U}_S$ and the value v_S are computed from $\mathcal{U}_{S'}$ and $v_{S'}$. Each cluster $C \in \mathcal{U}_{S'}$ is at one of following three statuses with respect to the string S:

- A cluster C is *satisfied* by a string S if all markers of C appear as a continuous block in S.
- A cluster C is *violated* by a string S if (1) all markers of C appear in S but they do not form a continuous block, or (2) there are two markers $m_i, m_j \in C$ such that there is a marker $m_l \notin C$ or a $\boxtimes$ appearing between m_i and m_j in S.[2]
- A cluster C is *uncertain* with respect to a string S if it is neither satisfied nor violated by S.

The details of the computation are given below in three cases.

Case 1 (Leaf bag). For the leaf bag B_1, we enumerate all possible simplified strings in T_1 as described in Lemma 3. Note that in a nice path decomposition, B_1 has only one marker m and $T_1 = \{m, \Box m, m\Box, \Box m\Box\}$. Recall that for each string S, v_S is the number of clusters that are violated by S and $\mathcal{U}_S$ is the set of clusters that are uncertain with respect to S. For each string S in T_1, we associate with it the set $\mathcal{U}_S$ and the value v_S. Note that for the leaf bag and for a given S, both $\mathcal{U}_S$ and v_S can be easily calculated.

Case 2 (Forget bag). Now let us consider a forget bag B_i, where a marker m is forgotten, i.e. $B_i = B_{i-1} \backslash \{m\}$. The table T_i is created from T_{i-1} as follows:

(a) For each string $S \in T_{i-1}$, replace the occurrence of m in S by a $\boxtimes$ and simplify S; then add the resulting string S' to T_i.
(b) For every string $S' \in T_i$, choose a string $S \in T_{i-1}$ with the minimum value v_s from among all the strings in T_{i-1} that yielded S', and associate $\mathcal{U}_{S'} = \mathcal{U}_S$ and $v_{S'} = v_S$ with S'.

Case 3 (Introduce bag). Finally, let us consider an introduce bag B_i, where a marker m is introduced, i.e., $B_i = B_{i-1} \cup \{m\}$. The table T_i is created from T_{i-1} as follows:

(a) For each string $S \in T_{i-1}$, replace every occurrence of $\Box$ in S by one of the following four choices: m, $\Box m$, $m\Box$, and $\Box m\Box$ and add the resulting strings to T_i.

---

[2] While it is clear that a cluster C is violated by a string if there are two markers $m_i, m_j \in C$ such that there is a marker $m_l \notin C$ appearing between m_i and m_j in the string, the same is true if m_l is replaced by a $\boxtimes$, because of the property of the path decomposition, as we will show in Lemma 4.

(b) For each string $S' \in T_i$ that resulted from $S \in T_{i-1}$ by (1), let $\mathcal{U}_{S'}$ be $\mathcal{U}_S$ minus the set of clusters in $\mathcal{U}_S$ that are satisfied or violated by S', and let $v_{S'}$ be v_S plus the number of clusters in $\mathcal{U}_S$ that are violated by S'.

(c) Purge every string $S' \in T_i$ with $v_{S'} > k$.

Finally, if T_t is non-empty, then the instance is a yes-instance, as justified in Lemma 7.

The following lemmas are needed for the correctness of algorithm.

Lemma 4. *If a marker m is forgotten at a bag B_i, then any cluster containing m is not uncertain with respect to any string in T_j where $j \geq i - 1$.*

Proof. Suppose that a marker m is forgotten by a bag B_i and C is a cluster containing m. Since the markers in C form a clique in the graph $G_\mathcal{M}$, there must exist a bag B_l, $l \leq i - 1$, such that all markers in C appear in B_l (see [21] for a proof). By the definition of the three statuses, every string in table T_l will either satisfy or violate C, and hence C is not uncertain for any string in B_l. According to the rules for creating the tables for bags B_i, $i > l$, C will not be uncertain for any string in the bags after B_l. The lemma follows. □

Lemma 5. *The simplification rules for strings are correct.*

Proof. The simplification Rule 1 is clearly correct by the definition of ⊠ and □.

Now suppose that a □ appears between two ⊠'s in a string S, as in ⊠□⊠. Suppose that a marker m is later introduced into the □ (if no marker is ever introduced into the □, then it can be safely removed) and m' is a forgotten marker in one of the two adjacent ⊠'s. Then m and m' cannot belong to the same cluster. This is because if m and m' belong to the same cluster, then by Lemma 4, m should have already been introduced before m' is forgotten. This means that the content of this □ is isolated from the contents of the two adjacent ⊠'s and from the rest of the string S. In other words, the content of this □ can be safely moved to the beginning or the end of the linear order. So Rule 2 is correct. □

Lemma 6. *Let B_i, B_{i-1}, m be as in Case 2 (Forget bag) of the algorithm. If more than one string $S_1, \ldots, S_q \in T_{i-1}$ yield the same string S at the forget bag B_i in Case 2(a), then $\mathcal{U}_{S_1} = \mathcal{U}_{S_2} = \ldots = \mathcal{U}_{S_q}$.*

Proof. Suppose that the statement is not true. Then there are two strings S_i and S_j, $1 \leq i, j \leq q$, such that $\mathcal{U}_{S_i} - \mathcal{U}_{S_j} \neq \emptyset$. Let $C \in \mathcal{U}_{S_i} - \mathcal{U}_{S_j}$ be a cluster. Since $C \in \mathcal{U}_{S_i}$, by Lemma 4, C does not contain m or any forgotten marker. On the other hand, $C \notin \mathcal{U}_{S_j}$, which means that C is satisfied or violated by S_j. Since C does not contain m or any forgotten marker, the fact that C is satisfied or violated by S_j can be determined by the markers in S. In other words, C is satisfied (or violated) by S_j if and only if it is satisfied (or violated) by S. For a similar reason, C is satisfied (or violated) by S_i if and only if it is satisfied (or violated) by S. Therefore, C is satisfied (or violated) by S_j if and only if it is satisfied (or violated) by S_i. This is a contradiction to $C \in \mathcal{U}_{S_i} - \mathcal{U}_{S_j}$. □

Lemma 7. *If the table T_t for the last bag B_t of the path decomposition is non-empty, then the instance is a yes-instance.*

Proof. After the last bag B_t of the path decomposition is processed, the status of every cluster C is decided. This is true because either at least one marker in C is forgotten and by Lemma 4 the status of C is decided, or all markers in C appear in B_t and by the definition of the three statuses the status of C is decided. Therefore if T_t is non-empty then there is a string $S \in T_t$ such that $\mathcal{U}_S = \emptyset$ (i.e. all clusters are not uncertain with respect to S) and $v_S \leq k$. This means that there is an ordering of the markers that violates at most k clusters and satisfies all other clusters, and hence the instance is a yes-instance. □

Theorem 2. *Let $(\mathcal{M}, \mathcal{C}, k)$ be an instance of RHMC_d. Given a nice path decomposition of $G_{\mathcal{M}}$ of width w and $n^{O(1)}$ bags (n is the number of clusters), the dynamic programming algorithm described above decides the instance $(\mathcal{M}, \mathcal{C}, k)$ in time $O^*(6^w w!)$.*

Proof. It is clear that Case 1 (Leaf bag) of the algorithm is correct because it enumerates all valid simplified strings for the bag B_1. Case 3 (Introduce bag) is correct because it enumerates all possible valid ways for inserting a marker m into a string S, and updates $\mathcal{U}_S$ and v_S accordingly. Case 2 (Forget bag) is justified because, by Lemma 6, all strings $S_1, \ldots, S_q \in T_{i-1}$ that yield the same string S have the same set of uncertain clusters $\mathcal{U}_{S_1} = \mathcal{U}_{S_2} = \ldots = \mathcal{U}_{S_q}$, and hence we can safely pick the string $S \in \{S_1, \ldots, S_q\}$ with the minimum value v_s. This proves the correctness of the algorithm.

By Lemma 3, the size of table T_i for every bag B_i is at most $6^{w+1} w!$. Since the number of bags is $n^{O(1)}$, and all other operations can be executed in polynomial time, the running time of the algorithm is $O^*(6^w w!)$. □

Combining Theorem 1 with Theorem 2, we get:

Corollary 2. *For any $d \in \mathbb{N}$, the RHMC_d problem is FPT.*

5   Conclusion

In this paper, we proved that the RHMC_d problem is fixed-parameter tractable for any $d \in \mathbb{N}$, answering a question posed by Zhang et al. [25]. Several interesting questions remain open. The obvious question is whether the (general) RHMC problem is fixed-parameter tractable or not (that is, with no upper bound on the cluster size). The approach used in the current paper does not seem to be applicable to the RHMC problem. In particular, the pathwidth of the auxiliary graph for an instance of RHMC may be unbounded. On the other hand, we point out that, due to the fact that one can control the resolution of the radiation dosage, in some applications d is quite small. For instance, Slonim et al. [24] only considered the case when $d = 3$.

Another interesting question that is worth pursuing is studying *kernelization* algorithms for the RHMC_d problem. A *kernelization* algorithm for a parameterized problem is a polynomial-time algorithm that maps an instance (x, k)

to an equivalent instance (x', k') such that both $|x'|$ and k' are bounded by a computable function of k. It is well known that a parameterized problem is fixed-parameter tractable if and only if the problem is kernelizable [12]. Therefore, the results in this paper, combined with [12], prove that the $RHMC_d$ problem is kernelizable; however, the upper bound on the kernel size is only of a theoretical interest. Zhang et al. [25] proved that $RHMC_3$ (and $RHMC_2$) has a linear-size kernel $(O(k))$. It is interesting to study whether or not $RHMC_d$, for any $d > 3$, admits a linear, quadratic, or a polynomial kernel.

Finally, it would be interesting to seek $O^*(c^k)$ parameterized algorithms for the problem. We leave those as open questions for future research.

References

1. Bodlaender, H.: A Linear-Time Algorithm for Finding Tree-Decompositions of Small Treewidth. SIAM Journal on Computing 25(6), 1305–1317 (1996)
2. Bodlaender, H., Fellows, M., Hallett, M., Wareham, H.: Parameterized complexity analysis in computational biology. Computer Applications in the Biosciences 11, 49–57 (1995)
3. Bodlaender, H., Fellows, M., Hallett, M., Wareham, H.: The parameterized complexity of the longest common subsequence problem. Theoretical Computer Science 147, 31–54 (1995)
4. Bodlaender, H.L., Thilikos, D.M.: Computing Small Search Numbers in Linear Time. In: Downey, R.G., Fellows, M.R., Dehne, F. (eds.) IWPEC 2004. LNCS, vol. 3162, pp. 37–48. Springer, Heidelberg (2004)
5. Cesati, M., Wareham, H.: Parameterized complexity analysis in robot motion planning. In: Proceedings of the 25th IEEE International Conference on Systems, Man and Cybernetics, pp. 880–885 (1995)
6. Chen, Z.-Z., Lin, G., Wang, L.: An approximation algorithm for the minimum co-path set problem. Algorithmica 60(4), 969–986 (2011)
7. Cheng, Y., Cai, Z., Goebel, R., Lin, G., Zhu, B.: The radiation hybrid map construction problem: recognition, hardness, and approximation algorithms (2008) (unpublished manuscript)
8. Cox, D., Burmeister, M., Price, E., Kim, S., Myers, R.: Radiation hybrid mapping: a somatic cell genetic method for constructing high resolution maps of mammalian chromosomes. Science 250, 245–250 (1990)
9. De Givry, S., Bouchez, M., Chabrier, P., Milan, D., Schiex, T.: Carh ta Gene: multi-population integrated genetic and radiation hybrid mapping. Bioinformatics 21(8), 1703 (2005)
10. Downey, R., Evans, P., Fellows, M.: Parameterized learning complexity. In: Proceedings of the 6th ACM Workshop on Computational Learning Theory, pp. 51–57 (1993)
11. Downey, R., Fellows, M.: Parameterized complexity. Springer, New York (1999)
12. Downey, R., Fellows, M., Stege, U.: Parameterized complexity: a framework for systematically confronting computational intractability. In: Roberts, F., Kratochvil, J., Nešetřil, J. (eds.) Contemporary Trends in Discrete Mathematics, AMS-DIMACS Proceedings, pp. 49–99. American Mathematical Society (1999)
13. Faraut, T., De Givry, S., Chabrier, P., Derrien, T., Galibert, F., Hitte, C., Schiex, T.: A comparative genome approach to marker ordering. Bioinformatics 23(2), e50 (2007)

14. Fellows, M., Hallett, M., Wareham, H.: DNA physical mapping: three ways of difficult. In: Lengauer, T. (ed.) ESA 1993. LNCS, vol. 726, pp. 157–168. Springer, Heidelberg (1993)

15. Fellows, M., Hallett, M., Wareham, H.: Fixed-parameter complexity and cryptography. In: Moreno, O., Cohen, G., Mora, T. (eds.) AAECC 1993. LNCS, vol. 673, pp. 121–131. Springer, Heidelberg (1993)

16. Fellows, M., Langston, M.: On search, decision and the efficiency of polynomial-time algorithms. In: Proceedings of the 21st ACM Symposium on Theory of Computing (STOC), pp. 501–512 (1989)

17. Fishburn, P.: Interval orders and interval graphs: A study of partially ordered sets. Wiley-Interscience Series in Discrete Mathematics, New York (1985)

18. Flum, J., Grohe, M.: Parameterized complexity theory. Springer-Verlag New York Inc. (2006)

19. Jiang, H., Zhu, B.: Weak Kernels. Electronic Colloquium on Computational Complexity (ECCC) 17, 5 (2010)

20. Kaplan, H., Shamir, R., Tarjan, R.: Tractability of parameterized completion problems on chordal, strongly chordal, and proper interval graphs. SIAM Journal on Computing 28, 880–892 (1999)

21. Niedermeier, R.: Invitation to fixed-parameter algorithms. Oxford University Press, USA (2006)

22. Rafiey, A.: Single Exponential FPT Algorithm for Interval Vertex Deletion and Interval Completion Problem, http://arxiv.org/pdf/1211.4629v1.pdf

23. Richard, C., Withers, D., Meeker, T., Maurer, S., Evans, G., Myers, R., Cox, D.: A radiation hybrid map of the proximal long arm of human chromosome 11 containing the multiple endocrine neoplasia type 1 (MEN-1) and bcl-1 disease loci. American Journal of Human Genetics 49(6), 1189 (1991)

24. Slonim, D., Kruglyak, L., Stein, L., Lander, E.: Building human genome maps with radiation hybrids. Journal of Computational Biology 4(4), 487–504 (1997)

25. Zhang, C., Jiang, H., Zhu, B.: Radiation hybrid map construction problem parameterized. In: Lin, G. (ed.) COCOA 2012. LNCS, vol. 7402, pp. 127–137. Springer, Heidelberg (2012)

26. West, D.: Introduction to graph theory. Prentice Hall Inc., Upper Saddle River (2006)

Reconstructing Ancestral Genomic Orders Using Binary Encoding and Probabilistic Models

Fei Hu[1,2], Lingxi Zhou[2], and Jijun Tang[1,2,*]

[1] School of Computer Science and Technology, Tianjin University, China
[2] Department of Computer Science & Engineering, Univ. of South Carolina, USA
jtang@cse.sc.edu

Abstract. Changes of gene ordering under rearrangements have been extensively used as a signal to reconstruct phylogenies and ancestral genomes. Inferring the gene order of an extinct species has the potential in revealing a more detailed evolutionary history of species descended from it. Current tools used in ancestral reconstruction may fall into parsimonious and probabilistic methods according to the criteria they follow. In this study, we propose a new probabilistic method called PMAG to infer the ancestral genomic orders by calculating the conditional probabilities of gene adjacencies using Bayes' theorem. The method incorporates a transition model designed particularly for genomic rearrangement scenarios, a reroot procedure to relocate the root to the target ancestor that is inferred as well as a greedy algorithm to connect adjacencies with high conditional probabilities into valid gene orders.

We conducted a series of simulation experiments to assess the performance of PMAG and compared it against previously existing probabilistic methods (InferCARsPro) and parsimonious methods (GRAPPA). As we learned from the results, PMAG can reconstruct more correct ancestral adjacencies and yet run several orders of magnitude faster than InferCARsPro and GRAPPA.

Keywords: ancestral genome, gene order, probabilistic method.

1 Introduction

1.1 Overview

Evolutionary biologists have had a long tradition in reconstructing genomes of extinct ancestral species. Mutations in a genomic sequence are made up not only at the level of base-pair changes but also by rearrangement operations on chromosomal structures such as inversions, transpositions, fissions and fusions [1]. Over the past few years, ancestral gene-order inference has brought profound predictions of protein functional shift and positive selection [2].

Methods for ancestral genome reconstruction either assume a given phylogeny that represents the evolutionary history among given species, known as the small

---

* Corresponding author.

Z. Cai et al. (Eds.): ISBRA 2013, LNBI 7875, pp. 17–27, 2013.
© Springer-Verlag Berlin Heidelberg 2013

phylogeny problem (SPP); or search the most appropriate tree along with a set of ancestral genomes to fit the observed data, called the big phylogeny problem (BPP). Most of parsimony methods (such as GRAPPA [3], MGR [4,5]) typically solve the SPP exactly by searching a set of ancestral gene orders to minimize the sum of the rearrangement distance over the entire edges of the phylogeny. Ma proposed another method for the SPP in the probabilistic framework (InferCARsPro [6]) by approximating the conditional probabilities for all possible gene adjacencies in an ancestral genome.

Current methods such as GRAPPA and InferCARsPro are capable to handle modern whole-genome data due to their intrinsic high complexity. In this paper, we propose a new probabilistic method called PMAG to reconstruct ancestral genomic orders given a phylogeny. We conducted extensive experiments to evaluate the performance of PMAG with other existing methods. According to the results, our new method can outperform all the other methods under study and still run at least hundreds of times faster than GRAPPA and InferCARsPro.

1.2   Genome Rearrangement

Given a set of n genes labeled as $\{1, 2, \cdots, n\}$, a genome can be represented by an *ordering* of these genes. Each gene is assigned with an orientation that is either positive, written i, or negative, written $-i$. Two genes i and j form an *adjacency* if i is immediately followed by j, or, equivalently, $-j$ is immediately followed by $-i$. An *breakpoint* of two genomes is defined as an adjacency appears in one but not in the other.

Genome rearrangement operations can change the ordering of genes. An *inversion* operation (also called *reversal*) reverses a segment of a chromosome. A *transposition* is an operation that swaps two adjacent segments of a chromosome. In the case of multiple chromosomes, *translocation* breaks a chromosome and reattaches a portion of it to another chromosome. Later Yancopoulos et al. [8] proposed a universal double-cut-and-join (DCJ) operation that accounts for all common events which resulted in a new genomic distance that can be computed in linear time.

1.3   Parsimony Methods for Ancestral Gene-Order Reconstruction

To find a solution for SPP, parsimony algorithms typically iterate over each internal node to solve for the median genomes until the sum of all edge distances (tree score) is minimized. The median problem can be formalized as follows: give a set of m genomes with permutations $\{x_i\}_{1 \le i \le m}$ and a distance measurement d, find another permutation x_t such that the median score defined as $\sum_{i=1}^{m} d(x_i, x_t)$ is minimized. GRAPPA and MGR (as well as their recently enhanced versions) are two widely-referenced methods that implemented a selection of median solvers for phylogeny and ancestral gene-order inference. However solving even the simplest case of median problem when m equals to three is NP-hard for most distance measurements.

Exact solutions to the problem of finding a median of three genomes can be obtained for the inversion, breakpoint and DCJ distances. Among all the median solvers, the best one is the DCJ median solver proposed by Xu and Sankoff (ASMedian [9]) based on the concept of adequate subgraph. Adequate subgraphs allow decompositions of an multiple breakpoint graph into smaller and easier graphs. Though the ASMedian solver could remarkably scale down the computational expenses of median searching, it yet runs very slow when the genomes are distant.

1.4 Reconstructing the Ancestral Gene Order in Probabilistic Frameworks

The probabilistic approach InferCARsPro proposed by Ma [6] is based on Bayes' theorem such that every possible predecessor and successor of a signed gene i denoted as X_i in the ancestral genome x, given D_x representing the observed data, can be expressed as

$$P(X_i\ in\ x | D_x) = \frac{P(D_x | X_i\ in\ x) P(X_i\ in\ x)}{\sum_{j=1}^{q} P(D_x | X_j\ in\ x) P(X_j\ in\ x)} = \frac{P(D_x | X_i\ in\ x)}{\sum_{j=1}^{q} P(D_x | X_j\ in\ x)}$$

where priors are assumed equal and the likelihood $P(D_x | X_i\ in\ x)$ can be calculated recursively in a post-order traversal fashion summed over q possible configurations. Its transition matrix is defined as an extension of the Jukes-Cantor model such that probability of transition from any character to any different character is always equal.

Let $s_x(\cdot)$ denote the successor of a gene and $p_x(\cdot)$ denote the predecessor of a gene, an adjacency pair $A_x(i, j)$ can be viewed as $s_x(i) = j$ and $p_x(j) = i$ simultaneously. After finishing the calculation of conditional probabilities for every successor and predecessor relationships, the conditional probability of an adjacency $A_x(i, j)$ in genome x can be approximated as

$$P(A_x(i, j) | D_x) = P(p_x(j) = i | D_x) \times P(s_x(i) = j | D_x)$$

Finally a fast greedy algorithm is adopted to connect adjacencies into contiguous ancestral regions. Although InferCARsPro showed good results and speedup over parsimonious methods, it is still too slow and inaccurate when dealing with even small number of distant genomes.

We investigated the following intrinsic characteristics of InferCARsPro that account for its difficulties in handling complex datasets, which in turn motivated us to propose our new method.

- InferCARsPro uses a neutral model accounting for all changes of adjacencies, however biased model for phylogeny reconstruction has been successfully applied for genome rearrangement scenarios [11].
- The total number of states for each gene is exactly equal to $2 \times n - 2$ where n is the number of genes. Thus computing the likelihood score on such excessive number of states clearly incurs huge computational burden.

- The conditional probability of an adjacency is approximated from the predecessor and successor relations. Although such approximation is intuitive, it is more desirable to directly calculate the conditional probability of an adjacency.
- InferCARsPro requires branch lengths of a given phylogeny as part of its inputs, but it is not always handy to obtain in practice.

2 Algorithm Detail

Given the topology of a model tree and a collection of gene orders at the leaves, our approach first encodes the gene orders into binary sequences and estimates the parameters in the transition model for adjacency changes. Ancestral nodes in the model tree are inferred independently and in each inference, we reroot the model tree to have the target ancestor as the root of a new tree. Then we utilize a probabilistic inference tool to compute the conditional probabilities of all the adjacencies encoded in the binary sequence of the target ancestor. At last we use a greedy algorithm as used in Ma's work to connect the adjacencies into contiguous regions. We call our new approach *Probabilistic Method of Ancestral Genomics (PMAG)*.

2.1 Encoding Gene Orders into Binary Sequences

A gene order can be expressed as a sequence of adjacency information that specifies presence or absence of all the adjacencies [10,11]. Denote the head of a gene i by i^h and its tail by i^t. We refer $+i$ as an indication of direction from head to tail ($i^h \rightarrow i^t$) and otherwise $-i$ as ($i^t \rightarrow i^h$). There are a total of four scenarios for two consecutive genes a and b in forming an *adjacency*: $\{a^t, b^t\}$, $\{a^h, b^t\}$, $\{a^t, b^h\}$, and $\{a^h, b^h\}$. If gene c is at the first or last place of a linear chromosome, then we have a corresponding singleton set, $\{c^t\}$ or $\{c^h\}$, called a *telomere*. A genome can then be expressed as a multiset of adjacencies and telomeres. For instance, a linear chromosome consists of four genes, $(+1,+2,-3,-4,)$ can be represented by the multiset of adjacencies and telomeres $\{\{1^h\}, \{1^t, 2^h\}, \{2^t, 3^t\}, \{3^h, 4^t\}, \{4^h\}\}$. We further write 1 (0) to indicate presence (absence) of an adjacency and we consider only those adjacencies and telomeres that appear at least once in the input genomes. Table 1 shows an example of encoding two artificial genomes into binary sequences.

Given a dataset D with m species and each of n genes, let k indicate the total number of linear chromosomes in D, then there are up to $\binom{2n+2}{2}$ distinct adjacencies and telomeres. However in reality if the length of the binary sequences extracted from D is l, then l is typically far smaller. In fact, in the extreme case when genomes in D share no adjacency and telomere, l equals at most to $n \times m + k$, and since m and k are commonly much smaller than n, thus the length of the binary sequences for a dataset is usually linear rather than quadratic to the number of genes.

Table 1. Example of encoding gene orders into binary sequences

$$G_1 : (1, \ 2, -3)$$
$$G_2 : (3, -2, \ 1)$$

(a) Two signed linear genomes

	$\{1^h\}$	$\{1^t, 2^h\}$	$\{2^t, 3^t\}$	$\{3^h\}$	$\{2^h, 1^h\}$	$\{1^t\}$
G_1	1	1	1	1	0	0
G_2	0	0	1	1	1	1

(b) Binary sequences

2.2 Estimating Transition Parameters

Since we are handling binary sequences with two characters, we use a general time-reversible framework to simulate the transitions from presence (1) to absence (0) and vice versa. Thus the rate matrix is

$$Q = \{q_{ij}\} = \begin{bmatrix} \cdot & a \\ a & \cdot \end{bmatrix} \begin{bmatrix} \pi_0 & 0 \\ 0 & \pi_1 \end{bmatrix}$$

The matrix involves 3 parameters: the relative rate a, and two frequencies π_0 and π_1.

Severl models have been proposed to probabilistically characterize the changes of gene adjacencies by common types of rearrangement operations such as inversion, transposition as well as DCJ [7,11]. In this study, we use the model that has been successfully applied for phylogeny reconstruction in the context of genome rearrangement as suggested in [11]. In particular, every DCJ operation breaks two random adjacencies uniformly chosen from the gene-order string and subsequently creates two new ones. Since each genome contains $n + O(1)$ adjacencies and telomeres where n is the gene number and $O(1)$ equals to the number of linear chromosomes in the genome, thus the probability that an adjacency changes from presence (1) to absence (0) in the sequence is $\frac{2}{n+O(1)}$ under one operation. Since there are up to $\binom{2n+2}{2}$ possible adjacencies and telomeres, the probability for an adjacency changing from absence (0) to presence (1) is $\frac{2}{2n^2+O(n)}$. Therefore we come to the conclusion that the transition from 1 to 0 is roughly $2n$ times more likely than that from 0 to 1.

2.3 Inferring the Probabilities of Ancestral Adjacencies for the Root Node

In principle, our probabilistic inference is categorized as marginal reconstruction which assigns characters to a single ancestral genome at a time. Once we have the tree topology and binary sequences encoding the input gene orders, we use the extended probabilistic approach for sequence data described by Yang [12] to infer the ancestral gene orders at the root node. In the binary sequences, each site represents an adjacency with character either 0 (absence) or 1 (presence) and for each site we seek to calculate the conditional probability of observing that adjacency. As the true branch lengths are not available, we take advantage of the widely-used maximum-likelihood estimation from the binary sequences at the leaves to estimate the branch length.

Suppose x is the root of a model tree, then the conditional probability that node x has the character s_x at the site, given D_x representing the observed data at the site in all leaves of the subtree rooted at x, is

$$P(s_x|D_x) = \frac{P(s_x)P(D_x|s_x)}{P(D_x)} = \frac{\pi_{s_x}L_x(s_x)}{\sum_{s_x}\pi_{s_x}L_x(s_x)}$$

where π_{s_x} is the character frequency for s_x. The conditional probability in the form of $L_x(s_x)$ is defined as the probability of observing the leaves that belong to the subtree rooted at x, given that the character at node x is s_x. It can be calculated recursively in a post-order traversal fashion suggested by Felsenstein [13] as:

$$L_x(s_x) = \begin{cases} 1 & \text{if } x \text{ is a leaf with character} = s_x \text{ at the site} \\ 0 & \text{if } x \text{ is a leaf with character} \neq s_x \text{ at the site} \\ \left[\sum_{s_f} p_{s_x s_f}(t_f)L_f(s_f)\right] \times \left[\sum_{s_g} p_{s_x s_g}(t_g)L_g(s_g)\right] & \text{otherwise} \end{cases}$$

where f and g are the two direct descendants of x. $p_{ij}(t)$ defines the transition probability that character i changes to j after an evolutionary distance t. Following the deduction of transition probability in [13], our transition-probability matrix can be written as

$$p_{ij}(t) = \pi_j + e^{-t}(\delta_{ij} - \pi_j)$$

Here the δ_{ij} is 1 if $i = j$, otherwise δ_{ij} is 0. In order to set up the $2n$ ratio, we simply set the rate a to 1 and add a direct assignment of the two frequencies in the code. For instance, if the character frequencies are $\pi_0 = 0.1$ and $\pi_1 = 0.9$, then the rate of 0 to 1 transitions is 10 times as high as the rate of transitions in the other direction under the same evolutionary distance.

RAxML [14,15] is one of the most widely used program for sequence-data analysis which implements the method for ancestral sequence inference developed by Yang [12]. In this study, we modified RAxML to infer the conditional probabilities of gene adjacencies at all sites. Once we obtain the conditional probability of every adjacency for the target ancestor x, we can construct an adjacency graph for x in which each gene i corresponds to two nodes, i^h and i^t, and each adjacency is connected by an edge with weight equal to the conditional probability of seeing that adjacency in x. The problem of searching the longest path in such a graph by visiting each gene's head and tail exactly once is indeed NP-hard as shown in Tang and Wang's study [16]. As a trade-off for time efficiency in dealing with large-scale datasets, we adopted the same greedy algorithm used in Ma's work [18] to connect adjacencies into contiguous ancestral regions.

2.4 Rerooting the Tree Topology

To infer the genomic order of a non-root ancestral node x, if x is taken as the root of the tree such that only the leaves in the subtree of x are considered into the recursive calculation of likelihood, potentially many good adjacencies in the outgroup of the subtree will be neglected and result in a loss of information. To minimize the influence, we incorporate the technique of rerooting so that original tree is rearranged and the target node x becomes the root of a new tree. The procedure of rerooting is a standard procedure implemented in many phylogenetic tools and it also has found to be useful for ancestral genome reconstruction in [6].

3 Experimental Results

3.1 Experimental Design

Since actual ancestors are rarely known for sure, it is difficult to evaluate ancestral reconstruction methods with real datasets. In order to carry out a complete evaluation over a group of methods under a wide range of configurations, we conducted a collection of simulation experiments following the standard steps of such tests that have been extensively adopted [17,11].

In particular, a group of tree topologies were firstly generated with edge lengths representing the expected number of evolutionary operations. An initial gene order was assigned at the root so it can evolve down to the leaves following the tree topology mimicking the natural process of evolution, by carrying out a number of predefined evolutionary events. In this way, we obtained the complete evolutionary history of the model tree and the whole set of genomes it has.

We utilized the simulator proposed by Lin et al. [20] to produce birth-death tree topologies. With a model tree, we were able to produce genomes of any size and difficulty by simply adjusting three main parameters: the number of genomes m, the number of genes n, and the tree diameter d.

Predicted ancestral genomes produced from a method were evaluated in terms of the total number of correctly inferred adjacencies (i.e. those also appear in the true ancestral genomes) divided by the total number of adjacencies in both true genome and predicted genome. In particular, if D represents the set of gene adjacencies in the real genome and D' the predicted genomes. We calculate C, the rate of `correct adjacency` by:

$$C = \frac{|D \cap D'|}{|D \cup D'|} \times 100\%$$

Errors are in two parts. If a gene adjacency in D is missing in D', such a gene adjacency is called a `false negative` (FN) adjacency. The false negative rate measures the proportion of false negative adjacencies with respect to the total number of gene adjacencies in D and D'. The `false positive` (FP) rate is defined similarly, by swapping D and D'.

3.2 Comparing the Performance with Existing Probabilistic Method

Though probabilistic methods of ancestral reconstruction for rearrangement data are relatively new, they have shown great potential in both scalability and efficiency. As we have mentioned, `InferCarsPro` and `PMAG` both aim to formulate the conditional probabilities of gene adjacencies, however due to excessive number of states `InferCarsPro` has to handle, it is much more computationally demanding than `PMAG`. In this section, we compared the performance of `PMAG` to `InferCarsPro`.

Figure 1 (left) shows the assessment result of the two methods using datasets of 10 genomes and each of 1000 genes. From the figure, `PMAG` achieved better accuracies than `InferCarsPro` in all tests, with about 5 percentage points of improvements. Given datasets containing more genomes and genes, `InferCarsPro` encountered great difficulty to finish, while `PMAG` scales well to handle them within a few hours of computation (Figure 1 right).

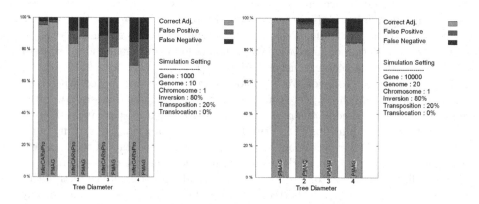

Fig. 1. Comparison between `PMAG` and `InferCARsPro`. X-axis represents the tree diameters from 1 to 4 times the number of genes.

3.3 Comparing the Performance with Parsimonious Methods

Parsimonious methods are in general time-consuming but very accurate. Their performances are sometimes referred as the upper bound of all methods [21], but such methods (`GRAPPA` for example) that directly optimize for the exact solution of the genome median problem suggested by Blanchette et al. [22] are NP-hard.

We compared the performance of `PMAG`, `InferCarsPro` and one direct optimization method `GRAPPA` with Xu's `ASMedian` solver [9] (`GRAPPA-DCJ`). Figure 2 shows the result of comparison. Because datasets are relatively easy, all methods can in average reconstruct more than 95% of true adjacencies and the differences among methods are not significant. However it is worth noting that `PMAG` receives less effect on tree diameters based on the observation that although `PMAG` performs sightly worse than `GRAPPA` methods under $0.6n$ tree diameter, it

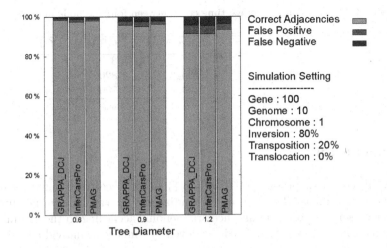

Fig. 2. Comparison among `PMAG`, `InferCARsPro` and `GRAPPA` with DCJ median solvers. X-axis represents the tree diameters that are 0.6, 0.9 and 1.2 times the number of genes.

can outperform the other methods at higher tree diameters. `InferCARsPro` is inferior to both `PMAG` and `GRAPPA` methods in the test which is consistent with the simulation results in Zhang et al.'s study [21].

3.4 Time Consumption

All tests were conducted on a workstation with 2.4Ghz CPUs and 4 GB RAM. We summarizes the running time of each method in the tests of Figure 1 and Figure 2 in Table 2 and Table 3 respectively. From table 2, we can see apparently `InferCARsPro` is computationally more demanding than `PMAG`, and hence restricted to handle small dataset. In table 3, both `InferCARsPro` and `GRAPPA` suffered significantly from high tree diameters, but tree diameter shows little impact on the running time of `PMAG`.

Table 2. Comparison of average time cost between two methods in seconds

Method	Genome#	Gene#	Tree Diameter			
			$1n$	$2n$	$3n$	$4n$
PMAG	10	1000	10	11	13	15
InferCARsPro	10	1000	5.4×10^3	1.4×10^4	2.9×10^4	7.2×10^4
PMAG	20	10000	2.4×10^3	3.6×10^3	5.7×10^3	9.5×10^3

Table 3. Comparison of average time cost between four methods in seconds

Tree Diameter	PMAG	InferCARsPro	GRAPPA-DCJ
0.6	1	300	8
0.9	1	1200	820
1.2	1	2600	7000

4   Conclusion

We introduced a new probabilistic method PMAG for ancestral gene-order inference. PMAG determines the state of each adjacency in the binary encoding to be either present or absent in an ancestral genome according to the conditional probability. Final ancestral genome is retrieved by connecting individual adjacencies into continuous regions. Experimental results show that ancestral genomes can be accurately inferred by PMAG. PMAG is also significantly faster in running time than InferCarsPro and parsimonious methods using direct optimization such as GRAPPA.

Much work remains to be done. In particular, we will try to extend our evolutionary model from rearrangements to a more general one in which other operations such as insertion (addition), duplication, or deletion (gene loss) are possible and hence introduce a new challenge to this study.

Acknowledgements. FH, LZ and JT are supported by grant NSF 0904179.

References

1. Kent, W., Baertsch, R., Hinrichs, A., Miller, W., Haussler, D.: Evolutions cauldron: duplication, deletion, andrearrangement in the mouse and human genomes. Proceedings of the National Academy of Sciences 100(20), 11484–11489 (2003)
2. Muller, K., Borsch, T., Legendre, L., Porembski, S., Theisen, I., Barthlott, W.: Evolution of carnivory in Lentibulariaceae and the Lamiales. Plant Biology 6(4), 477–490 (2008)
3. Moret, B., et al.: A New Implmentation and Detailed Study of Breakpoint Analysis. In: Pacific Symposium on Biocomputing (2001)
4. Guillaume, B., Pevzner, P.: Genome-scale evolution: reconstructing gene orders in the ancestral species. Genome Research 12(1), 26–36 (2002)
5. Max, A., Pevzner, P.: Breakpoint graphs and ancestral genome reconstructions. Genome Research 19(5), 943–957 (2009)
6. Ma, J.: A probabilistic framework for inferring ancestral genomic orders. In: 2010 IEEE International Conference on Bioinformatics and Biomedicine (BIBM). IEEE (2010)
7. Sankoff, D., Blanchette, M.: Probability models for genome rearrangement and linear invariants for phylogenetic inference. In: Proceedings of the Third Annual International Conference on Computational Molecular Biology. ACM (1999)

8. Sophia, Y., Attie, O., Friedberg, R.: Efficient sorting of genomic permutations by translocation, inversion and block interchange. Bioinformatics 21(16), 3340–3346 (2005)
9. Xu, A.W., Sankoff, D.: Decompositions of multiple breakpoint graphs and rapid exact solutions to the median problem. In: Crandall, K.A., Lagergren, J. (eds.) WABI 2008. LNCS (LNBI), vol. 5251, pp. 25–37. Springer, Heidelberg (2008)
10. Hu, F., et al.: Maximum likelihood phylogenetic reconstruction using gene order encodings. In: 2011 IEEE Symposium on Computational Intelligence in Bioinformatics and Computational Biology (CIBCB). IEEE (2011)
11. Lin, Y., Hu, F., Tang, J., Moret, B.: Maximum likelihood phylogenetic reconstruction from high-resolution whole-genome data and a tree of 68 eukaryotes. In: Proc. 18th Pacific Symp. on Biocomputing, PSB 2013, pp. 285–296 (2013)
12. Yang, Z., Sudhir, K., Masatoshi, N.: A new method of inference of ancestral nucleotide and amino acid sequences. Genetics 141(4), 1641–1650 (1995)
13. Felsenstein, J.: Evolutionary trees from DNA sequences: a maximum likelihood approach. Journal of molecular evolution 17(6), 368–376 (1981)
14. Stamatakis, A.: RAxML-VI-HPC: maximum likelihood-based phylogenetic analyses with thousands of taxa and mixed models. Bioinformatics 22(21), 2688–2690 (2006)
15. Stamatakis, A.: New standard RAxML version with marginal ancestral state computationas, https://github.com/stamatak/standard-RAxML
16. Tang, J., Wang, L.: Improving genome rearrangement phylogeny using sequence-style parsimony. In: Fifth IEEE Symposium on Bioinformatics and Bioengineering, BIBE 2005, pp. 137–144. IEEE (2005)
17. Jahn, K., Zheng, C., Kováč, J., Sankoff, D.: A consolidation algorithm for genomes fractionated after higher order polyploidization. BMC Bioinformatics 13(suppl. 19), S8 (2012)
18. Ma, J., Zhang, L., Suh, B., Raney, B., Burhans, R., Kent, W., Blanchette, M., Haussler, D., Miller, W.: Reconstructing contiguous regions of an ancestral genome. Genome Research 16(12), 1557–1565 (2006)
19. Lin, Y., Rajan, V., Moret, B.: Bootstrapping phylogenies inferred from rearrangement data. BMC Algorithms for Molecular Biology 7, 21 (2012)
20. Lin, Y., Rajan, V., Moret, B.: Fast and accurate phylogenetic reconstruction from high-resolution whole-genome data and a novel robustness estimator. J. Computational Biology 18(9), 1131–1139 (2011) (special issue on RECOMB-CG 2010)
21. Zhang, Y., Hu, F., Tang, J.: A mixture framework for inferring ancestral gene orders. BMC Genomics 13(suppl. 1), S7 (2012)
22. Blanchette, M., Bourque, G., Sankoff, D.: Breakpoint phylogenies. Genome Informatics 8, 25–34 (1997)

Computational Methods for the Parallel 3D Simulation of Biochemical Kinetics at the Microscopic Scale

Laurent Crépin[1,2], Fabrice Harrouet[1], Sébastien Kerdélo[1,2],
Jacques Tisseau[1], and Pascal Redou[1]

[1] Lab-STICC, UMR 6285 CNRS, UEB/ENIB/CERV, France
[2] Diagnostica Stago, Gennevilliers, France
crepin@enib.fr

Abstract. This work takes place in the context of biochemical kinetics simulation for the understanding of complex biological systems such as hæmostasis. The classical approach, based on the numerical solving of differential systems, cannot satisfactorily handle local geometrical constraints, such as membrane binding events. To address this problem, we propose a particle-based system in which each molecular species is represented by a three-dimensional entity which diffuses and may undergo reactions. Such a system can be computationaly intensive, since a small time step and a very large number of entities are required to get significant results. Therefore, we propose a model that is suitable for parallel computing and that can especially take advantage of recent multicore and multiprocessor architectures. We present our particle-based system, detail the behaviour of our entities, and describe our parallel computing algorithms. Comparisons between simulations and theoretical results are exposed, as well as a performance evaluation of our algorithms.

Keywords: Particle-based system, parallel computing, biochemical kinetics, microscopic scale simulation.

1 Introduction

A common way to improve the understanding of complex biological systems is to run experiments on models in order to test hypotheses. The results are then extrapolated from the models to the real systems they represent. One usually distinguishes *in vivo* experimentation, taking place in a living organism, from *in vitro* experimentation, run in test tubes. One challenge for the upcoming years would be to save the various expenses of some of these experiments by running reliable computer simulations of predictive numerical models.

Complex biological systems necessarily involve the occurence of biochemical events such as chemical reactions; numerical simulations of these complex systems thus imply the modelisation of their kinetics. Biochemical kinetics can be simulated in many ways, but this is originally done by implementing empirical

Z. Cai et al. (Eds.): ISBRA 2013, LNBI 7875, pp. 28–39, 2013.
© Springer-Verlag Berlin Heidelberg 2013

laws, e.g. *mass-action* or *Henri-Michaelis-Menten* [1], into a set of ordinary differential equations that is solved using a numerical method [2]. The presence of biological material, such as membranes, can induce heterogeneity in the medium where the biochemical events occur. Partial differential equations are thereby required to take into account this spatial heterogeneity. The differential equation approach relies on the hypothesis that the medium is large enough to provide significant results. When it comes to small volumes, e.g. inside a biological cell, this assumption fails, which makes this approach only suitable for macroscopic scale simulations. Gillespie [3, 4] thus introduced *stochastic simulation algorithms* to mimic accurately the behaviour of the solution of the *chemical master equation* that describes biochemical kinetics at the mesoscopic scale. This approach takes into account the discrete and stochastic aspects of biochemical reactions [5] and makes the numerical simulation of biochemical kinetics possible, whatever the volume. The case of heterogeneous media has also been treated [6].

It is now widely accepted that, in addition to the discrete and stochastic aspects, spatial constraints must also be taken into account in order to simulate accurately biochemical kinetics [7]. As an example, the blood coagulation cascade [8] involves enzymatic reactions that take place both in solution and on a membrane surface. Moreover, as the clotting process goes on, a fibrin mesh is formed and the medium becomes insoluble: classical enzyme kinetics no longer applies. Simulations of such complex phenomena are typically achieved using *particle-based methods* that intend to describe biochemical kinetics at the microscopic scale [9]. These methods track individual molecules, named particles or entities, as they diffuse in three dimensions, collide and react. The main drawbacks with particle-based methods stand in the small time steps and the very large number of particles required to obtain significant results, which makes the simulations computationaly intensive. Although other particle-based simulations of biochemical kinetics at the microscopic scale already exist [10–13], none of them, to our knowledge, focused on a noteworthy gain of computational performances.

This work addresses this issue using a particle-based method for the simulation of biochemical kinetics at the microscopic scale that is suitable for parallel computing and that can especially take advantage of recent multicore and multiprocessor architectures. We present our model in two steps. First, we introduce our particle-based system and detail the behaviour of our entities, as well as their specific scheduling scheme. Then, we describe the cache-aware simulation engine and the parallel computing algorithms that we developed for performance purpose. This description is followed by a validation section in which we illustrate our approach on both a reversible and an enzymatic reactions and compare the results of our method with those obtained with the classical approach. As our work intends to improve the performances of the simulations, we then show the computational gain that our algorithms offer. Finally, we discuss our choices and give some perspectives for our work.

2 Model

2.1 Particle-Based System

Our method uses a particle-based approach, illustrated on Fig. 1, to simulate the kinetics of biochemical systems with spatial and stochastic details [14]. Two kinds of biochemical species can be represented: the `Species3D` entities which diffuse in solution (in a `Volume`) and the `Species2D` entities which diffuse along a physiological membrane (on a `Surface`). Each molecular species is represented by its geometrical shape, i.e. an ellipsoid for the 3D species in solution and a disc for the 2D species bound to a membrane.

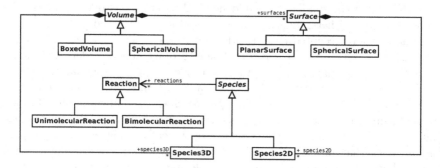

Fig. 1. UML class diagram of our model – The `Species` class represents an entity diffusing in a `Volume` or on a `Surface`, and which may undergo a `Reaction`.

Each entity diffuses in the reactional volume according to Brownian motion, with a diffusion coefficient computed from the entity's radii, the volume temperature and viscosity. It can also undergo two main biochemical reactions: unimolecular ones or bimolecular ones. These reactions are responsible for the creation or the destruction of other entities and, as a result, they govern the variations of the chemical concentrations in the system. The whole life cycle of an entity is detailed in Fig. 2 algorithm. The following sections describe each step of this algorithm.

Unimolecular Reactions. Unimolecular reactions are phenomena which can transform a biochemical species (the reagent) into one or more products. A reaction R converting a molecular species C in a couple A and B is represented by the scheme:

$$R : C \xrightarrow{k} A + B \tag{1}$$

where k is the reaction rate characterising the velocity of the phenomenon.

A molecular species can take part in one or many reactions. In order to simulate a system of n reactions, it is necessary to compute the probability of each

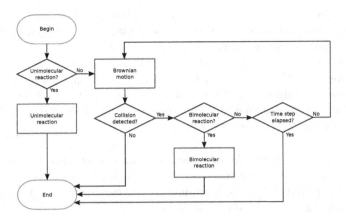

Fig. 2. Life cycle of an entity – This diagram summarises the global behaviour of an entity during one simulation step. First, the entity checks if an unimolecular reaction should happen. If not, the entity diffuses in the environment according to Brownian motion. Then, if a collision happens, the entity tries to react with the collided one and undergoes a bimolecular reaction if necessary. This process is repeated until the time step allocated to the entity is elapsed.

one, i.e. the probability $P(R_i)$ of the event "the reaction R_i occurs on a given time step Δt" and the probability $P(\overline{R})$ of the event "no reaction occurs". These probabilities, also detailed in [15], are given by:

$$
\begin{cases}
P(R_i) = \dfrac{k_i}{\displaystyle\sum_{j=1}^{n} k_j} \times \left(1 - \exp(-\sum_{j=1}^{n} k_j \times \Delta t)\right) \\[2em]
P(\overline{R}) = \exp(-\sum_{j=1}^{n} k_j \times \Delta t) \ .
\end{cases}
\tag{2}
$$

This system determines which reaction will occur during the current time step Δt thanks to uniform random sampling.

Brownian Motion and Collision Detection. Brownian motion characterises the erratic motion of an entity within a fluid. It arises from the collisions between the solvent particles and the entity itself which seems to move randomly under these impacts [16]. The theory associated with the brownian motion of ellipsoidal molecules is studied in [17, 18]. At the microscopic scale, this phenomenon is usually modeled using three-dimensional random walks, i.e. Markov processes defined by a succession of random elementary steps. The future position and orientation of the entity solely depend on its current location, which means the process is memoryless. This succession of elementary steps can be approximated by random Gaussian sampling. As a matter of fact, during a time step Δt, each three-dimensional translation and rotation are computed from six random variables following a Gaussian law with a zero mean and a standard deviation of $\sqrt{2D\Delta t}$, where D is the diffusion coefficient of the entity.

During its diffusion step, an entity may encounter one of its close neighbours. To determine if a collision occured between two molecular species, we use the collision detection algorithms for ellipsoids presented in [19]. Then we use a dichotomy algorithm to manage the collision and avoid the overlap of the entities. Such collisions can lead to bimolecular reactions.

Bimolecular Reactions. As opposed to unimolecular ones, bimolecular reactions can only occur when collisions happen in the system. As a result, two reagents are necessary to create a product. A reaction R between two species A and B, producing C, can be represented as follows:

$$R : A + B \xrightarrow{k} C \tag{3}$$

where k is the reaction rate. To simulate such reactions, the probability $P(R)$ that a reaction occurs upon a collision between the species A and B has to be determined. This probability is maximal, i.e. $P(R) = 1$, when all the collisions between the reagents lead to the creation of a product. The corresponding maximum rate constant k_{max} can easily be computed by using our simulator and by fitting the concentrations with the solution of the mass action law equation. It is worth noting that the values we obtain for different simulations are rather consistent with Von Smoluchowski theory [1] which states that the maximum rate constant is $k_{max} = 4\pi(D_A + D_B)(r_A + r_B)N_A$, where D_A and r_A are respectively the diffusion coefficient and the radius of the species A (likewise for B), and N_A is the Avogadro number. Once k_{max} has been determined, the probability $P(R)$ is given by the ratio $\frac{k}{k_{max}}$ where k is the rate constant that we want to simulate. A reaction R occurs only if the value of a random variable X, following a standard uniform distribution, is less than $P(R)$.

However, if two entities do not react the first time they collide, there is a very high probability that they will during the next simulation steps. This is due to the fact that the entities are still very close to each other after their first encounter. To avoid the bias introduced by this recollision problem, we decide to consider only the first collision in a sequence of encounters.

Scheduling. The interactions between our entities are not predetermined and the overall behaviour of the system is unknown. Because these entities are not just numerical equations which results could be added, as in a synchronous system, they introduce concurrency in the simulation. When two entities collide and react in one place, a third one cannot pretend having reacted with one of them (which could have just disappeared) in another place during the same time step; the state of the system has been irreversibly changed by the preceding reaction and all the following actions have to consider this new state. Therefore we chose to use an asynchronous and chaotic iteration scheme to schedule the

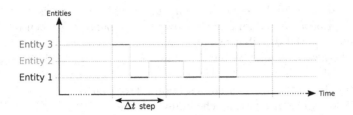

Fig. 3. Asynchronous and chaotic scheduling – This figure shows the execution order of three entities scheduled according to the asynchronous and chaotic iteration scheme. Three simulation steps are represented. The scheduler executes every entity exactly once in a time step, randomly reordering the sequence at each new step.

entities [20]. Although a common time step ensures the temporal consistency of the entities, the asynchronous scheduler executes every entity one after another inside this time step in order to take into account every single event. However, since every entity is affected by the previous ones actions, a fixed scheduling order would have implied an unwanted priority between them. As shown on Fig. 3, the chaotic scheduler gets rid of this artifact with a random reordering at each new step. The convergence and stability of such a scheduling scheme in the context of differential system solving were strictly validated in [21].

2.2   Parallel Asynchronous Scheduler

According to the law of large numbers, a particle-based biochemical kinetics simulation requires many entities (approximately 10^5) to be significant. Moreover, the microscopic scale implies the use of very short time steps (approximately 10 nanoseconds) and therefore, many iterations are required to compute the whole simulation. To speed up such intensive computations, we developed algorithms suitable for parallel computing on multicore and multiprocessor architectures.

Background: Cache-Aware Simulation Engine. As detailed in [22], we previously designed a simulation engine that can harness the full potential of all the Central Processing Units (CPUs) (would they be processors, physical cores, or logical cores) in a parallel computer.

To prevent cache-memory trashing, the whole set of entities to schedule is subdivided in as many subsets as there are CPUs. Since our simulations imply a common repetitive time step, some work-stealing [23] is used at the end of each step to dynamically balance the CPU workloads, keeping all of them busy until the end. This stealing relies on the knowledge of the cache-memory hierarchy to minimise trashing. The assignment of every entity to the CPU in charge of its execution is arbitrary at first but is dynamically adjusted: each executed entity keeps track of its neighbours' current CPU and moves to the most represented one for the next simulation step. It is then very likely for an entity and its neighbours to be run on the same CPU, and thus to find their respective data already up to date in the same cache-memory. This solution is more generic than a spatial partitioning and offers a better load balancing when it comes to heterogeneous spatial distributions or gregarious behaviours.

This cache-aware simulation engine shows a very good scalability related to the number of CPUs used [22]. However, it was formerly dedicated to synchronous simulations and, as stated in section 2.1, our particle-based biochemical model relies on an asynchronous scheduling scheme.

From Synchronous to Asynchronous Parallel Scheduling. The first issue we have to deal with consists in keeping the consistency of any entity when accessed by many CPUs simultaneously. Each entity has its own reader-writer-lock [23] which is locked for writing (one at a time) when its behaviour is executed. When an entity collects informations from its neighbours, it locks them for reading (many reader-lock operations are allowed at the same time on a given lock). In case it needs to modify a neighbour, the reader-lock is promoted to writer-lock (one at a time). Even though the consistency is now guaranteed for concurrent accesses, the main drawback with overlapped lock operations stands in deadlock situations: several entities having to lock themselves and one another, thus waiting endlessly for these locks to become free.

To prevent this new issue from happening, we turn the locking operations into attempts that may immediately fail if the lock is not free. This requires that the behaviour of the entities has to be written so that all the decision making takes place in local variables; the entities are finally modified only if all the chain of locking operations succeeds. When a locking failure occurs, the currently executed behaviour is simply given up as if it has never been started; it will be rescheduled later in the same time step. Nevertheless, when approaching the end of the step, a live-lock situation may occur: the rescheduled entities which interact with one another on different CPUs are probably the only remaining ones and will forever miss their locking attempts.

To get around this situation, as soon as an entity fails twice to be scheduled in the same time step, we postpone its execution to the next step. As the scheduling follows a random order, it is very unlikely for these concurrent entities, amongst many other ones now, to be scheduled simultaneously one more time. Of course, a postponed entity has to be scheduled twice in this next step to ensure a long term temporal consistency between entities.

Although these locking attempts, rescheduling and postponing decisions may seem to raise the computation workload, they actually do not so much. Since our scheduler tends to assign the entities to the CPUs according to their neighbourhood, most of the time a whole set of interacting entities is scheduled by only one CPU. Consequently, the locking attempts are serialised and mostly succeed, thus it is scarcely ever necessary to reschedule or even postpone some entities.

3   Results

This section presents some results which validate our algorithms on simple simulations. We achieve this validation by comparing the simulation results obtained with our approach with the ones determined by solving the differential equations of the mass action law. Then, we study how simulation speed scales related to the number of CPUs involved.

3.1 Validation

We chose to illustrate our approach on a reversible reaction and an enzymatic reaction as they are the most frequent in biochemical kinetics. These reactions are modeled by one bimolecular reaction and one or two unimolecular reactions. As an example for the simulation of a reversible reaction, our approach is illustrated on the interaction of blood coagulation factor Xa with its tight-binding inhibitor, tissue factor pathway inhibitor (TFPI). The whole biochemical description of this interaction can especially be found in [24]. Then, we illustrate our method on the activation of prothrombin (II) into thrombin (IIa) by an enzymatic complex (PT_{Ptex}) that can be found in the venom of the Australian snake *Pseudonaja textilis*. This interaction is fully detailed in [25]. Table 1 gathers all the data we used to set up our validation simulations.

Table 1. Validation parameters – This table presents the validation conditions of our model. It gathers every data necessary to reproduce our results. It should be noted that the shape of factors Xa and II are assumed to be prolate ellipsoids whereas the other species have a spherical shape.

Reversible reaction			Enzymatic reaction			
k_{on} $\text{Xa} + \text{TFPI} \leftrightarrows \text{Xa} \cdot \text{TFPI}$ k_{off}			$K_{\text{M}} \qquad\qquad k_{\text{cat}}$ $\text{PT}_{\text{Ptex}} + \text{II} \leftrightarrows \text{PT}_{\text{Ptex}} \cdot \text{II} \rightarrow \text{PT}_{\text{Ptex}} + \text{IIa}$			
$k_{\text{on}} = 0.9 \times 10^6\,\text{M}^{-1} \cdot \text{s}^{-1}$ $k_{\text{off}} = 3.6 \times 10^{-4}\,\text{s}^{-1}$			$K_{\text{M}} = 1.83 \times 10^{-6}\,\text{M}$ $k_{\text{cat}} = 5.87\,\text{s}^{-1}$			
Initial concentrations						
$[\text{Xa}]_0 = 170\,\text{nM}$ $[\text{TFPI}]_0 = 2.5\,\text{nM}$			$[\text{PT}_{\text{Ptex}}]_0 = 100\,\text{nM}$ $[\text{II}]_0 = 1.4\,\mu\text{M}$			
Dimensions						
Xa	TFPI	Xa·TFPI	PT_{Ptex}	II	$\text{PT}_{\text{Ptex}} \cdot \text{II}$	IIa
$r_{\text{x}} = 26.0\,\text{Å}$ $r_{\text{y}} = 26.0\,\text{Å}$ $r_{\text{z}} = 51.5\,\text{Å}$	$r = 22.5\,\text{Å}$	$r = 35.9\,\text{Å}$	$r = 39.3\,\text{Å}$	$r_{\text{x}} = 22.5\,\text{Å}$ $r_{\text{y}} = 22.5\,\text{Å}$ $r_{\text{z}} = 60.0\,\text{Å}$	$r = 45.0\,\text{Å}$	$r = 25.0\,\text{Å}$

The results computed from our simulations are illustrated on Fig. 4, and are consistent with the ones coming from reference laws: the mean relative errors at steady state are about 1% for both reversible and enzymatic reactions. These two validations are essential before addressing more complex simulations. One may notice that we do not deal with membrane binding events in these examples. This point will be developed in the perspective section.

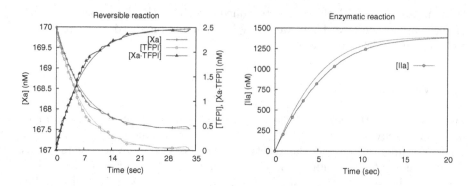

Fig. 4. Reversible and enzymatic reaction validation – These curves illustrate the validation of our model on two kinds of reaction. Smooth curves are computed from mass action law whereas noisy ones are the results of our simulations.

3.2 Performances

We ran many simulations, similar to the preceding ones, on a single computer with two Intel® Xeon® X5680 processors at 3.33 GHz clockspeed and a 12-Mbyte level-3 cache each. Thanks to the 2-way SMT technology, the twelve CPU cores provided here can be seen as twenty four logical CPUs running simultaneously. Figure 5 reports the computational frequency, i.e. the number of simulation steps per second, as well as the number of simulated entities running at a given rate, depending on the number of CPUs used. Both curves seem to scale linearly but with a slight change in the slopes around twelve CPUs. When using only the first dozen of CPUs, the physical cores fully exploit their respective hardware resources and cache hierarchy, but when it comes to the second dozen of CPUs, the 2-way SMT technology is involved and implies the sharing of the same resources for twice as much workload. This explains the slightly lower efficiency.

These results show that, as long as the CPUs come along with some cache memory, the simulation performances scale linearly with the number of CPUs involved. This lets us foresee that the cache-aware design of our simulator would enable even bigger simulations when using over twenty four CPUs.

4 Discussions and Perspectives

4.1 Alternatives for Computing Power

Intensive computations can be handled by several technologies involving many computing units. Some of these parallel solutions are local, taking place in a single computer, such as multicore and multiprocessor computers or graphical processing units (GPUs), whereas others are distributed, implying communication through a network (dedicated or not), such as computer clusters, grid computing or cloud computing.

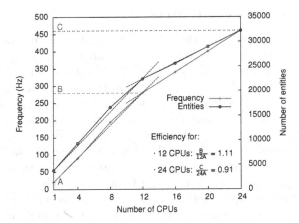

Fig. 5. Performance scaling – One curve illustrates the scaling of the computational frequency, depending on the number of CPUs used, for a simulation with 22438 entities. The other one shows the number of entities that our simulator can handle at a given computational frequency (280 Hz).

While distributed solutions are flexible and may provide a huge amount of computing units, the communication delays make them only suitable for computational problems which can be naturally subdivided into many independent subproblems requiring very little or even no synchronisation. Our simulations are made of multiple recurrent interactions which cannot be split in fixed and independent subsets beforehand; they require intensive synchronisations between the computing units. Because the hardware of general purpose parallel computers ensures very efficiently the cache-memory coherency, we decided to focus on this technology in the design of our simulation engine. Even in this favourable local context, our experiments showed that the slightest clumsiness in synchronisation and cache usage impacted significantly the overall computing performance; this definitely discouraged us from investigating distributed solutions.

GPUs are local to a computer and provide much more computing units than CPUs do. Our experiments highlighted that, not only the data transfer delay with the GPUs was far from being negligible compared to the computation duration, but also the programming model associated with this technology [26] was well-suited for a synchronous simulation scheme in which thousands of entities can compute simultaneously an identical behaviour without taking care of any unexpected data change in their neighbourhood. Unfortunately, as stated in section 2.2, the asynchronous simulation scheme associated with our model implies a careful synchronisation strategy. Although some synchronisation primitives are available on GPUs, their usage totally contradicts the programming model of these devices and tends to ruin dramatically the raw computing performance they are capable of. Consequently, we prefer saving this technology for synchronous simulations of physical phenomena which could however interact with our particle-based asynchronous simulations.

4.2 Perspectives and Future Work

Due to the fact that we chose the microscopic scale, our diffusion algorithm requires a very short time step in order not to miss collisions (approximately 10 nanoseconds). This constraint somewhat limitates the performances of the simulations. As a matter of fact, two main parameters govern these performances: the number of entities executed during each simulation step and the length of this time step. Parallel computing enables us to increase the first parameter by using more computational units. However, it has no impact on the second one. We are currently working on a statistical method to address this problem. For the time being, we recommend the use of our simulator for short time applications, i.e. biochemical processes lasting no longer than ten seconds.

Beside this time step optimisation, a part of our model still needs to be thoroughly validated. As stated previously in section 2.1, the molecular species in solution may bind with membranes. Altough such bindings (and unbindings) are fully implemented, we still have to ensure the consistency of our results with experimental ones. In the present situation, our simulator provides nothing more than the classical approach. Nevertheless, this method is the only one capable of handling local geometrical constraints in biochemical systems. We will therefore focus our future work on modeling and validating these membrane binding events.

References

1. Purich, D., Allison, R.: Handbook of biochemical kinetics. Academic Press (2000)
2. Alves, R., Antunes, F., Salvador, A.: Tools for kinetic modeling of biochemical networks. Nature Biotechnology (6), 667–672 (2006)
3. Gillespie, D.T.: A general method for numerically simulating the stochastic time evolution of coupled chemical reactions. Journal of Computational Physics 22(4), 403–434 (1976)
4. Gillespie, D.T.: Exact stochastic simulation of coupled chemical reactions. The Journal of Physical Chemistry 81(25), 2340–2361 (1977)
5. Van Kampen, N.: Stochastic processes in physics and chemistry. North Holland (March 2007)
6. Stundzia, A.B., Lumsden, C.J.: Stochastic simulation of coupled reaction-diffusion processes. Journal of Computational Physics 127(1), 196–207 (1996)
7. Resat, H., Costa, M.N., Shankaran, H.: Spatial aspects in biological system simulations. In: Johnson, M.L., Brand, L. (eds.) Computer Methods, Part C. Methods in Enzymology, vol. 487, pp. 485–511. Academic Press (2011)
8. Marder, V.J., Aird, W.C., Bennett, J.S., Schulman, S., White, G.C.: Hemostasis and Thrombosis: Basic Principles and Clinical Practice. Lippincott Williams & Wilkins (November 2012)
9. Tolle, D.P., Le Novère, N.: Particle-based stochastic simulation in systems biology. Current Bioinformatics 1(3), 315–320 (2006)
10. Andrews, S.S., Addy, N.J., Brent, R., Arkin, A.P.: Detailed simulations of cell biology with smoldyn 2.1. PLoS Computational Biology 6(3) (March 2010)
11. Stiles, J.R., Bartol, T.M.: 4. In: Monte Carlo Methods for Simulating Realistic Synaptic Microphysiology Using MCell. CRC Press (2001)

12. Plimpton, S.J.: Chemcell: a particle-based model of protein chemistry and diffusion in microbial cells. Technical report, Sandia National Laboratories (December 2003)
13. Van Zon, J., Ten Wolde, P.: Green's-function reaction dynamics: a particle-based approach for simulating biochemical networks in time and space. The Journal of Chemical. Physics 123, 234910, 1–16 (2005)
14. Kerdélo, S.: Méthodes informatiques pour l'expérimentation in virtuo de la cinétique biochimique - Application à la coagulation du sang. PhD thesis, Université de Rennes 1 (January 2006)
15. Andrews, S.S., Bray, D.: Stochastic simulation of chemical reactions with spatial resolution and single molecule detail. Physical Biology 1(3), 137 (2004)
16. Berg, H.C.: Random Walks in Biology. Princeton University Press (September 1993)
17. Perrin, F.: Mouvement brownien d'un ellipsoïde (i): Dispersion diélectrique pour des molécules ellipsoidales. Journal de Physique et le Radium 5(10), 497–511 (1934)
18. Perrin, F.: Mouvement brownien d'un ellipsoïde (ii): Rotation libre et dépolarisation des fluorescences. translation et diffusion des molécules ellipsoïdales. Journal de Physique et le Radium 7(1), 1–11 (1936)
19. Wang, W., Wang, J., Kim, M.S.: An algebraic condition for the separation of two ellipsoids. Computer Aided Geometric Design 18(6), 531–539 (2001)
20. Harrouet, F.: oRis: s'immerger par le langage pour le prototypage d'univers virtuels à base d'entités autonomes. PhD thesis, Université de Bretagne Occidentale (December 2000)
21. Redou, P., Gaubert, L., Desmeulles, G., Béal, P.A., Le Gal, C., Rodin, V.: Absolute stability of chaotic asynchronous multi-interactions schemes for solving ode. CMES: Computer Modeling in Engineering & Sciences (December 2010)
22. Harrouet, F.: Designing a multicore and multiprocessor individual-based simulation engine. IEEE Micro. 32(1), 54–65 (2012)
23. Padua, D.A.: Encyclopedia of Parallel Computing. Springer (2011)
24. Baugh, R.J., Broze, G.J., Krishnaswamy, S.: Regulation of extrinsic pathway factor Xa formation by tissue factor pathway inhibitor. Journal of Biological Chemistry 273(8), 4378–4386 (1998)
25. Bos, M.H.A., Boltz, M., St. Pierre, L., Masci, P.P., de Jersey, J., Lavin, M.F., Camire, R.M.: Venom factor V from the common brown snake escapes hemostatic regulation through procoagulant adaptations. Blood 114(3), 686–692 (2009)
26. NVIDIA: NVIDIA CUDA C best practice. Technical report, NVIDIA (October 2012)

A Tool for Non-binary Tree Reconciliation

Yu Zheng and Louxin Zhang

Department of Mathematics, National University of Singapore, Singapore 119076

Abstract. Tree reconciliation has been widely used to study the important roles of gene duplication and loss, and to infer a species tree from gene trees in evolutionary biology. Motivated by the fact that both reference species trees and real gene trees are often non-binary, we develop a novel computer program to reconcile two non-binary trees. Such a program extends the usefulness of tree reconciliation greatly, as it can be used for gene duplication inference and species tree inference.

1   Introduction

Genes are usually gained through duplication and are lost via deletion or pseudogenization throughout evolution. Because of gene duplication and loss, the evolutionary history of a gene family – the gene tree – is often not concordant with the evolutionary history of the species – the species tree – in which the genes are present. Hence, a gene tree and the corresponding species tree are often compared using a procedure known as tree reconciliation to study the roles of gene duplication and loss, and to infer the species tree from gene trees in evolutionary biology. Gene tree and species tree reconciliation are parsimonious approaches that formalize the following intuition: If the offspring of a gene tree node is distributed in the same set of species as that of one of its direct descendants, then the node corresponds to a gene duplication event.

Tree reconciliation was originally proposed for binary trees [10,18]. Real gene trees, however, often contain non-binary nodes or weakly supported branches, as there is not enough signal in the gene sequence data to date gene divergence events. On top of ambiguity in a gene tree, there are also uncertainties in the species tree. The reference species trees in the NCBI taxonomy database are frequently non-binary due to the unresolved order of species divergence. Therefore, it is important and much more challenging to study the reconciliation of non-binary trees [3,9,21]. The reconciliation problem is solvable in linear time if the input gene and species trees are all binary [6,22], solvable in polynomial time for binary species trees [8,4], and NP-hard for non-binary species trees as shown here.

We shall study the general reconciliation problem in the binary refinement model [9]. The binary refinement model was first used for non-binary tree reconciliation in [3]. However, their tool, Softparsmap, has limitations. For example, Softparsmap may overestimate the number of gene loss events for a non-binary gene tree node. In the present paper, we present a novel approach for reconciling non-binary trees. Our approach has been implemented into a computer tool call TxT, for which the online server is on http://phylotoo.appspot.com.

Z. Cai et al. (Eds.): ISBRA 2013, LNBI 7875, pp. 40–51, 2013.
© Springer-Verlag Berlin Heidelberg 2013

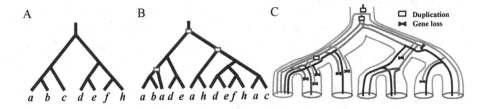

Fig. 1. (A) A binary species tree S over species a, b, c, d, e, f, h. (B) A binary gene tree G of a gene family, which contains four genes present in a, two genes in d, e, f, and one gene in b, c, f. (C) A duplication history of three duplication and eight loss events is inferred from λ_G^S.

2   Basic Concepts and Notation

2.1   Gene Trees and Species Trees

Gene or species trees are rooted graphs in which there is exactly one distinguished 'root' node such that there is a unique path from it to any other node. A species tree represents the evolutionary history of a set of modern species. Its leaves are nodes with degree one and labeled uniquely by modern species. A leaf or the branch incident to it represents the labeling species. Non-leaf nodes are internal nodes. A branch between two internal nodes represents an ancestral species. Here, a species tree also contains a 'branch' entering the root to represent the common ancestor of all the extant species (Fig 1A).

A gene tree represents the evolutionary relationships of a gene family, in which a leaf represents a family member. Here, we assume that the multiple gene family members that are present in a species are products of gene duplication. Gene tree and species tree reconciliations are used to infer the duplication history of the corresponding gene family. In the study of tree reconciliation, a gene tree leaf is labeled with the species in which the corresponding gene is present. Note that the leaves of the gene tree of a multiple-gene family are not uniquely labeled.

Given a gene or species tree T, $r(T)$ denotes its root; $LV(T)$, $\overset{\circ}{V}(T)$, $V(T)$, and $E(T)$ denote the sets of its leaves, internal nodes, all the nodes and all the branches, respectively. For $u, v \in V(T)$, v is said to be an *ancestor* of u and u is an *descendant* of v if v is in the unique path from $r(T)$ to u; v is the parent of u and u is a child of v if v is an ancestor of u and $(v, u) \in E(T)$. We use $p(u)$ to denote the parent of u if u is a non-root node. The induced subtree of T of all the descendants of u is written $T(u)$. The subset of the labels of the leaves in $T(u)$ is called a *cluster* of T and denoted by $L(u)$.

Each $u \in \overset{\circ}{V}(T)$ has two or more children. It is *binary* if it has two children. A tree is *binary* if all its internal nodes are binary and *non-binary* otherwise.

2.2   Tree Reconciliation

Consider the gene tree G of a gene family and the corresponding species tree S such that $L(G) \subseteq L(S)$. The *least common ancestor (lca) reconciliation* λ_G^S is a map from $V(G)$ to $V(S)$ defined as:

$$\lambda_G^S(u) = \begin{cases} v \text{ such that } L(v) = L(u) & \text{if } u \in \mathrm{LV}(G), \\ \mathrm{lca}\{\lambda_G^S(x) : x \in \mathrm{LV}(G(u))\} & \text{if } u \in \overset{\circ}{V}(G), \end{cases} \tag{1}$$

where $\mathrm{lca}\{\cdots\}$ denotes the most recent common ancestor of the species in the set. Clearly, for $g \in V(G)$ with k children g_i, $\lambda_G^S(g) = \mathrm{lca}\{\lambda_G^S(g_i) : i \leq k\}$. We shall write λ when no confusion is likely to arise after G and S are dropped.

Assume that both G and S are binary. A node $u \in \overset{\circ}{V}(G)$ is a duplication node if $\lambda_G^S(u) = \lambda_G^S(u')$ for some child u' of u. If we assume that the duplication event corresponding to a duplication node u occurs in the ancestral species corresponding to $(p(\lambda_G^S(u)), \lambda_G^S(u))$, we obtain a hypothetical duplication history of the gene family [10] (Fig 1). The number of inferred duplication nodes is defined as the duplication cost of λ_G^S, denoted by $D(G, S)$ [18]. Additionally, gene losses usually have to be assumed in the inferred duplication history. For instance, for $u \in V(G)$ with a child u', if $\lambda_G^S(u) \neq \lambda_G^S(u')$, we have to assume the corresponding gene has been lost in each branch off the lineage path from $\lambda(u)$ to $\lambda(u')$. Overall, we have to assume the following number of gene loss events:

$$L(G, S) = \sum_{u \in V(G)} [l(u, u') + l(u, u'') + 2(I_u - 2)], \tag{2}$$

where u' and u'' are the children of u, I_u is 1 if u is a duplication node and 0 otherwise, and $l(u, x) = |\{v \in V(S) : L(\lambda(x)) \subseteq v \subseteq L(\lambda(u))\}|$ for $x = u', u''$. $L(G, S)$ is called the gene loss cost of λ_G^S. In fact, any duplication history of the gene family contains at least $D(G, S)$ duplication and $L(G, S)$ loss events [5,11]. Thus λ_G^S induces a parsimonious duplication history of the corresponding gene family.

We define the affine duplication cost as the weighted sum of the duplication and gene loss cost, given the weights of duplication and loss events. Affine costs have been used in recent studies of tree reconciliation [7,8].

3   Non-binary Tree Reconciliation

For non-binary gene and species trees, the lca reconciliation between them does not necessarily induce a duplication history that has the fewest duplication and loss costs [3]. Different models have been proposed to study how to reconcile non-binary trees for gene duplication inference [4,8,9]. In this work, we study non-binary tree reconciliation in the binary refinement model [9].

A (binary) tree T is a (binary) refinement of T' if every cluster of T' is also a cluster of T or, equivalently, if T' can be obtained from T via a series of branch contractions. Formally, we shall investigate the following problem:

General Reconciliation

INPUT: A species tree S, a set of gene trees G_i ($1 \leq i \leq k$), and a reconciliation cost function $c(\ )$.

OUTPUT: The binary refinements $\hat{S}$ and $\hat{G}_i$ of S and G_i ($1 \leq i \leq k$), respectively, such that $\sum_{1 \leq i \leq k} c(\hat{G}_i, \hat{S})$ is minimized.

The unique binary refinement of a fully binary tree is itself. Hence, the traditional reconciliation problem is a special case of the above problem. In addition, the species tree inference problem is also a special case of it, as every binary tree is a binary refinement of the corresponding star tree (in which all non-root nodes are leaves). The NP-hardness of the species tree inference proved in [15] (see also [1]) leads to the following complexity result, which is stronger than the NP-hardness result given in [3].

Theorem 1. *The general reconciliation problem under the duplication cost is NP-hard even for one binary gene tree when the input species tree is non-binary.*

The full proof of this theorem is omitted due to space limits and can be found in the full version of this work [23].

4   A Heuristic Algorithm for Binary Refinement of A Species Tree

Consider a gene tree G and the corresponding species tree S such that $L(G) \subseteq L(S)$. To find the best binary refinement of S, we resolves its non-binary nodes one by one. This is because the gene duplication and loss events associated with each gene tree node are counted independently.

Let $s \in V(S)$ have the children $s_1, s_2, \cdots, s_{n(s)}$, where $n(s) \geq 3$. We define $\lambda^-(s) = \{g \in V(G) : \lambda(g) = s\}$. We have the following facts:

- For each $g \in \lambda^-(s)$, there are at least two children s_i and s_j of s such that
$$L(g) \cap L(s_i) \neq \phi, \ L(g) \cap L(s_j) \neq \phi.$$
- For each $g \in \lambda^-(s)$ and a child g' of g, $\lambda(g') \in S(s_j)$ for some $j \leq n(s)$ if $g' \notin \lambda^-(s)$.

To refine s, we need to replace the star tree consisting of s and its children with a rooted binary tree T_s with root s and $n(s)$ leaves each labeled uniquely by some s_i, $1 \leq i \leq n(s)$. It is well known that T_s is equivalent to a partial partition subset $\mathcal{P}(T_s) = \{[L(u_1), L(u_2)] : u_1$ and u_2 are siblings in $T_s\}$ over $\{s_1, s_2, \cdots, s_{n(s)}\}$. The partition corresponding to the children of $r(T_s)$ is called the **first partition** of the tree. Hence, we can refine s by recursively solving the **minimum duplication bipartition** (MDB) problem [17]:

INPUT: A gene tree G.

OUTPUT: A bipartition $[P_1, P_2]$ of $L(G)$ that minimizes the number of internal nodes g such that we have:

$$\exists g' \in C(v) \text{ such that } L(g') \cap P_1 \neq \phi \text{ and } L(g') \cap P_2 \neq \phi, \tag{3}$$

where $C(g)$ is the set of the children of g.

We take this approach for two purposes. First, it may reduce the overall duplication cost. Second, pushing duplication down in the species tree may reduce the gene loss cost even if the resulting reconciliation does not have the smallest duplication cost.

Unfortunately, it is an open problem whether the MDB problem is NP-hard or polynomial-time solvable. In the rest of this section, we present a novel heuristic method for it. Consider a gene tree G. Let $[P_1, P_2]$ be a bipartition of $L(G)$. For an internal node $g \in \overset{o}{V}(G)$ with children g_1, g_2, g does not satisfy Eqn. (3) if and only if, for every $i = 1, 2$, we have:

$$L(g_i) \subseteq P_1 \text{ or } L(g_i) \subseteq P_2. \tag{4}$$

In the rest of this discussion, for clarity, we call $L(g_1)|L(g_2)$ a split rather than a partial partition for any gene tree node g. Motivated by this fact, we require that the partition $[P_1, P_2]$ does not cut a gene tree split $L(g_1)|L(g_2)|\cdots|L(g_k)$ if and only if for every i,

$$L(g_i) \cap P_1 = \phi, \text{ or } L(g_i) \cap P_2 = \phi. \tag{5}$$

Our proposed algorithm for the MDB problem is called the First Partition (FP) algorithm and summarized below. It attempts to maximize the split $L(g_1)|L(g_2)|\cdots|L(g_k)$ that satisfies the condition given in Eqn. (5) instead of Eqn. (3).

The FP algorithm is illustrated by an example in Fig. 2, where the computation flow of the FirstExtension($\{c\}$, ϕ) is outlined. In the example, we try to find the first partition for a gene tree with the leaf labels a, b, c, d, e, f. The gene tree splits are used in Step 1 of FirstExtension( ) and SplitExtension( ), but are not listed explicitly. After the 'partial' partition $[\{c\}, \{f\}]$ is obtained, the subprocedure SplitExtension( ) is called to extend $[\{c\}, \{f\}]$ into a partition $[\{c, e, b, d\}, \{f, a\}]$. Since the computation of FirstExtension() is heuristic, the partition $[\{c, e, b, d\}, \{f, a\}]$ obtained from $[\{c\}, \{f\}]$ might not be the optimal solution. Hence, FirstExtension() is called on $[\{a, f\}, \phi]$ to obtain good partitions in the case that $[\{c\}, \{f\}]$ does not lead to the optimal first partition. As such, FirstExtension() is recursively called during computation. Overall, the subprocedure FirstExtension() is recursively called five times, outputting the following partial partitions (in red):

$$[\{c\}, \{f\}], [\{c, f\}, \{b\}], [\{c, f, b\}, \{d\}], [\{c, f, b, d\}, \{a\}], [\{c, f, b, d, a\}, \{e\}].$$

SplitExtension() is called on these partial partitions to produce the five partitions listed at the bottom (in green). Finally, the algorithm selects the best partition over all the partitions obtained.

We generated 8,000 random datasets divided into eight groups to evaluate the FP algorithm. For each dataset, we checked whether it outputted a partition with the maximum number of splits satisfying Eqn. (5) or not. Here, the maximum number of splits not cut by a partition was obtained by an exhaustive search for evaluation purposes. We compared the FP algorithm with the algorithm reported

---

First Partition Algorithm

Input: A gene tree G.

$\mathcal{S} = \phi$; /* This is used to keep bipartitions */
For each i
    FirstExtension($[\{i\}, \phi], \mathcal{S}$);
Output the best partition in $\mathcal{S}$;

FirstExtension($[P, \phi], \mathcal{S}$) {
    1. For each $i \not\in P$
        Compute $n(i)$, the # of the gene tree splits not cut by $[P, \{i\}]$;
    2. Select j such that $n(j) = \max_i n(i)$;
    3. If $P \cup \{j\} \neq L(G)$ **do** {
        SplitExtension($[P, \{j\}], \mathcal{S}$); FirstExtension($[\{j\} \cup P, \phi], \mathcal{S}$);
    } **else**
        Add $[P, \{j\}]$ into $\mathcal{S}$;
} /* End of FirstExtension */
SplitExtension($[P_1, P_2], \mathcal{S}$) {
    1. For each $i \not\in P_1 \cup P - 2$
        Compute $n_1(i)$, the # of the gene tree splits not cut by $[P_1, P_2 \cup \{i\}]$;
        Compute $n_2(i)$, the # of the gene tree splits not cut by $[P_1 \cup \{i\}, P_2]$;
    2. Select j such that $\max\{n_1(j), n_2(j)\} = \max_i\{n_1(i), n_2(i)\}$;
    3. If $(P_1 \cup P_2 \cup \{j\} \neq L(G))$ **do** {
        SplitExtension($[\{j\} \cup P_1, P_2], \mathcal{S}$) if $n_1(j) \geq n_2(j)$;
        SplitExtension($[P_1, P_2 \cup \{j\}], \mathcal{S}$) if $n_2(j) > n_1(j)$;
    } **else** {
        Add $[\{j\} \cup P_1, P_2]$ into $\mathcal{S}$ if $n_1(j) \geq n_2(j)$;
        Add $[P_1, P_2 \cup \{j\}]$ into $\mathcal{S}$ if $n_2(j) > n_1(j)$;
    }
} /* End of SplitExtension */

---

in [17]. We call the latter the HC algorithm, as it is derived from an algorithm for the hypergraph min cut problem given in [16]. Our tests showed that the FP algorithm outperformed the HC algorithm for most datasets (Table 1).

The FP algorithm takes $O(k^3)$ set operations in the worst case for refining a non-binary species tree node with k children. Putting all the refinements of non-binary nodes together, we obtain a good binary refinement $\hat{S}$ of the species tree. It takes less than a minute to resolve a non-binary species tree over 100 species, as the degrees of non-binary nodes are usually small in a species tree.

5   Tool Implementation

In a separate work [23], we presented a linear-time algorithm for reconciling an arbitrary gene tree and a binary species tree. On an input gene tree G and the corresponding binary species tree S, the algorithm finds a binary refinement G' of G such that $D(G', S) = \min_{R \in \mathcal{BF}} D(R, S)$ and $L(G', S) = \min_{R \in \mathcal{BF}:D(R,S)=D(G',S)} L(R, S)$, where $\mathcal{BF}$ is the set of all the binary refinements of G. Such an optimality criterion is used as a gene lose event is often attributed to an absence in the species when many species have sparse sampling of genes [3].

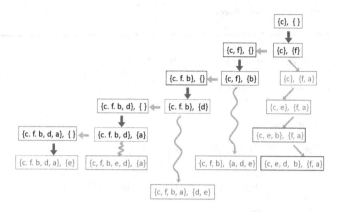

Fig. 2. An illustration of the execution of FirstExtension($\{c\}$, ϕ). Here, the input gene trees are over the genes present in six species a, b, c, d, e, f. FirstExtension( ) is recursively executed five times (indicated with black vertical arrows), generating partial partitions (in red). Light horizontal arrows denote the four recursive calls of FirstExtension( ). SplitExtension() is called on each of these partial partitions to produce the five partitions shown in green, where a diagonal arrow denotes a call and a curved arrow a series of calls to it. The gene tree topology is irrelevant and thus is not given.

By integrating such an algorithm and the FP algorithm, we obtain a novel tool named TxT for reconciling non-binary trees. As illustrated in Fig. 3, on a gene tree G and the corresponding species tree S, TxT computes a binary refinement $\hat{S}$ of S based on the splits of G using the FP algorithm. It then executes the reconciliation algorithm reported in [23] to compute a binary refinement $\hat{G}$ of G based on $\hat{S}$. Finally, it outputs a duplication history for the corresponding gene family based on $\hat{G}$ and $\hat{S}$. TxT, implemented in Python and available on http://phylotoo.appspot.com, has the following features:

Table 1. Comparison of the HC [17] and FP algorithms. In each test case, there are 1,000 datasets each consisting of c_s random splits over a fixed set of c species. An algorithm made an error if it did not output an optimal partition that gave the smallest number of splits satisfying the condition in Eqn. (3). Each entry in the last two rows indicates how many times the corresponding algorithm did not output an optimal partition in the 1,000 tests of the corresponding group.

Test cases	I	II	III	IV	V	VI	VII	VIII
# of species (c)	5	5	10	10	10	15	15	15
# of splits (c_s)	5	10	5	10	20	7	15	20
# of errors made by FP	7	0	0	1	0	0	0	0
# of errors made by HC	15	18	4	2	0	3	1	1

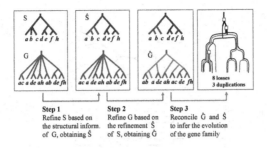

Fig. 3. A schematic view of the proposed method for reconciliation of a gene tree G and the corresponding species tree S.

1. It provides information on whether an inferred duplication is compulsory or weakly supported hypothetical, as in [21].
2. For a gene family, it outputs a set of solutions with the same reconciliation cost.
3. By taking a set of gene trees and a star species tree as input, it outputs a binary tree over the species. Hence, it can be used for species tree inference.
4. It also has a command-line version to allow for automated analysis of a large number of gene trees.

6   Validation Tests

6.1   Inference of Tor Gene Duplications

The target of rapamycin (Tor) is a eukaryotic gene responsible for sensing nutrients. In mammals, the unique mTor governs cellular processes via two distinct complexes: Tor Complex1 (TorC1) and TorC2. In fungal species, however, there are two Tor genes. Hence, how fungal Tor homologs were produced is an interesting question.

Shertz *et al.* investigated the evolution of the Tor family in the fungal kingdom [19]. They reconstructed the Tor tree over 13 fungal species (redrawn in Fig. 4A) and inferred the duplication events that probably were responsible for the two Tor genes present in the fungal species. Their work suggests that a whole-genome duplication event, occurring about one 100 million years (MYS) ago, produced the two Tors in *Saccharomyces cerevisiae*, *Saccharomyces paradoxus*, and several other species, whereas three independent lineage-specific duplications were responsible for the Tor genes in *Schizosaccharomyces pombe*, *Batrachochytrium dendrobatids* and *Pleurotus ostreatus*. When TxT was applied to the Tor tree and the non-binary species tree downloaded from the NCBI taxonomy database (drawn in Fig. 4B), the same duplication events were inferred.

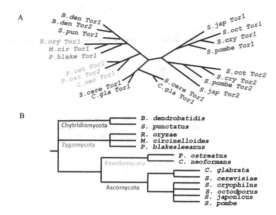

Fig. 4. (A) A Tor gene tree over 13 fungal species, redrawn based on the Tor gene tree reported in [19]. (B) The corresponding reference species tree of the species downloaded from the NCBI taxonomy database.

6.2   Simulation Data

To study gene duplication in *Drosophila* species, Hahn *et al.* obtained 13,376 gene trees over 12 species and the following corresponding species tree [12] (in Newick format):

```
((D.gri,(D.vir,D.moj)),(D.wil,((D.pse,D.per),(D.ana,((D.ere,D.yak),(D.mel,(D.sim,D.sec))))))).
```

We called it the *Drosophila* tree. The selected species have evolved from their most recent common ancestor for the past roughly 63 MYS. In the *Drosophila* tree, we randomly generated 1,000 gene families in the birth–death model by setting both duplication and loss rates to 0.002 per MYS, as estimated in [12]. For each random gene family, we obtained its true gene tree from the recorded duplication events and derived two more approximate gene trees by respectively contracting branches that were shorter than 2 and 3 MYS in the true tree. We inferred duplication events by reconciling each of the three trees and the species tree using TxT.

   As a reconciliation method, TxT tends to overestimate duplication events in 'deep' branches that are close to the root [13]. In our experiment, it correctly inferred the true duplication history for all except for one gene family. When approximate trees were used, however, our program frequently overestimated duplications. Nevertheless, it still inferred duplication events on all branches with high accuracy. The accuracy statistics are omitted here and can be found in the full version of this work [23].

6.3   Accuracy of Species Tree Inference for *Drosophila* Species

We evaluated the accuracy of TxT in binary refinement of species tree using the same simulated datasets. We obtained the following non-binary species tree (in Newick format):

$$((\texttt{D.gri,D.moj,D.vir}),\texttt{D.wil},(\texttt{D.pse,D.per}),(\texttt{D.mel,D.sec,D.sim,D.ere,D.yak,D.ana})). \qquad (6)$$

by contracting the branches that were shorter than 10 MYS in the *Drosophila* tree. We then evaluated the accuracy of TxT by counting how many times we obtain the *Drosophila* tree as the binary refinement of the tree given in (6) when the latter and a set of gene trees were given as input. We observed the following facts:

1. When true gene trees were used, TxT outputted the *Drosophila* tree..
2. When contracted gene trees were used, TxT still performed well. For example, with more than 15 contracted gene trees with three or less non-binary nodes, it outputted the tree in about 97% cases (Table 2).

Table 2. Accuracy of TxT, given as a percentage of the tests for which it outputted the *Drosophila* tree as the binary refinement of the non-binary species tree given in (6). Each test case includes 100 tests in which N inputted gene trees were obtained by contracting random branches in randomly selected gene trees. RE: the average number of branches that were contracted in each gene tree; MD: the maximum degree of a non-binary gene tree node; A: the accuracy of TxT.

Cases	I	II	III	IV	V	VI	VII	VIII	IX	X	XI	XII	XIII	XIV	XV
N	5	10	15	20	30	5	10	15	20	30	5	10	15	20	30
RE	0.97	0.99	1.03	0.99	0.99	2.91	2.95	2.90	2.95	2.99	4.83	5.00	4.94	4.91	5.02
MD	2.73	2.75	2.75	2.72	2.73	3.73	3.78	3.75	3.77	3. 80	4.96	5.14	5.09	5.01	5.08
A(%)	90	100	100	100	100	72	90	97	99	100	27	65	66	76	90

Table 3. Accuracy of inferring the unrooted *Drosophila* tree form unrooted gene trees. Accuracy0: The accuracy of inferring the tree from the original gene trees in [13]; accuracyX: The accuracy of the inference with the non-binary gene trees obtained from the original gene trees via branch contraction with a cut-off value X of 60 or 90.

No. of gene trees	Accuracy0 (%)	Accuracy60 (%)	Accuracy90 (%)
5	21	35	34
10	45	72	54
20	61	87	68
30	76	92	84

The accuracy of TxT in inferring an unrooted species tree was again evaluated by using the *Drosophila* species tree. We considered the original gene trees as well as the two classes of rooted non-binary gene trees obtained by contracting the branches with cut-off support values of 60 and 90. Our results (summarized in Table 3) suggest that contracting the weakly supported edges (with support value below 60) improves the accuracy of the inference of unrooted species trees. It also indicates that contraction of strongly supported branches reduces the inference accuracy. Which cut-off value achieves the highest accuracy for species tree inference is worthy for further study.

7   Conclusion

The general reconciliation problem is an important and challenging problem in phylogenetic analysis. Only special cases of it have been studied in different models [4,8,21]. Here, we have investigated it using the binary refinement model [9]. By proposing a novel method for refining non-binary species tree nodes based on the input gene trees, we developed a novel tool for the general reconciliation problem. This tool can be used for the study of gene duplication as well as for species tree inference.

Because of low taxon sampling or the long branch attraction phenomena, deep branches in both gene and species trees are often reconstructed with low support value [14]. Any error occurring in deep branch estimation might lead to overestimation of duplications on an incorrectly inferred deep branch. In its application to gene duplication inference, TxT attempts to reduce error by reconciling non-binary gene and species trees. For species tree inference, it is worth pointing out that contracting weakly supported gene tree branches improves the inference of the corresponding species tree from gene trees as indicated by our validation test.

TxT has several strengths. First, it reconciles two non-binary trees. The current version of NOTUNG requires either the gene or the species tree to be binary [6]. SoftParsmap can take two non-binary trees as input but does not refine species trees [3]. Secondly, our tool can be applied to both gene duplication inference and species tree inference.

Horizontal gene transfer (HGT) and incomplete lineage sorting (ILS) are two other mutational events that may cause the discordance of a gene tree and the corresponding species tree. Recently, simultaneous inference of HGT, ILS, gene duplication and loss events have been studied [2,20]. As a future research topic, we shall investigate how to reconcile non-binary trees for detecting HGT, ILS, gene duplication and loss events.

Acknowledgment. LX Zhang would like to thank David A. Liberles for useful comments on the preliminary draft of this paper. This work was supported by Singapore MOE Tier-2 grant R-146-000-134-112.

References

1. Bansal, M.S., Shamir, S.: A note on the fixed parameter tractability of the gene-duplication problem. IEEE-ACM Trans. Comput. Biol. Bioinform. 8, 848–850 (2010)
2. Bansal, M.S., Alm, E.J., Kellis, M.: Efficient algorithms for the reconciliation problem with gene duplication, horizontal transfer and loss. Bioinform. 28, i283–i291 (2012)
3. Berglund-Sonnhammer, A., et al.: Optimal gene trees from sequences and species trees using a soft interpretation of parsimony. J. Mol. Evol. 63, 240–250 (2006)
4. Chang, W.-C., Eulenstein, O.: Reconciling gene trees with apparent polytomies. In: Chen, D.Z., Lee, D.T. (eds.) COCOON 2006. LNCS, vol. 4112, pp. 235–244. Springer, Heidelberg (2006)

5. Chauve, C., El-Mabrouk, N.: New perspectives on gene family evolution: Losses in reconciliation and a link with supertrees. In: Batzoglou, S. (ed.) RECOMB 2009. LNCS, vol. 5541, pp. 46–58. Springer, Heidelberg (2009)

6. Chen, K., Durand, D., Farach-Colton, M.: NOTUNG: a program for dating gene duplications and optimizing gene family trees. J. Comput. Biol. 7, 429–447 (2000)

7. Doyon, J.-P., Ranwez, V., Daubin, V., Berry, V.: Models, algorithms and programs for phylogeny reconciliation. Briefings Bioinform. 12, 392–400 (2012)

8. Durand, D., Halldorsson, B., Vernot, B.: A hybrid micro- macroevolutionary approach to gene tree reconstruction. J. Comput. Biol. 13(2), 320–335 (2005)

9. Eulenstein, O., Huzurbazar, S., Liberles, D.: Reconciling phylogenetic trees. In: Dittmar, K., Liberles, D. (eds.) Evolution After Duplication, pp. 185–206. Wiley-Blackwell, New Jersey, USA (2010)

10. Goodman, M., et al.: Fitting the gene lineage into its species lineage, a parsimony strategy illustrated by cladograms constructed from globin sequences. Syst. Zool. 28, 132–163 (1979)

11. Górecki, P., Tiuryn, J.: DLS-trees: a model of evolutionary scenarios. Theoret. Comput. Sci. 359, 378–399 (2006)

12. Hahn, M.W., et al.: Estimating the tempo and mode of gene family evolution from comparative genomic data. Genome Res. 15, 1153–1160 (2005)

13. Hahn, M.W.: Bias in phylogenetic tree reconciliation methods: implications for vertebrate genome evolution. Genome Biol. 8(7), R141 (2007)

14. Koonin, E.V.: The origin and early evolution of eukaryotes in the light of phylogenomics. Genome Biol. 11, 209 (2010)

15. Ma, B., Li, M., Zhang, L.X.: From gene trees to species trees. SIAM J. Comput. 30, 729–752 (2000); also in Proc. RECOMB 1998, pp. 182–191 (2000)

16. Mak, W.-K.: Faster min-cut computation in unweighted hypergraphs/circuit netlists. In: Proc. 2005 IEEE Int'l. Symp. VLSI, Automation and Test, pp. 67–70 (2005)

17. Ouangraoua, A., Swenson, K., Chauve, C.: A 2-approximation for the minimum duplication speciation problem. J. Comput. Biol. 18, 1041–1053 (2011)

18. Page, R.: Maps between trees and cladistic analysis of historical associations among genes, organisms, and areas. Syst. Biol. 43, 58–77 (1994)

19. Shertz, C.A., Bastidas, R.J., Li, W., Heitman, J., Cardenas, M.E.: Conservation, duplication, and loss of the Tor signaling pathway in the fungal kingdom. BMC Genomics 11, 510 (2010)

20. Stolzer, M., Lai, H., Xu, M., Sathaye, D., Vernot, B., Durand, D.: Inferring duplications, losses, transfers and incomplete lineage sorting with nonbinary species trees. Bioinform. 28, 409–415 (2012)

21. Vernot, B., Stolzer, M., Goldman, A., Durand, D.: Reconciliation with non-binary species trees. J. Comput. Biol. 15(8), 981–1006 (2008)

22. Zhang, L.X.: On a Mirkin–Muchnik–Smith conjecture for comparing molecular phylogenies. J. Comput. Biol. 4, 177–187 (1997)

23. Zheng, Y., Wu, T., Zhang, L.X.: Reconciliation of gene and species trees with polytomies, arXiv:1201.3995, arxiv.org (2012)

Patterns of Chromatin-Modifications Discriminate Different Genomic Features in *Arabidopsis*

Anuj Srivastava[1], Xiaoyu Zhang[2], Sal LaMarca[3], Liming Cai[1,3], and Russell L. Malmberg[1,2]

[1] Institute of Bioinformatics, University of Georgia, Athens, GA 30602-7229
[2] Department of Plant Biology, University of Georgia, Athens, GA 30602-7404
[3] Department of Computer Science, University of Georgia, Athens, GA 30602

Abstract.

Motivation: Dynamic regulation and packaging of genetic information is achieved by the organization of DNA into chromatin. Nucleosomal core histones, which form the basic repeating unit of chromatin, are subject to various post-translational modifications such as acetylation, methylation, phosphorylation and ubiquitinylation. These modifications have effects on chromatin structure and, along with DNA methylation, regulate gene transcription. The goal of this study was to determine if patterns in modifications were related to different categories of genomic features, and, if so, if the patterns had predictive value.

Results: In this study, we used publically available data (ChIP-chip) for different types of histone modifications (methylation and acetylation) and for DNA methylation for *Arabidopsis thaliana* and then applied a machine learning based approach (a support vector machine) to demonstrate that patterns of these modifications are very different among different kinds of genomic feature categories (protein, RNA, pseudogene and transposon elements). These patterns can be used to distinguish the types of genomic features. DNA methylation and H3K4me3 methylation emerged as features with most discriminative power. From our analysis on *Arabidopsis*, we were able to predict 33 novel genomic features, whose existence was also supported by analysis of RNA-seq experiments. In summary, we present a novel approach which can be used to discriminate/detect different categories of genomic features based upon their patterns of chromatin modification and DNA methylation.

Keywords: Chromatin Modification, DNA Methylation, Support Vector Machine, Machine Learning, *Arabidopsis*.

1 Introduction

In eukaryotic nuclei, DNA associates with proteins to form chromatin. The structure of chromatin plays an essential role in organization of genome, transcriptional activity and developmental state memory (Bernstein, et al., 2002). The basic unit is the nucleosome in which 146 base pairs of DNA are wrapped around an octamer of four

Z. Cai et al. (Eds.): ISBRA 2013, LNBI 7875, pp. 52–63, 2013.
© Springer-Verlag Berlin Heidelberg 2013

core histone proteins (H2A, H2B, H3 and H4) (Luger, et al., 1997). The structures of core histone protein are predominantly globular with the exception an unstructured amino-terminal 'tail' of 25-40 residues. A variety of post-translational modifications (acetylation, phosphorylation and methylation) occurs on these unstructured tails (Zhang and Reinberg, 2001) and have effects on gene expression. These changes are referred to as epigenetic modifications as changes in gene expression are caused by mechanisms other than changes in the underlying DNA sequence.

A second type of epigenetic modification is the addition of methyl groups to the DNA (DNA methylation), primarily at CpG sites, to convert cytosine to 5-methylcytosine. Cytosine DNA methylation is a conserved epigenetic silencing mechanism involved in many important biological processes including defense against transposon proliferation, heterochromatin formation, control of genome imprinting, regulation of endogenous gene expression, and silencing of transgenes (Bender, 2004; Paszkowski and Whitham, 2001; Zhang, et al., 2006). Another type of epigenetic data (but not a modification) is the enrichment of RNA Pol II in different genic regions (Chodavarapu, et al., 2010). Relating the multitude of epigenetic modifications to their regulatory effects poses a complex and fascinating challenge.

In recent years, the use of modification-specific antibodies in chromatin immune-precipitations (ChIP) coupled to gene array technology (ChIP on CHIP) has become an important experimental tool to determine these modifications (Kouzarides, 2007). An advance on ChIP-chip technology is ChIP-Seq which involves chromatin immune-precipitations followed by sequencing. ChIP-Seq offers greater coverage, less noise and higher resolution than its predecessor ChIP-chip, owing largely to advances in next generation sequencing technology (Park, 2009).

In this research, we used chromatin modification data to test their ability to be markers to discriminate or detect different classes of genomic features (Protein coding, RNA, Pseudogene and Transposable elements). The two main questions that we asked here are: 1) Are there differences between the epigenetic modification patterns of different genomic feature types? 2) Can these patterns be used to find the new instances of these features from the un-annotated regions of genome? To perform this analysis, we gathered data for different kinds of epigenetic modifications (DNA methylation, H3 methylation and H4 acetylation at different lysine residue) and also RNA Pol II occupancy of *Arabidopsis thaliana* and then used a machine learning based approach (support vector machine) to distinguish/detect different genomic features.

Support vector machines (SVMs) (Vapnik, 1995) are machine learning techniques widely used to solve classification problems (Barutcuoglu, et al., 2006; Bhardwaj, et al., 2005; Hoglund, et al., 2006). In bioinformatics, the SVM is a widely used classification method in studies such as prediction of DNA-binding proteins (Bhardwaj, et al., 2005), gene function (Barutcuoglu, et al., 2006) and protein subcellular localization (Hoglund, et al., 2006). We implemented an SVM which found that there are substantial differences between modification patterns of genomic feature types which can be readily used to distinguish them. We also showed that these patterns can be used to classify novel genomic features from the genomic background whose existence was then confirmed by RNA-seq experiment.

2   Methods

Datasets. We obtained data for 7 different types of chromatin modifications for *Arabidopsis thaliana*. These datasets were generated using biochemical methods in combination with whole-genome tiling microarrays at 35bp resolution (Chodavarapu, et al., 2010; Costas, et al., 2011; Kong, et al., 2007; Zhang, et al., 2006; Zhang, et al., 2009). The datasets came in the form of probabilities of modification of particular genomic region. These probabilities were obtained by a two-state hidden Markov model (HMM) based on probe-level t statistics by tool tilemap (Ji and Wong, 2005). Prior to analysis, we converted the 35 bp region probability values in the dataset, into base specific probabilities. The region ± 20bp were assigned the same modification probabilities and values in adjacent overlapping regions were averaged together. The detailed methodology used in obtaining datasets can be found in these articles (Chodavarapu, et al., 2010; Costas, et al., 2011; Kong, et al., 2007; Zhang, et al., 2006; Zhang, et al., 2009).

In addition, we also had expression information obtained by an RNA-seq experiment for *Arabidopsis thaliana* (obtained from 2-week-old seedlings). RNA-seq was performed as previously described by Lister, et al., 2008. Image analysis and base calling were performed with the standard Illumina pipeline (Firecrest v1.3.4 and Bustard v.1.3.4). The resulting reads were aligned to the *Arabidopsis* genome (TAIR9) using Tophat (version 1.0.13)/Bowtie (version 0.12.3) with the following commands: --solexa1.3-quals -F 0 -g 1 -I 5000 (Langmead, et al., 2009; Trapnell, et al., 2009).

Obtaining the Genomic Features. Prior to obtaining the genomic features, we adjusted the coordinates of the epigenetic modification probabilities (which was based on TAIR5) using the assembly update information file obtained from TAIR database (ftp://ftp.arabidopsis.org/home/tair/Software/UpdateCoord/). This enabled us to use feature coordinates based upon the latest TAIR release i.e. TAIR10. We obtained General feature format (GFF) file containing coordinates of genomic features (protein, RNA, pseudogene and transposable element gene) of *Arabidopsis* from the TAIR database (TAIR10) and assigned epigenetic modification probabilities to genomic features. Overlapping genomic features and features with less than 30% of the regions covered with modification probabilities were ignored. This cut-off was decided by plotting the number of features against different spanning thresholds.

Feature Selection. We used the feature selection tool provided in LIBSVM (Chang and Lin, 2001) to determine which of the initially considered epigenetic features are actually useful in discriminating different genomic features. An F-score (Chang and Lin, 2001) was used to measure the discriminating power of each feature value to our classification problem in different categories. The code used and information regarding the F-score can be found at the LIBSVM website (http://www.csie.ntu.edu.tw/~cjlin/libsvmtools/#feature_selection_tool).

Creating Datasets for SVM Classification. The 24,872 proteins, 806 RNAs, 803 pseudogenes, and 3006 transposons with their corresponding 7 epigenetic feature modification probabilities were used to create datasets for 5-fold cross-validation experiments, for 2-class and 4-class SVMs. The datasets for 5-fold cross-validation for the 4-class SVMs were created by randomly shuffling the data for proteins, RNAs, pseudogenes, and transposons. After the random shuffling, the first 20% of proteins, RNAs, pseudogenes, and transposons were extracted and a union of these 4 extractions was used to create the first validation set; the union of the remaining 80% from each class was used to create the first training set. Similarly, the second 20% of each class was extracted and combined for the second validation dataset, and the remaining 80% were used as the second training dataset, and this was continued until there were 5 independent validation sets and 5 training sets for the 5-fold cross-validation experiments for the 4-class SVM.

To equalize the number of features of each type, oversampling was applied by copying all RNAs (30 times), pseudogenes (30 times) and transposons (8 times) in each dataset, to roughly equalize the number of proteins, RNAs, pseudogenes, and transposons in each validation and training dataset for the 4-class SVMs. Thus, 5 validation and 5 training oversampled datasets were created that had the properties that the validation sets were independent of their corresponding training sets, and each set had roughly an equal number of data points consisting of each class. The "all" dataset for the 4-class SVM was created by taking the union of the 5 validation sets. The training sets were used for training 4-class SVMs to predict which of the 4 classes a set of 7 epigenetic feature modifications belong to for 5-fold cross-validation experiments, the validation sets were used as an unbiased testing set for the trained 4-class SVMs for 5-fold cross-validation experiments, and the "all" dataset was used to train the 4-class SVM on 100% of the oversampled data.

For the 2-class SVMs, and the 4-class SVM training and validation, all oversampled datasets were split into 6 subsets consisting of 6 binary combinations of the 4 different classes: {protein, RNA}, {protein, pseudogene}, {protein, transposon}, {RNA, pseudogene}, {pseudogene, transposon}, and {RNA, transposon}. For example, the {protein, RNA} validation datasets consisted of subsets of the 4-class SVM validation datasets that contained all of the protein and RNA data points, but no data points from the other classes. This was also used to create the training and "all" datasets for the 2-class SVMs that were used to predict if a sequence is protein or RNA. This process was repeated to create validation, training, and "all" datasets for the other binary combinations.

SVM Classification Experiments: Six, 2-class SVM classifiers were created using the LIBSVM package (Chang and Lin, 2001) that was trained on the 2-class training datasets for each of the 6 binary combinations for 5-fold cross-validation experiments. The radial basis kernel function and SVM probability estimates were used in the LIBSVM package. The 4-class SVM was built in a similar fashion using LIBSVM, but it used the majority vote multi-label class SVM that splits the problem of 4-classification into a 6 part, 2-classification problem.

The training of the 4-class SVM-based classifier was performed using a standard procedure provided in LIBSVM (Chang and Lin, 2001) to find values of two parameters C and γ, where C controls the trade-off between training errors and classification margins, and γ determines the width of the radial basis kernel (Chang and Lin, 2001). A grid search using 5-fold cross-validation was used on the training and validation sets for the 4-class SVM to find the optimal parameters of $C = 1$ and $\gamma = 2$ that yielded the lowest average error (1 – average accuracy) on their independent validation sets without showing signs of over-fitting as shown in Figure 1. The optimal parameters of C and γ were used for training the 2-class and 4-class SVMs with LIBSVM using the radial basis kernel and modification probability estimates aforementioned.

Prediction of Novel Genomic Features: The coordinates and sequences of intergenic regions were obtained from the TAIR database. Intergenic sequences were divided into two parts - intergenic transcribed/non-transcribed, based on RNA-seq expression data. The intergenic regions with at least 2 RNA-seq reads covering at least 20% of the total intergenic region length were considered to be transcribed. This threshold was decided after plotting different combinations i.e. number of reads/spanning length against the number of sequences. We further ignored the intergenic regions which are less than 200bp and also ignored the sequences ±50bp from both ends of the intergenic region.

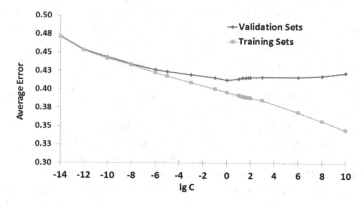

Fig. 1. Fivefold cross-validation results (for the 4- class SVM with γ = 2) indicate the lack of over fitting which might occur due to over-sampling of the data. At the optimal parameters, the average error (1- average accuracy) is same for the training and validation sets. These parameters were used during feature discrimination and detection.

Afterwards, we extracted the chromatin modification probabilities of the intergenic regions and used them to identify novel features. Potential novel features were identified by using genomic feature probabilities and intergenic region probabilities as training and testing datasets during classification, respectively. The multi-level classifier provides the probability estimates for a test data instance of it belonging to each of four feature classes. We chose a probability threshold of 0.70 (determined after plotting the distribution of values of each feature classes) to assign the data instance to particular feature type. To make our prediction more reliable, we took the sequences of intergenic regions and checked the coding potential of predicted features by coding potential calculator (CPC) (Kong, et al., 2007). The regions predicted as protein coding genes and also predicted as coding by CPC were considered as protein coding and vice-versa. A similar analysis was performed for both intergenic transcribed/non-transcribed categories.

3   Results

We gathered datasets on different types of epigenetic modifications (DNA methylation, H3 methylation and H4 acetylation at different lysine residue and RNA Pol II occupancy) for *Arabidopsis thaliana*; these were in the form of probabilities for each type of modification for regions of the *Arabidopsis* chromosomes. Scripts were developed which convert the experimentally determined feature probabilities for regions of the genome to basepair coordinates to match the coordinate system of the features in the *Arabidopiss* GFF files. We performed two analyses: first determining the level of bias in chromatin modification pattern which exists between different feature classes using an SVM, and then second predicting novel genomic features by applying the SVM classification method to regions currently labeled as intergenic.

3.1   Binary and Multi-level Classifiers

Binary classifiers (two-class SVMs) were used to determine discrimination in epigenetic modification patterns of different genomic features and multiple way classifiers (four-class SVMs) were used to assign the intergenic regions to different feature classes. We performed 6 different comparisons among 4 different feature classes (protein, RNA, pseudogenes and transposon element genes) using two-class SVMs. We obtained modification probabilities for 24,872, 806, 803 and 3006 protein coding, RNA, pseudogenes and transposable element genes, respectively. In each classification, SVM based classifiers were used to separate two feature classes and an F-score was calculated to determine the discrimination power of each epigenetic feature in every comparison (Table 1). Over-sampling was used by the SVMs to balance the number of data points for the modification probabilities for the four classes.

In general, the larger an F-score, the more discriminative the corresponding feature is. Based upon F-scores, DNA methylation and H3K4me3 emerged as the features with most discriminative power (Figure 2). The average value of each type of epigenetic modifications in every feature class determined using their genomic coordinates is shown in Table 2, and the results of the two-class SVMs, showing the

Table 1. F-score result for each binary classification category

Epigenetic Feature	Protein RNA	Protein Pseudo	Protein Trans	RNA Pseudo	RNA Trans	Pseudo Trans
H3K27me3	0.000945	0.000305	0.011745	0.021394	0.015270	0.075851
H4K5ac	0.000098	0.000042	0.000689	0.000639	0.000642	0.001500
DNA methylation	0.003920	0.015529	0.632277	0.142858	0.601012	0.213082
RNA Pol II	0.001991	0.001688	0.014568	0.031097	0.071851	0.006889
H3K4me1	0.008662	0.004281	0.033849	0.013567	0.001405	0.025655
H3K4me2	0.002330	0.005418	0.03183	0.007523	0.032187	0.010474
H3K4me3	0.000388	0.013048	0.062806	0.135458	0.285946	0.019144

Table 2. Average value ± SEM (standard error of mean) of each feature obtained using their genomic coordinates

Epigenetic feature	Protein	RNA	Pseudogene	Transposon
H3K27me3	0.0728 ± 0.00003	0.0748 ± 0.0003	0.1206 ± 0.0003	0.0161 ± 0.00004
H4K5ac	0.0018 ± 0.000003	0.0013 ± 0.00002	0.0014 ± 0.00002	0.0006 ± 0.000006
DNA methylation	0.1231 ± 0.00004	0.0723 ± 0.0003	0.2467 ± 0.0004	0.6569 ± 0.0001
RNA Pol II	0.0778 ± 0.00003	0.1083 ± 0.0004	0.0460 ± 0.0002	0.0175 ± 0.00004
H3K4me1	0.1414 ± 0.00004	0.0364 ± 0.0002	0.0384 ± 0.0001	0.0116 ± 0.00003
H3K4me2	0.1164 ± 0.00003	0.1000 ± 0.0004	0.0546 ± 0.0002	0.0196 ± 0.00005
H3K4me3	0.2051 ± 0.00005	0.2532 ± 0.0006	0.0473 ± 0.0002	0.0094 ± 0.00003

Table 3. Two-class SVM results with $C = 1$ and $\gamma = 2$. Protein (P), RNA (R), Pseudogene (Ps), Transposon (T), Accuracy (Acc), Precision (Prec), Sensitivity (Sens), Specificity (Spec).

+	-	Dataset	TP	FP	FN	TN	Acc	Prec	Sens	Spec	MCC
P	R	Validation	18352	6030	6520	18150	0.7441	0.7527	0.7379	0.7506	0.4884
		Training	73603	20670	25885	76050	0.7627	0.7807	0.7398	0.7863	0.5265
		All	18723	5580	6149	18600	0.7609	0.7704	0.7528	0.7692	0.5220
P	Ps	Validation	16873	5160	7999	18930	0.7312	0.7658	0.6784	0.7858	0.4665
		Training	67654	17160	31834	79200	0.7498	0.7977	0.6800	0.8219	0.5064
		All	17007	4260	7865	19830	0.7524	0.7997	0.6838	0.8232	0.5113
P	T	Validation	23199	3312	1673	20736	0.8981	0.8751	0.9327	0.8623	0.7977
		Training	92853	12880	6635	83312	0.9003	0.8782	0.9333	0.8661	0.8019
		All	23217	3232	1655	20816	0.9001	0.8778	0.9335	0.8656	0.8016
Ps	R	Validation	15270	5130	8820	19050	0.7110	0.7485	0.6339	0.7878	0.4269
		Training	64140	18210	32220	78510	0.7388	0.7789	0.6656	0.8117	0.4826
		All	16050	4620	8040	19560	0.7377	0.7765	0.6663	0.8089	0.4802
Ps	T	Validation	16440	4144	7650	19904	0.7550	0.7987	0.6824	0.8277	0.5156
		Training	67740	16152	28620	80040	0.7675	0.8075	0.7030	0.8321	0.5396
		All	16860	4032	7230	20016	0.7660	0.8070	0.6999	0.8323	0.5369
T	R	Validation	21184	1650	2864	22530	0.9064	0.9277	0.8809	0.9318	0.8138
		Training	85144	5460	11048	91260	0.9144	0.9397	0.8851	0.9435	0.8302
		All	21304	1380	2744	22800	0.9145	0.9392	0.8859	0.9429	0.8303

discrimination between different feature classes, are given in Table 3, where MCC denotes Matthews correlation coefficient. The Table 1 and Table 2 values are also shown in the form of bar plots in Figure 2 and Figure 3.

Our testing strategy used a 5-fold cross-validation scheme where the validation dataset was independent of the training dataset. The "validation" row in Table 3 shows the results from 5 independent validation datasets used in 5-fold cross-validation on the SVMs trained on their corresponding 5 training datasets; the rows containing "training" show the results on the training datasets used in 5-fold cross-validation. The rows labeled as "all" are the results from the SVM trained on all of the oversampled data. The two-class SVM has the highest accuracy (0.90) and MCC

(0.81) on its validation sets for transposon/ncRNA classification. The protein/transposon and protein/ncRNA classification have accuracies of 0.90, 0.74 and MCC of 0.80, 0.49, respectively and finally, the ncRNA/pseudogenes classification has the lowest accuracy (0.71) and MCC (0.43) on its validation sets.

We developed a multi-level classifier by combining the binary SVMs after exploring the optimum parameters for doing this. We first tested the ability of the SVMs to discriminate between genomic features using a cross-validation approach similar to that used previously. The results from the 4-class SVM, which is built from 6, 2–class SVMs, are shown in Table 4. In the 4-class SVM, for each data point, each of the 6, 2-class SVMs makes a prediction and the final prediction from the 4-class is the majority vote from the 6, 2-class SVMs. In the case of a tie, the 4-class SVM will predict the class with the highest probability (Wu, et al., 2004).

Table 4 contains three results of the 4-class SVMs in confusion matrix form i.e. validation sets, training sets results (in 5-fold cross-validation) and the 4-class SVM results (trained on all oversampled data). As shown in Table 4, the 4-class SVM (trained on 5 training sets) on the validation datasets has an accuracy of 0.587, 0.709, 0.219 and 0.832 for protein, RNA, pseudogenes and transposon, respectively. The overall accuracy on the 4-class SVM's on validation set was 0.587, which is 0.337 above randomly classifying one class out of the four classes (i.e. 0.25). The 4-class SVMs had the highest and lowest accuracies/reliabilities for predicting transposons and pseudogenes, respectively.

3.2   Novel Feature Prediction

We used the multi-level classifier to detect novel genomic features by analysis of the chromatin modification probabilities of Arabidopsis genomic regions currently annotated as intergenic in the data set. Based upon the number of RNA-seq reads covering the region, the intergenic data for Arabidopsis was divided into two parts i.e. intergenic transcribed (617 sequences) and non-transcribed (25331sequences). Afterwards, these two datasets were used as a testing set in multi-level classification. The predicted features from the multi-level classification were further filtered by checking their potential for coding by a coding potential calculator (CPC) (Kong, et al., 2007). The consensus results of SVM prediction and CPC were included in the final prediction. In all, we were able to identify 4 protein, 21 ncRNA, 1 pseudogene, and 7 transposons, respectively in intergenic transcribed category (supplementary file 1) and 15 protein, 479 ncRNA, 8 pseudogenes and 734 transposons, respectively in the intergenic non-transcribed category (supplementary file 2).

4   Discussion

We used data for different types of epigenetic modification from *Arabidopsis* and then used binary SVM classifiers to discriminate the patterns of epigenetic modification among different genomic features.

Table 4. Four-class SVM results ($C = 1$ and $\gamma = 2$) showing the confusion matrices resulting from a 5-fold cross-validation from the over-sampled datasets

		Predicted by SVM				
		Protein	RNA	Pseudogene	Transposon	Accuracy
Validation dataset	Protein	14600	5409	3539	1324	0.587
	RNA	3720	17160	2280	1020	0.709
	Pseudogene	4020	7440	5280	7350	0.219
	Transposon	1024	1776	1232	20016	0.832
	Reliability	0.624	0.539	0.428	0.673	
	Avg. Class Accuracy: 0.505		Avg. Class Reliability: 0.566		Overall Accuracy: 0.587	
Training dataset	Protein	58591	21508	14118	5271	0.588
	RNA	13260	70620	8670	4170	0.730
	Pseudogene	13560	28500	25290	29010	0.262
	Transposon	3976	6960	5056	80200	0.833
	Reliability	0.655	0.553	0.476	0.675	
	Avg. Class Accuracy: 0.527		Avg. Class Reliability: 0.590		Overall Accuracy: 0.603	
All data	Protein	14715	5380	3464	1313	0.591
	RNA	3360	17640	2130	1050	0.729
	Pseudo gene	3420	7140	6240	7290	0.259
	Transposon	1008	1736	1232	20072	0.834
	Reliability	0.653	0.553	0.477	0.675	
	Avg. Class Accuracy: 0.526		Avg. Class Reliability: 0.590		Overall Accuracy: 0.603	

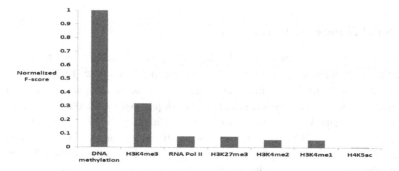

Fig. 2. The power of each epigenetic feature in discrimination (normalized F-score) different genomic classes. In the normalized F-score plot, DNA methylation has the maximum F-score so it is defined to be 1.0 and other feature F-scores were divided by the DNA methylation values.

For protein/RNA binary classification, the feature with the most discriminative power is H3K4me1. From the determination of probabilities within the feature region (Table 2), we found that average modification probabilities for H3K4me1 are higher for protein coding genes compared to ncRNA genes. Previously, it has been found that H3K4me1 modification occurs predominantly in the transcribed region of genes and has positive correlation with length of the genes (Zhang, et al., 2009). The low value of H3K4me1 for RNA genes could be due to their length as in our dataset 65% of the RNAs are less than 200bp in length, compared to 2% of protein coding genes being this short.

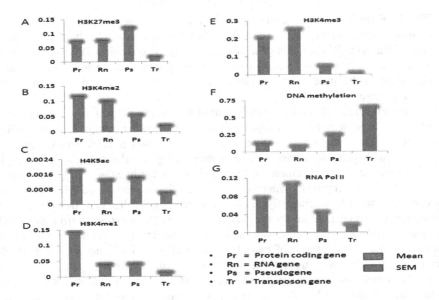

Fig. 3. The average value of each epigenetic feature in protein, RNA, pseudogene and transposon element regions. The overall mean and SEM (standard error of mean) values were obtained using the co-ordinate of genomic features. The standard error is indicated by a fuzzy area at the top of the bar.

In the protein/transposon comparison, DNA methylation emerged as the feature with the most discriminative power (Table 1) and was associated with transposons (Figure 3F). A strong pattern of methylation is known to be associated with transposable element genes (Bender, 2004; Paszkowski and Whitham, 2001; Zhang, et al., 2006) and it serves as a defense mechanism against proliferation of transposons in the genome. In the protein/pseudogene classification, DNA methylation was the top feature with high F-score (Table 1). Pseudogenes also have overall high DNA methylation value comparing to protein coding genes (Figure 3F). Similar to transposable element genes, their high value of DNA methylation is related to transcriptional silencing of pseudogenes (Zhang, et al., 2006). However, unlike transposable element genes which are methylated to prevent their deleterious effects, pseudogenes might be methylated to prevent the cost of transcription of a non-functional unit of genome.

A strong DNA methylation pattern associated with transposons has also the most discriminative power in ncRNA/transposon classification (Table 1). The second best feature, H3K4me3, also has a high F-score and was associated with ncRNA (Figure 3E). Several categories of ncRNA genes (tRNA, miRNA, snoRNA) were previously shown to have higher H3K4me3 methylation compared to DNA methylation and the H3K4me2 type of modifications in rice (Li, et al., 2008); sixty-nine percent of the ncRNA in our dataset were comprised of these 3 types of RNA which explains the high value of H3K4me3.

The H3K4me3 type of modification is found in genes known to be highly expressed (Zhang, et al., 2009) and these ncRNA genes are likely to be highly expressed. In the ncRNA/pseudogene comparison, the two features DNA methylation and H3K4me3 have the most discriminative power and, based on average modification probabilities, they are found to be associated with pseudogenes and RNA, respectively (Table 2).

For pseudogenes/transposons, DNA methylation also emerged as a feature with most power in separating two classes, similarly to the other comparisons involving transposons. The second best feature in this classification is H3K27me3, which is associated with gene silencing in *Arabidopsis* (Kong, et al., 2007), and has high average modification probabilities (along with DNA methylation) for pseudogenes (Figure 3A). To determine, whether DNA methylation and H3K27me3 occur in tandem or mutually exclusively in pseudogenes, we calculated the correlation coefficient (Spearman) value for the pseudogenes from data extracted from the SVM feature file and found that these two modifications are inversely related ($r = -0.20$, p-value < 0.01). The existence of two alternate mechanisms of gene silencing which occur largely exclusively suggests the importance of silencing pseudogenes.

In the normalized F-score plot in Figure 2, H4K5ac has the lowest value. This epigenetic feature also has the lowest average value compared to other epigenetic modifications in all four feature classes (Figure 3C). Acetylation patterns are positively correlated with gene expression and in particular H4K5ac modification are elevated in transcribed regions of active genes in human (Wang, et al., 2008); there is also enrichment of this modification at origin of replications (Costas, et al., 2011). The lower average values indicate that this modification is not frequent as compared to others and particularly is rare in transposons which makes sense as genomes in general try to silence transposon not activate them.

We predicted novel genomic features from epigenetic modification patterns of intergenic using the multi-level SVM. Data from an RNA-seq experiment and CPC was used to further verify the predicted features. The higher number of ncRNA genes in the intergenic transcribed dataset (RNA-seq reads present) makes biological sense as protein coding genes are already well annotated in *Arabidopsis* and therefore transcribed reads has more likelihood to be associated with ncRNA. In the non-transcribed class (RNA-seq reads absent), the number of ncRNA genes is second to transposon element genes. This is reasonable due to the abundance of transposons in genomes and the lack of transcription evidence in the RNA-seq data.

In conclusion, we provided support for distinctive patterns of chromatin modifications being associated with different kinds of genomic features, and we demonstrated a novel approach for discriminating/detecting different genomic features based upon these modifications. We did not predict many new features in *Arabidopsis* as it has already being extensively studied. However, with the continuous progress in the field of high-throughput sequencing generating this kind of data is become simpler and cheaper, and this approach might be used to discriminate/detect novel features in many newly sequenced plant species such as *Populus* and *Vitis*.

Acknowledgements. This work was supported by National Science Foundation (IIS 0916250); and The University of Georgia Franklin College of Arts & Science's research fund.

References

1. Barutcuoglu, Z., Schapire, R.E., Troyanskaya, O.G.: Hierarchical multi-label prediction of gene function. Bioinformatics 22, 830–836 (2006)
2. Bender, J.: DNA methylation and epigenetics. Annu. Rev. Plant Biol. 55, 41–68 (2004)
3. Bernstein, B.E., et al.: Methylation of histone H3 Lys 4 in coding regions of active genes. Proc. Natl. Acad. Sci. U.S.A. 99, 8695–8700 (2002)
4. Bhardwaj, N., et al.: Kernel-based machine learning protocol for predicting DNA-binding proteins. Nucleic Acids Res. 33, 6486–6493 (2005)
5. Chang, C.-C., Lin, C.-J.: LIBSVM: a library for support vector machines (2001)
6. Chodavarapu, R.K., et al.: Relationship between nucleosome positioning and DNA methylation. Nature 466, 388–392 (2010)
7. Costas, C., et al.: Genome-wide mapping of Arabidopsis thaliana origins of DNA replication and their associated epigenetic marks. Nat. Struct. Mol. Biol. 18, 395-400 (2011)
8. Hoglund, A., et al.: MultiLoc: prediction of protein subcellular localization using N-terminal targeting sequences, sequence motifs and amino acid composition. Bioinformatics 22, 1158–1165 (2006)
9. Ji, H.K., Wong, W.H.: TileMap: create chromosomal map of tiling array hybridizations. Bioinformatics 21, 3629–3636 (2005)
10. Kong, L., et al.: CPC: assess the protein-coding potential of transcripts using sequence features and support vector machine. Nucleic Acids Res. 35, W345–W349 (2007)
11. Kouzarides, T.: Chromatin modifications and their function. Cell 128, 693–705 (2007)
12. Langmead, B., et al.: Ultrafast and memory-efficient alignment of short DNA sequences to the human genome. Genome Biol. 10, R25 (2009)
13. Li, X.Y., et al.: High-resolution mapping of epigenetic modifications of the rice genome uncovers interplay between DNA methylation, histone methylation, and gene expression. Plant Cell 20, 259–276 (2008)
14. Lister, R., et al.: Highly integrated single-base resolution maps of the epigenome in Arabidopsis. Cell 133, 523–536 (2008)
15. Luger, K., et al.: Crystal structure of the nucleosome core particle at 2.8 A resolution. Nature 389, 251–260 (1997)
16. Park, P.J.: ChIP-seq: advantages and challenges of a maturing technology. Nat. Rev. Genet. 10, 669–680 (2009)
17. Paszkowski, J., Whitham, S.A.: Gene silencing and DNA methylation processes. Curr. Opin. Plant Biol. 4, 123–129 (2001)
18. Trapnell, C., Pachter, L., Salzberg, S.L.: TopHat: discovering splice junctions with RNA-Seq. Bioinformatics 25, 1105–1111 (2009)
19. Vapnik, N.V.: The Nature of Statistical Learning Theory. Springer (1995)
20. Wang, Z., et al.: Combinatorial patterns of histone acetylations and methylations in the human genome. Nat. Genet. 40, 897–903 (2008)
21. Wu, T.F., Lin, C.J., Weng, R.C.: Probability estimates for multi-class classification by pairwise coupling. J. Mach. Learn. Res. 5, 975–1005 (2004)
22. Zhang, X., et al.: Genome-wide high-resolution mapping and functional analysis of DNA methylation in Arabidopsis. Cell 126, 1189–1201 (2006)
23. Zhang, X.Y., et al.: Genome-wide analysis of mono-, di- and trimethylation of histone H3 lysine 4 in Arabidopsis thaliana. Genome Biol. 10 (2009)
24. Zhang, Y., Reinberg, D.: Transcription regulation by histone methylation: interplay between different covalent modifications of the core histone tails. Genes. Dev. 15, 2343–2360 (2001)

Inferring Time-Delayed Gene Regulatory Networks Using Cross-Correlation and Sparse Regression

Piyushkumar A. Mundra[1], Jie Zheng[1,5], Niranjan Mahesan[2],
Roy E. Welsch[3,4], and Jagath C. Rajapakse[1,3,6]

[1] BioInformatics Research Centre, School of Computer Engineering,
Nanyang Technological University, Singapore 639798
[2] School of Electronics and Computer Science, University of Southampton,
Southampton, United Kingdom
[3] Computation and Systems Biology, Singapore-MIT Alliance,
Nanyang Technological University, Singapore 637460
[4] Sloan School of Management, Massachusetts Institute of Technology, Cambridge,
MA 02142, USA
[5] Genome Institute of Singapore, Biopolis Street, Singapore 138672
[6] Department of Biological Engineering, Massachusetts Institute of Technology, USA

Abstract. Inferring a time-delayed gene regulatory network from microarray gene-expression is challenging due to the small numbers of time samples and requirements to estimate a large number of parameters. In this paper, we present a two-step approach to tackle this challenge: first, an unbiased cross-correlation is used to determine the probable list of time-delays and then, a penalized regression technique such as the LASSO is used to infer the time-delayed network. This approach is tested on several synthetic and one real dataset. The results indicate the efficacy of the approach with promising future directions.

Keywords: LASSO, gene regulation, time-delayed interactions, microarray analysis, cross-correlation.

1 Introduction

The collection of high-throughput molecular data using advanced technology has enabled researchers to reverse engineer the dynamics of the underlying complex biological system. The inferences of mechanisms are generally achieved by building the gene regulatory networks (GRNs). Using time-series gene expression data, gathered by microarray chips, a typical yet simple GRN is inferred and it consists of interconnected nodes (genes) and edges that demonstrate how a particular gene is regulated by a set of genes.

With microarrays, it is possible to measure the expressions of thousands of genes simultaneously. However, the data are only gathered over a few time-samples for several reasons, such as cost of experiments, availability of subjects, etc. Sometimes this is also because the biological state we are interested in

Z. Cai et al. (Eds.): ISBRA 2013, LNBI 7875, pp. 64–75, 2013.
© Springer-Verlag Berlin Heidelberg 2013

cannot be known precisely. For example, while studying development of fruitfly embryos, hundreds of embryos are bred and gene expressions are measured at different points in time/stages of development. This results in a variability along the time axis because not all embryos are going to grow at the same speed [1]. From a computational view, while modeling a GRN, it is assumed that gene expressions at a given time point only depend on the immediate previous time point [2–4]. Such an assumption leads to GRN with first order (or delay or lag). In reality, in many cases, the regulation of one gene by another gene may occur only after a number of time points, resulting in an invalid first order assumption. However, modeling higher order GRN is very challenging due to the significant increase in numbers of parameters which need to be estimated and the reduction in numbers of available time samples.

In the past, several approaches have been presented to build a first order GRN using time-series gene-expression data. These approaches include a Bayesian framework, Dynamic Bayesian Networks (DBN), Boolean Networks and their probabilistic approaches, ordinary differential equations (ODE), linear or non-linear regression approaches, information theory based models, etc. The readers are referred to [2, 3, 5–7] for excellent reviews on this topic. With respect to time-delayed GRNs, a decision tree with delayed correlation was used to discover the time-delayed regulations between the genes [8]. A first order DBN model is extended to a higher-order DBN where mutual information has been used to determine the best time-delay of an interaction [9, 10]. In another approach, the ARACNE (Algorithm for the Reconstruction of Accurate Cellular Networks) model has been extended to TimeDelay-ARACNE by using a stationary Markov Random Field [11]. Using protein-protein interaction and microarray data, a skip chain model was introduced to obtain a GRN [12]. Recently, based on the mutual information and minimum description length principles, a novel scoring metric was proposed to infer time delayed GRN [13]. Although several DBN based approaches were proposed for inference of time-delayed GRN and show the importance of inferring time-delay edges, these methods are only applicable to small networks due to high computational cost. In [14], a simple time delay Boolean networks framework was presented to tackle the computational complexity. However, many of these approaches need discretization of data to infer the GRN and hence, possibly suffer from loss of information.

Using continuous data, sparse regression based approaches have been developed for inferring first-order GRNs [4,15,16]. However, to the best of our knowledge, such regression approaches for inference of higher-order GRN are not yet developed. In this paper, we propose a simple yet effective solution to model a higher-order GRN under a sparse linear regression framework. In a two-step method, we first determine a probable order of regulation using cross-correlation, and then, a LASSO (least absolute shrinkage and selection operator) regression in a multivariate autoregression (MVAR) framework is applied to infer a time-delayed GRN. The efficacy of this approach is tested on both synthetic datasets with varying numbers of genes and numbers of time points and a real dataset.

The rest of the paper is organized as follows: In the next subsection, we propose the two-step cross-correlation based methodology to infer time-delayed GRN. Next, details on synthetic as well as real datasets, parameter estimations, and performance evaluations metrics are presented. Finally, key results, discussion and future directions are discussed.

2   Methods

Let $X = \{x_i(t)\}_{i=1,t=1}^{I,T}$ denote expressions of I genes gathered over equally-spaced T time samples. Here, $x_i(t)$ denotes expression of gene i at time t. We also assume that the gene expressions of all genes at time t are represented by the vector $x(t) = (x_i(t))_{i=1}^{I}$. A higher order fully-connected network of these I variables (genes) could be derived by using an r-th order multivariate vector autoregressive (MVAR) model:

$$x(t) = \sum_{\tau=1}^{r} \beta^{\tau} x(t - \tau) + \varepsilon(t) \tag{1}$$

where $\beta^{\tau} = \{\beta_{i,j}^{\tau}\}_{i=1,j=1}^{I,I}$ represents the strength of interactions (i.e. regression coefficients) between all the pairs of genes for a model of order τ, and $\varepsilon(t) = (\varepsilon_i(t))_{i=1}^{I}$ denotes residuals that are assumed to follow a Gaussian distribution with zero mean and are independently and identically distributed (i.i.d.). For an r-th order model, $I^2 r$ coefficients ($\beta^1, \beta^2, \ldots, \beta^r$) need to be estimated from the given data. This could be easily achieved by using a standard regression formulation [16].

The above mentioned MVAR model needs to be modified for inference related to biological networks, such as time-delayed gene regulatory networks for the following reasons: (1) it is generally assumed that expression time-series are stationary and no multiple regulation edges with different time lags exist between two genes; (2) GRNs are sparse in nature while a standard formulation derives a fully connected network; (3) in a typical gene-expression time-series data, the numbers of genes whose expressions are measured are far higher than numbers of time samples. Hence a standard regression technique to derive strength of connections is inapplicable. In the following, we propose a method to tackle these challenges. First, we fix the time-delay by using cross correlation and then a sparse regression technique is used to infer a time-delayed network.

Mathematically, the assumption of a single time-delayed regulation (out of possible r lags) between two genes i and j implies that for $\exists \tau$ if $\beta_{i,j}^{\tau} > 0$, then $\beta_{i,j}^{\tau} = 0$ for all other τ. This requirement could be achieved by using the cross-correlation between two genes and using the lag that gives maximum absolute cross-correlation. If gene j regulates gene i, the unbiased cross-correlation is given by [17],

$$\widehat{C}(x_i, x_j, \tau) = \frac{1}{T - |\tau|} \sum_{t=1}^{T-\tau-1} x_i(t + \tau) x_j(t) \qquad \tau \geq 1 \tag{2}$$

Here, $\widehat{C}(x_i, x_j, \tau)$ is an estimated unbiased cross-correlation for regulation of gene i by gene j. Such cross-correlation was computed after normalising expressions of a gene to have zero mean and a standard deviation of one. These values are computed for all $\tau = 1, 2, \ldots, r$ and the maximum of the absolute cross-correlation denotes the probable time lag regulation. Let C_{ij} denotes the maximum absolute value of $\widehat{C}(x_i, x_j)$ vector and it corresponds to a time-lag e_{ij}.

Once the probable time lag is fixed, the next step is to identify the relevant regulators for gene i from all the possible I genes. This can be obtained by employing sparse linear regression techniques like the LASSO.

Let's assume that gene j regulates gene i with the time lag k_{ij} which is estimated using cross-correlation. Let $y(t) = (x_i(t))_{i=1,t=r+1}^{I,T}$ denote a gene expression vector of i-th gene at time t and $z(t) = (x_i(t))_{i=1,t=r-k_{ij}+1}^{I,T-k_{ij}}$ denote the vector of gene expression at the corresponding time lag k_{ij} for each gene. Then, using the multivariate vector autoregressive model, the strength of the time delayed regulation by each of the genes could be estimated by,

$$y^t = z^t \beta^* + \varepsilon^t \tag{3}$$

β^* is regulation strength (regression coefficients) matrix of size $I \times I$, and $\varepsilon^t = [\varepsilon_1(t), \varepsilon_2(t), \ldots \varepsilon_I(t)]$ the corresponding innovations. If we assume that the t-th row of matrices Y, Z, and E, are y^t, z^t, and ε^t respectively, Eq. (3) could be written as $Y = Z\beta + E$ and the parameters could be estimated using a standard least square procedure,

$$\hat{\beta} = (Z^T Z)^{-1} Z^T Y \tag{4}$$

Considering that GRNs are sparse and more importantly, the number of time samples are significantly smaller than the number of genes in a typical gene-expression dataset, Eq. (4) can not determine the strength of regulatory connections. However, by using sparse regression techniques, these inherent constraints could be solved. By treating each of the genes independently to identify its potential regulators, the LASSO loss function is given by

$$L(\beta_i, \alpha_i) = ||y_i - Z\beta_i||^2 + \alpha_i |\beta_i|_1 \tag{5}$$

where α_i is a regularization parameter.

The solution provided by Eq. (5) gives only a few non zero β_i coefficients which denote regulation of i-th gene by a very few genes. Using cross-correlation and LASSO regression, we obtain a sparse time-delayed linear GRN.

Algorithm 1 describes the complete approach to derive a time-delayed GRN. This is basically a two-step procedure. Starting with time-series gene-expression data and a fixed maximum time-delay, for a given gene, cross-correlation is used to determine the probable time lags of regulations by other genes. In the second step, LASSO regression is used to derive the regulators. By repeating the same process for all the genes, a complete time-delayed GRN is derived.

Algorithm 1. Time-delayed Gene Regulatory Network with LASSO regression

---

Begin
Time-series gene expression data X; Maximum possible time-delay r; Final time
delay matrix $k = [\ ]$
for Each gene i **do**
   A temperory vector $k^* = [\ ]$
   for All other gene j **do**
      Compute the cross-correlation $\widehat{C}(x_i, x_j)$ between i-th gene and j-th genes using
      Eq. (2)
      Determine the probable time-delay e_{ij} based on maximum absolute cross-
      correlation C_{ij}
   end for
   Store all e_{ij} values in temperory vector k^*
   Derive dependent variable matrix y_i and independent variable matrix Z based on
   probable time-delays e_{ij}
   Using five-fold cross validation, Determine α_i parameter for LASSO regression
   Determine the LASSO regression coefficients $(\beta_{i,.})$ using the best α_i value
   if $\beta_{i,j} = 0$ **then**
      $k_j^* = 0$
   end if
   Append time-delay information matrix $k = [k\ k^*]$
end for
Output: β (and hence gene regulatory network as a non-zero β denotes an edge)
and time-lag information k for each edge

---

3 Experiments

The performance of the proposed method was tested using both simulated and
real time-series gene expression datasets. To generate simulated datasets, we extracted sub-networks of size 20, 50, or 100 genes by using gene net weaver (GNW)
software [18]. These networks are in fact extracted from a global Saccharomyces
cerevisiae network, and hence, the extracted network topologies resemble actual
regulatory networks.

Once the network is extracted, each of the regulatory edges is randomly assigned a time-delay. In reality, the maximum time-delay information is unknown.
In the worst case scenario, the longest delayed response can be expected to be
$T - 1$ time points. However, as discussed earlier, this will make the estimation
of parameters (β_{ij}) intractable. Hence, in this study, the maximum time-delay
(r) was fixed at either 3 or 5.

3.1 Simulating Synthetic Data

For a given network topology, the regression coefficients corresponding to no
interactions among genes were set to zero. For all the edges with respective
τ values, MVAR coefficients $(\beta_{i,j}^\tau)$ were obtained by drawing samples from a
uniform distribution on the interval $[0.8, 1]$. Coefficients for all other time-lags

$(\tau^* \in r$ where $\tau^* \neq \tau)$ were set to zero, i.e., $\beta_{i,j}^{\tau^*} = 0$. For example, if the j-th gene regulates the i-th gene with 2nd order time delay and $r = 5$, then $\beta_{i,j}^2 \in [0.8, 1]$ and $\beta_{i,j}^1 = \beta_{i,j}^3 = \beta_{i,j}^4 = \beta_{i,j}^5 = 0$.

The initial gene expression values at $t = 0, 1, ..., r$ were drawn from a uniform distribution on the interval $[0, 1]$. For successive time points, expressions were generated using a higher-order MVAR model with added i.i.d. Gaussian random noise $\Sigma = \mathbf{I}$. The first 10,000 samples were discarded. The numbers of time points were varied from 20, 30, or 40 and, for each combination of network size and number of time points, we generated 100 time-series datasets by randomly initializing the gene expressions.

3.2 Parameter Estimation and Performance Evaluation

In both synthetic and real datasets, expressions of a gene were normalized to have zero mean and one standard deviation. In the proposed algorithm, LASSO regression was used to identify regulatory edges and to generate sparse time-delayed GRNs. The network topology is essentially achieved by I separate LASSO regressions. The LASSO solutions were achieved by using the GLMNET package [19] which can generate the whole solution path for α_i. For each such regression, the penalty parameter α_i was chosen by using five-fold cross-validation.

We evaluated the performance of the proposed approach over a hundred simulated datasets for each combination of number of genes and number of timepoints. In generating simulated datasets, the network topology was extracted from GNW software and each regulatory edge was randomly assigned a timedelay. Hence, the true information (ground truth) of regulatory connection and their delay was available. Using this information, we employed precision, recall and F-measure as performance metrics. Let TP, FP, TN, and FN denotes true positive, false positive, true negative, and false negative between the generated network and ground truth. TP were computed for exact time delays while FPs were computed by counting all instances when a false edge (of any time order) is detected. The precision, recall, and F-measure are defined below:

$$Precision = \frac{TP}{TP + FP} \tag{6}$$

$$Recall = \frac{TP}{TP + FN} \tag{7}$$

$$F - measure = 2 \times \frac{Precision \times Recall}{Precision + Recall} \tag{8}$$

We further defined order identification accuracy (OIA) as the number of edges which were identified with true time-delays divided by total number of identification of true edges irrespective of time order, i.e., $OIA = \frac{TP}{w}$ where w denotes the number of all true edges.

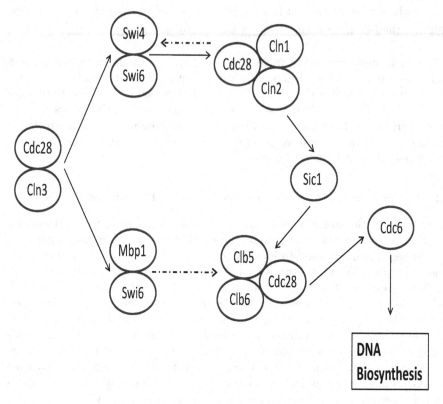

Fig. 1. *S. cerevisiae* KEGG pathway in G1 phase. Dotted line represents indirect regulation.

3.3   Real Dataset

We selected the *Saccharomyces cerevisiae* (yeast) cell cycle dataset to test the performance of the proposed method. Spellman et al. have identified 800 differentially expressed genes for cell-cycle regulation covering four phases (G1,S,G2 and M) of yeast development [20]. For our analysis, eleven genes (Cln3, Cdc28, Swi4, Swi6, Clb5, Clb6, Cln1, Cln2, Cdc6, Sic1, Mbp1) were specifically selected from the cdc28 experiment of G1-phase resulting in dataset with 11 genes and 17 time points. As suggested in [11], the first time point is excluded as it is related to the M step. This dataset is used in two recent studies and is available with TDARACNE package [11].

In the proposed method, the α parameter plays an important role in determining regulators of a particular gene. To avoid errors due to parameter estimation with a five-fold cross validation, we repeated the complete process for 100 times and used edge stability of 0.75 to infer the final single network structure [16]. An edge stability of 0.75 implies that an edge is derived at least 75% of the time.

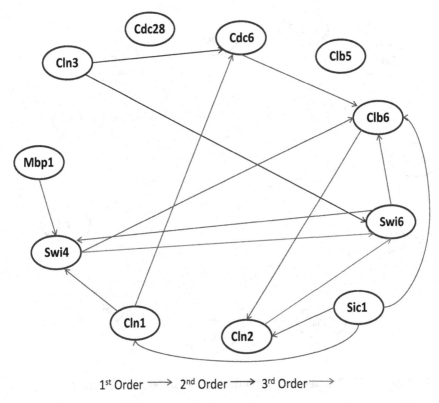

Fig. 2. An inferred time-delayed gene regulatory network of 11 genes of *S. cerevisiae*. The maximum time delay is set to 3.

4   Results and Discussion

Inferring a time-delayed GRN from the gene-expression data is an important step to understand the dynamics of the underlying gene regulation. In this paper, we have proposed a two-step approach to infer such a network using cross-correlation and sparse regression. To evaluate the efficacy of this approach, several synthetic datasets with varying time points and numbers of genes were generated. By fixing the maximum delay to 3 or 5, the performances of the proposed approach are shown in Table 1 and Table 2, respectively.

The results on synthetic datasets show that increase in number of genes and decrease in length of time series reduces precision, recall and F-measure. The results also show that within truly identified edges, the correct delay is also generally identified with a high accuracy. At the same time, by fixing the lower value of the maximum possible time delay, the performance could be improved, because a single point increase in maximum delay (r) increases the number of parameters to be estimated by I^2. Moreover, the available number of time samples is reduced by one. Hence, it is important to not choose too high a

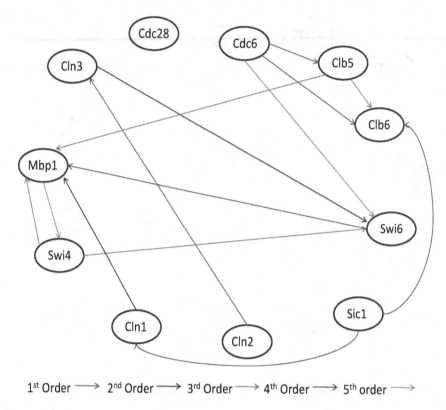

Fig. 3. An inferred time-delayed gene regulatory network of 11 genes of *S. cerevisiae*. The maximum time delay is set to 5.

value of the maximum possible time-delay to get any meaningful results by the proposed computational algorithm.

As a true underlying network of *S. cerevisiae* is unknown yet, we use the KEGG pathway to validate the reconstructed GRN (Figure 1). In yeast cell cycle progression, G1 and G2 phases are gaps between DNA replication (S phase) and mitosis (M phase). As per KEGG pathway and [21], in G1 phase, an association between Cln3 and Cdc28 is needed to initiate the start of the cycle . After reaching a certain threshold of the Cln3/Cdc28 complex, two transcription factors SBF and MBF are activated. Swi4 and Swi6 form the SCB complex with SBF which results in activation of Cln1 and Cln2 genes [22] while Mbp1 and Swi6 form a complex with MBF to promote transcription of other genes required for S-phase progression. Cln1 and Cln2 interacting with Cdc28 promote the activation of B-type cyclin associated CDK, which drives DNA replication and entry into mitosis. Further, Clb1 and Clb2 are associated with Cdc28 and this complex represses Sic1, which in turn represses the Clb5/Clb6/Cdc28 complex.

Table 1. The performance of the proposed method over 100 simulated datasets with $r = 3$

Network Size	Time Points	Precision	Recall	F-measure	OIA
20	20	0.35	0.31	0.32	0.76
	30	0.45	0.50	0.47	0.82
	40	0.52	0.62	0.56	0.84
50	20	0.24	0.36	0.29	0.92
	30	0.30	0.66	0.41	0.96
	40	0.35	0.82	0.49	0.97
100	20	0.18	0.14	0.16	0.90
	30	0.22	0.31	0.26	0.92
	40	0.26	0.47	0.33	0.93

Table 2. The performance of the proposed method over 100 simulated datasets with $r = 5$

Network Size	Time Points	Precision	Recall	F-measure	OIA
20	20	0.21	0.22	0.21	0.57
	30	0.28	0.40	0.33	0.67
	40	0.36	0.53	0.42	0.74
50	20	0.17	0.25	0.20	0.86
	30	0.22	0.57	0.31	0.92
	40	0.26	0.75	0.39	0.94
100	20	0.13	0.11	0.12	0.86
	30	0.16	0.26	0.20	0.87
	40	0.18	0.42	0.25	0.88

The GRNs inferred by the proposed method are shown in Figure 2 and Figure 3. As can be seen, several true gene-gene interactions have been recovered. For example, in Figure 2, we find interaction between (1) Cln3 and Swi6, (2) Clb6 and Cdc6, (3) interaction of Sic1 with Cln1, Clb6 and Cln2, (4) Swi4 and Swi6, and (5) interaction of Swi4 and Swi6 with Cln1 and Cln2. However, we also note that there are few wrong directions of regulation. Further, comparison between Figure 2 and 3 reveals that few new edges are formed and few are not recovered. Such phenomenon could be attributed to loss of time samples and increase in parameter space.

As discussed earlier, building a time-delayed GRN is a very challenging problem and several future directions may lead to better solutions. In our earlier work, we proposed a bootstrapping technique for short time-series datasets with a first-order assumption [23]. Developing such techniques for higher order models and integrating stability criteria is a promising possible extension of this work. In the current two-step procedure, cross-correlation is used to determine the probable time lags. Since cross-correlation may suffer due to small sample size, developing a robust technique with possibly a single step procedure would be another interesting extension of this work. Last but not least, an extension of the data integration approach for first-order GRN inference [24,25] to higher order may help in deriving a highly accurate time-delayed GRN.

Acknowledgments. This work is supported by a AcRF Tier 2 grant MOE2010-T2-1-056 (ARC 9/10), Ministry of Education, Singapore. The support provided by the Singapore-MIT Alliance to Roy E. Welsch is also acknowledged.

References

1. Pisarev, A., Poustelnikova, E., Samsonova, M., Reinitz, J.: Flyex, the quantitative atlas on segmentation gene expression at cellular resolution. Nucleic Acid Research 37, D560–D566 (2009)
2. Huang, Y., Tienda-Luna, I., Wang, Y.: Reverse engineering gene regulatory networks. IEEE Signal Processing Magazine 26(1), 76–91 (2009)
3. Kim, S., Imoto, S., Miyano, S.: Inferring gene networks from time series microarray data using dynamic bayesian networks. Briefings in Bioinformatics 4(3), 228–235 (2003)
4. Fujita, A., Sato, J., Garay-Malpartida, H., Yamaguchi, R., Miyano, S., Sogayar, M., Ferreira, C.: Modeling gene expression regulatory networks with the sparse vector autoregressive model. BMC Systems Biology 1, 39 (2007)
5. Chima, C., Hua, J., Jung, S.: Inference of gene regulatory networks using time-series data: A survey. Current Genomics 10, 416–429 (2009)
6. de Jong, H.: Modeling and simulation of genetic regulatory systems: A literature review. Journal of Computational Biology 9(1), 67–103 (2002)
7. Fogelberg, C., Palade, V.: Machine learning and genetic regulatory networks: A review and a roadmap. In: Hassanien, A.-E., Abraham, A., Vasilakos, A.V., Pedrycz, W. (eds.) Foundations of Computational, Intelligence 1. SCI, vol. 201, pp. 3–34. Springer, Heidelberg (2009)
8. Li, X., Rao, S., Jiang, W., Li, C., Xiao, Y., Guo, Z., Zhang, Q., Wang, L., Du, L., Li, J., Li, L., Zhang, T., Wang, Q.: Discovery of time-delayed gene regulatory networks based on temporal gene expression profiling. BMC Bioinformatics 7, 26 (2006)
9. Chaitankar, V., Ghosh, P., Perkins, E., Gong, P., Zhang, C.: Time lagged information theoretic approaches to the reverse engineering of gene regulatory networks. BMC Bioinformatics 11(suppl. 6), S19 (2010)
10. Chaturvedi, I., Rajapakse, J.C.: Detecting robust time-delayed regulation in mycobacterium tuberculosis. BMC Genomics 10(suppl. 3), S28 (2009)
11. Zoppoli, P., Morganella, S., Ceccarelli, M.: TimeDelayed-ARACNE: Reverse engineering of gene networks from time-course data by an information theoretic approach. BMC Bioinformatics 11, 154 (2010)
12. Chaturvedi, I., Rajapakse, J.C.: Building gene networks with time-delayed regulations. Pattern Recognition Letters 31(14), 2133–2137 (2010)
13. Morshed, N., Chetty, M., Vinh, N.: Simultaneous learning of instantaneous and time-delayed genetic interactions using novel information theoretic scoring technique. BMC Systems Biology 6, 62 (2012)
14. Chueh, T.H., Lu, H.: Inference of biological pathway from gene expression profiles by time delay boolean networks. PLOS ONE 7(8), e42095 (2012)

15. Shimamura, T., Imoto, S., Yamaguchi, R., Fujita, A., Nagasaki, M., Miyano, S.: Recursive regularization for inferring gene networks from time-course gene expression profiles. BMC Systems Biology 3, 41 (2009)
16. Rajapakse, J.C., Mundra, P.A.: Stability of building gene regulatory networks with sparse autoregressive models. BMC Bioinformatics 12(suppl. 13), S17 (2011)
17. Orfanidis, S.: Optimum Signal Processing. An Introduction. Prentice-Hall (1996)
18. Marbach, D., Schaffter, T., Mattiussi, C., Floreano, D.: Generating realistic in silico gene networks for performance assessment of reverse engineering methods. Journal of Computational Biology 16(2), 229–239 (2009)
19. Friedman, J., Hastie, T., Tibshirani, R.: glmnet: Lasso and elastic-net regularized generalized linear models
20. Spellman, P., Sherlock, G., Zhang, M., Iyer, V., Anders, K., Eisen, M., Brown, P., Botstein, D., Futcher, B.: Comprehensive identification of cell cycle regulated genes of the yeast saccharomyces cerevisiae by microarray hybridization. Molecular Biology of the Cell 9(12), 3273–3297 (1998)
21. Nasmyth, K.: Control of the yeast cell cycle by the cdc28 protein kinase. Current Opinion in Cell Biology 5(2), 166–179 (1993)
22. Siegmund, R., Nasmyth, K.: The saccharomyces cerevisiae start-specific transcription factor Swi4 interacts through the ankyrin repeats with the mitotic Clb2/Cdc28 kinase and through its conserved carboxy terminus with Swi6. Molecular Biology of the Cell 16(6), 2647–2655 (1996)
23. Mundra, P.A., Welsch, R.E., Rajapakse, J.C.: Bootstrapping of short time-series multivariate gene-expression data. In: Colubi, A., Fokianos, K., Gonzalez-Rodriguez, G., Kontaghiorghes, E. (eds.) Proceedings of 20th International Conference on Computational Statistics(COMPSTAT 2012), pp. 605–616 (2012)
24. Chen, H., Maduranga, D., Mundra, P., Zheng, J.: Integrating epigenetic prior in dynamic bayesian network for gene regulatory network inference. In: IEEE Symposium on Computational Intelligence in Bioinformatics and Computational Biology (accepted, 2013)
25. Hecker, M., Lambeck, S., Toepfer, S., van Someren, E., Guthke, R.: Gene regulatory network inference: Data integration in dynamic models: A review. Biosystems 96, 86–103 (2009)

A Simulation of Synthetic *agr* System in *E.coli*

Xiangmiao Zeng, Ke Liu, Fangping Xie, Ying Zhang, Lei Qiao, Cuihong Dai,
Aiju Hou, and Dechang Xu

School of Food Science & Engineering
Harbin Institute of Technology, P.R.C
dcxu@hit.edu.cn

Abstract. *Staphylococcus aureus* (*S.aureus*) is an important human pathogen.
Its strong infection ability benefits from the quorum-sensing system *agr*
(accessory gene regulator). In order to eliminate *S.aureus* from the
environment, an engineered *E.coli* was designed. It can sense the extracellular
AIP (auto-inducing peptide) and then, as a response, produce Lysostaphin to
kill *S.aureus*. To characterizing how *E.coli* sense *S.aureus* and secrete
Lysostaphin, a mathematical model was developed. According to the model, it
is at least 2.5 hours for the system to sense the AIP (*S.aureus*) and then produce
enough Lysostaphin to kill the *S.aureus*, and therefore keep the AIP
concentration at a relative low condition.

Keywords: *S.aureus*, *E.coli*, *agr*, quorum-sensing system, simulation.

1 Introduction

Staphylococcus aureus (*S.aureus*) is an important human pathogen. Compared to
single symptom pathogen such as Influenza Virus [1,2], many diseases for example,
osteomyelitis, infective endocarditis, septic arthritis and metastatic abscess
formation[3] are caused by *S.aureus* infection which are major causes of morbidity
and mortality in community and hospital settings. Many strains have the resistance to
spectrum of antibiotics, making elimination often difficult. Its strong infection
ability benefits from the quorum-sensing system *agr* (accessory gene regulator)[4].
Quorum-sensing (QS) is the ability of bacteria to communicate and coordinate
behavior via signaling molecules. This signal could be, for example, temperature,
pressure, pH, or the bacterial population density (in the *agr* system of *S.aureus*). The
indicator of population density in *S.aureus* is called auto-inducing peptide (AIP).
When the concentration of the indicator reaches a threshold, bacteria make some
changes in its population as a response to the environment they live in.

The quorum-sensing system mentioned above comprises a sensor protein in the
membrane of the cell and a response regulator within the cell cytoplasm[5]. The
receptor detects the signal in the environment and then triggers auto-phosphorylation
causing the phosphate transfers to the response regulator protein. The phosphorylated
proteins usually show a higher affinity for the relevant DNA binding sites than
un-phosphorylated one. As mentioned above, the *agr* system like other QS systems is
a cell-to-cell communication mechanism which is usually considered to facilitate the
coordination of gene expression at the population level and control the production of
virulence factors while infection.

Z. Cai et al. (Eds.): ISBRA 2013, LNBI 7875, pp. 76–86, 2013.
© Springer-Verlag Berlin Heidelberg 2013

In *S.aureus*, the *agr* operon consists of two transcriptional units, termed RNAII and RNAIII[6], transcribed from the divergent promoters, *agrP2* and *agrP3*. The *agrP2* locus contains four genes *agrB*, *agrD*, *agrC*, *agrA*. *agrC* and *agrA* encode the sensor and the response regulator, respectively, while *agrB* and *agrD* are responsible for the synthesis of the AIP. In broth cultures, the *agr* system is activated in mid- to post-exponential growth, when AIP concentration reaches a threshold. AIP binds to an extracellular domain of AgrC resulting in auto-phosphorylation at a conserved cytoplasmic histidine residue[7]. The phosphate group is then transferred to AgrA, which have a higher affinity for the agrP2 and agrP3 than un-phosphorylated AgrA. Phosphorylated AgrA trigger transcription from both P2 and P3[8].

In order to eliminate *S.aureus* from the environment, an engineered *Escherichia coli* (*E.coli*) was invented. The *E.coli* carries a plasmid containing *agrC*, *agrA*, *P2*, *P3* GFP (green fluorescence protein) and Lysostaphin[9] which degrades the cell wall of *Staphylococcus aureus* and then make them inactive. To better understanding the dynamics of the *agr* system in *E.coli*, a mathematical model is described below to characterize the *agr* system in *E.coli*.

2   Materials and Methods

Plasmids containing agrA, agrC, agrP2, agrP3 and GFP (Lysostaphin) were constructed with protocols used at iGEM headquarters. Gene agrA, agrC, agrP2, agrP3, Lysostaphin were modified with iGEM standard prefixes and suffixes. DNA assembly and transformation are following protocols of 3A assembly and transformation, respectively.

Cultures of the *E.coli* cells were grown in 50ml LB supplemented with 50μg/ml ampicillin at 37℃ (160r/min) to exponential phase (A_{600}=0.2). *S.aureus* were cultured in 30ml tryptic soy broth to steady phase (A_{600}=0.4). Then, the *E.coli* (A_{600}=0.2) medium was treated with 5ml supernatant of *S.aureus* (A_{600}=0.4). After 24 hours of treatment, cultures were placed under a fluorescence microscope (×1000) to observe the expression of GFP.

3   Model

The mathematical model is developed to describe how the concentration of extracellular AIP affects the *agr* system in *E.coli*. To be able to calculate the steady-state levels of Lysostaphin (or GFP) in cells at different extracellular AIP concentrations, a non-growing homogenous bacterial population is assumed.

The model assumptions are as follows.

(1)The *E.coli* population contain a number (n) of cells, the volume of a cell's cytoplasm is v. The extracellular AIP concentration is P. Neglecting inhibiting AIP.

(2) Concentration of a substance in the cytoplasm is assumed to be homogeneous. For molecules in the membrane, the concentration by area is scaled to an amount per cytoplasm volume since total membrane area due to cell division is proportional to total cytoplasm volume.

(3) All reactions are following mass action principle and saturation kinetics.

(4) The level of translation product is assumed to be proportional to mRNA levels.

(5) The agrP2 promoter activity is high enough to make stochastic effects negligible.

(6) There is a basal amount of AgrA and AgrC.

(7) There is no delay in synthesis of either component or delay because of protein transportations.

(8) The variables and parameters are described in Table 1 and 2.

(9) The values of variables and parameters are partly obtained from the present publication on *agr* system, and partly assumed reasonably.

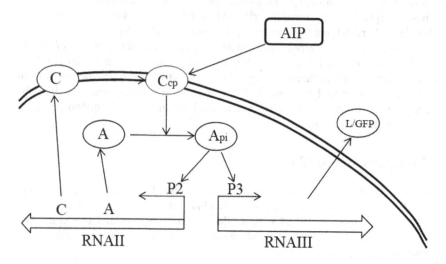

Fig. 1. A Schematic View of the Joint Metabolic and Gene Regulation Network

Table 1. Variables Used in the Model

Variable	Dimension	Description	Initial value
C	μmol	Total amount of un-complex AgrC in the population	0.1
C_{CP}	μmol	Total amount of complex between AgrC and AIP	0
A	μmol	Total amount of un-phosphorylated AgrA in the population	0.1
A_{pi}	μmol	Total amount of phosphorylated AgrA in the population	0
RII	μmol	Total amount of RNAII in the population	0
RIII	μmol	Total amount of RNAIII in the population	0
L	μmol	Total amount of Lysostaphin in the population	0
		Degradation of the substance	0

Table 2. Parameters Used in the Model

Parameter	Dimension	Description	Value
n		Number of cells in the *E.coli* population	
v	μm^3	Volume of a *E.coli* cell	1
P	$\mu mol \cdot \mu m^{-3}$	Concentration of AIP in the environment	
k_1	$\mu mol^{-1} \cdot \mu m^3 \cdot hour^{-1}$	Association rate of complex C_{CP}	1
k_2	$hour^{-1}$	Dissociation rate of complex C_{CP}	0.1
k_3	$\mu mol^{-1} \cdot \mu m^3 \cdot hour^{-1}$	Phosphorylation rate of AgrA	10
k_4	$hour^{-1}$	Dephosphorylation rate of AgrA	1
D_1	$hour^{-1}$	Degradation rate of complex C_{CP}	2
D_2	$hour^{-1}$	Degradation rate of phosphorylated AgrA	2
K_{Api}	$\mu mol \cdot hour^{-1}$	Maximal AgrA dependent transcription rate of P2/P3	10
k_{Api}	$\mu mol \cdot \mu m^{-3}$	Concentration of phosphorylated AgrA required for half-maximal AgrA dependent transcription rate of P2/P3	1
K_b	$\mu mol \cdot hour^{-1}$	Basal transcription rate of P2	0.1
D_{R2}	$hour^{-1}$	Degradation rate of RNAII	2
D_{R3}	$hour^{-1}$	Degradation rate of RNAIII	2
α_A		Effective factor of AgrA protein synthesis	1
D_A	$hour^{-1}$	Degradation rate of un-phosphorylated AgrA	2
α_C		Effective factor of AgrC protein synthesis	1
D_C	$hour^{-1}$	Degradation rate of un-complex AgrC	2
α_L		Effective factor of Lysostaphin protein synthesis	1
D_L	$hour^{-1}$	Degradation rate of Lysostaphin	2

3.1   Interaction between Sensor AgrC and AIP

We assume that the *agr* system activates at a low concentration of AIP. AIP (P) binds to AgrC (C) forming a complex (C_{CP}) at a rate k_1 and dissociates at a rate k_2. The complex C_{CP} is assumed to degrade at a rate D_1 . Thus, reaction (1) and (2) are

$$C <=> C_{CP} : R_1 = k_1CP - k_2C_{CP} \tag{1}$$

and

$$C_{CP} => : R_2 = D_1C_{CP} \tag{2}$$

3.2   Phosphorylation of AgrA

If the phosphate available is plenty enough, the phosphorylation rate depends on the amount of complex C_{CP} and un-phosphorylated AgrA and the stability of AgrA. Complex C_{CP} activates the phosphorylation of AgrA at a rate k_3. The AgrA-phosphate complex dissociates at a rate k_4 and degrades at a rate D_2. Reaction (3) and (4) are

$$A \Longleftrightarrow A_{pi} : R_3 = \frac{k_3}{nv} C_{CP} A - k_4 A_{pi} \tag{3}$$

and

$$A_{pi} \Longrightarrow : R_4 = D_2 A_{pi} \tag{4}$$

3.3   Transcription of RNAII

In the model, we assume that phosphorylated AgrA (Api) influences the transcription of RNAII. The activity of AgrA-dependent agrP2 promoter follows saturation kinetics. A maximal AgrA-dependent transcription rate of P2/P3 is K_{Api} and half-maximum as the concentration of phosphorylated AgrA equals k_{Api}. The AgrA-independent transcription rate of agrP2 is K_b. Reaction of transcription of RNAII is

$$\Longleftrightarrow RNAII : R_5 = nK_{Api} \frac{\dfrac{A_{pi}}{nv}}{k_{Api} + \dfrac{A_{pi}}{nv}} + nK_b - D_{R_{II}} R_{II} \tag{5}$$

3.4   Translation of AgrC and AgrA

The translation of AgrA and AgrC from RNAII is assumed to proceed with an efficiency of α_A and α_C. Since AgrA and AgrC are located in the same operon, the synthesis rate of AgrA and AgrC is assumed to be the same. And AgrA and AgrC degrade at the rate D_A and D_C, respectively. So reactions are

$$\Longleftrightarrow A : R_6 = \alpha_A nK_{Api} \frac{\dfrac{A_{pi}}{nv}}{k_{Api} + \dfrac{A_{pi}}{nv}} + \alpha_A nK_b - D_A A \tag{6}$$

and

$$\Longleftrightarrow C : R_7 = \alpha_C nK_{Api} \frac{\dfrac{A_{pi}}{nv}}{k_{Api} + \dfrac{A_{pi}}{nv}} + \alpha_C nK_b - D_C C \tag{7}$$

3.5   Transcription of GFP and Lysostaphin

Transcription of GFP and Lysostaphin is similar to RNAII. But here we assumed P3 all depended on phosphorylated AgrA. So the transcription reaction is

$$<=> RNAIII : R_8 = nK_{Api} \frac{\frac{A_{pi}}{nv}}{k_{Api} + \frac{A_{pi}}{nv}} - D_{R_{III}} R_{III} \tag{8}$$

3.6   Translation of GFP and Lysostaphin

The efficiency of RNA translating to Lysostaphin is assumed to be α_L, and the translation product of agrP3 degrades at a rate D_L.

$$<=> L : R_9 = \alpha_L nK_{Api} \frac{\frac{A_{pi}}{nv}}{k_{Api} + \frac{A_{pi}}{nv}} - D_L L \tag{9}$$

The whole model described above is summarized below

$$\frac{dA}{dt} = -R_3 + R_6 = -\frac{k_3}{nv} C_{CP} A + k_4 A_{pi} + \alpha_A nK_{Api} \frac{\frac{A_{pi}}{nv}}{k_{Api} + \frac{A_{pi}}{nv}} + \alpha_A nK_b - D_A A$$

$$\frac{dC}{dt} = -R_1 + R_7 = -k_1 CP + k_2 C_{CP} + \alpha_C nK_{Api} \frac{\frac{A_{pi}}{nv}}{k_{Api} + \frac{A_{pi}}{nv}} + \alpha_C nK_b - D_C C$$

$$\frac{dA_{pi}}{dt} = +R_3 - R_4 = \frac{k_3}{nv} C_{CP} A - k_4 A_{pi} - D_2 A_{pi}$$

$$\frac{dC_{CP}}{dt} = +R_1 - R_2 = k_1 CP - k_2 C_{CP} - D_1 C_{CP}$$

$$\frac{dL}{dt} = +R_9 = \alpha_L nK_{Api} \frac{\frac{A_{pi}}{nv}}{k_{Api} + \frac{A_{pi}}{nv}} - D_L L$$

$$\frac{dR_{II}}{dt} = +R_5 = nK_{Api}\frac{\dfrac{A_{pi}}{nv}}{k_{Api}+\dfrac{A_{pi}}{nv}} + nK_b - D_{R_{II}}R_{II}$$

$$\frac{dR_{III}}{dt} = +R_8 = nK_{Api}\frac{\dfrac{A_{pi}}{nv}}{k_{Api}+\dfrac{A_{pi}}{nv}} - D_{R_{III}}R_{III}$$

4   Results

In experimental research, we have managed to construct *E.coli* that can sense the extracellular AIP and successfully observed the green fluorescence after 24 hours treatment. By varying the AIP concentration, we got a series points of GFP expression state level, and found the threshold time of *agr* system changing with AIP concentration. When changed in number (n) of cells, the substance increase proportionally with n. When AIP concentration is at a high level, the GFP expression is about 100 times than at a low level.

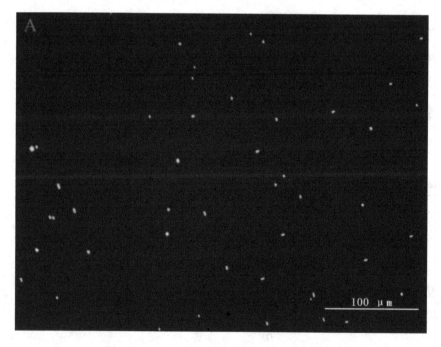

Fig. 2. Fluorescence in *E.coli* treated *S.aureus*

A sharp increase of GFP expression can be seen in Figure 3 when treated with different concentration of AIP. The threshold of 30 hours after treating AIP is about 1.145 (Table 3). When AIP concentration is higher than 1.2, the time for GFP expression reached a high level is less than 25 hours. While the AIP concentration is at a high level, for example 100 or more the reaction time of the *agr* system is at least 2.5 hours (Table 4).

The steady-state level of GFP expression at different AIP concentration is calculated using SBToolbox2[10] for MatLab.(Figure 5 and 6).

Table 3. State Value of GFP 30 hours after treating AIP

AIP	0	0.25	0.5	0.75	1.0	1.1	1.11	1.12	1.13
GFP	0	0.53	0.117	0.204	0.37	0.64	0.745	0.96	1.6
AIP	1.14	1.145	1.15	1.16	1.17	1.2	1.4	1.6	2.0
GFP	6.5	20	32.5	37	38	38	38	38	38

The fluorescence observed in the *E.coli* ($A_{600}=0.2$) medium treating with 5ml supernatant of *S.aureus* ($A_{600}=0.4$) after 24 hours using fluorescence microscope ($\times 1000$)

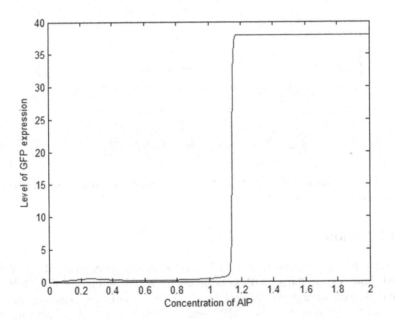

Fig. 3. State Value of GFP 30 hours after treating AIP

We changed the parameter P from 0 to 2 and obtained the response of L at the time 30.

Table 4. Threshold time changing with AIP

AIP	1.08	1.085	1.095	1.1	1.115	1.13	1.145	1.16	1.175
Time	125	86	59	52	41	34	30	27	25
AIP	1.20	1.3	1.5	2	5	10	50	100	200
Time	22	16	11	7	4	3.5	2.7	2.5	2.5

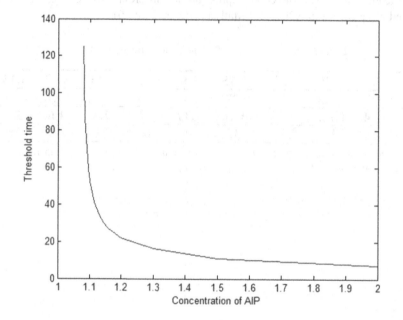

Fig. 4. Threshold time changing with AIP

We changed the parameter P from 1.08 to 200 to figure out the time that the system switches on as a response to the changing AIP.

5   Discussion

We have developed a mathematical model of *agr* system in an engineered *E.coli*, and described how extracellular AIP concentration would influence the GFP (or Lysostaphin) level in a homogeneous bacterial cell population. The system is a positive feedback system. The AIP bind to AgrC and then activate the phosphorylation of AgrA. The phosphorylated AgrA then triggers the production of AgrA, AgrC and Lysostaphin. As a result, there will be more and more Lysostaphin. If it is effective enough to kill the *S.aureus*, the Lysostaphin would reduce the population of *S.aureus* which results in the decrease of AIP. According to the model, it is at least 2.5 hours for the system to sense the AIP (*S.aureus*) and then produce enough Lysostaphin to kill the *S.aureus*, and therefore keep the AIP concentration at a relative low condition.

6 Appendix

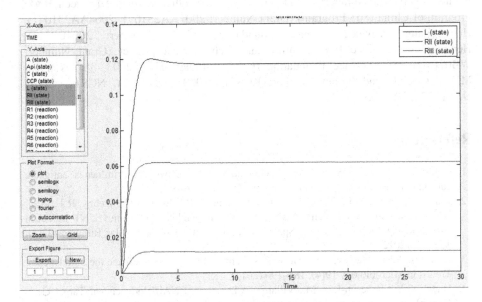

Fig. 5. Level of Lysostaphin RNAII and RNAIII when AIP=0.5 in MatLab GUI

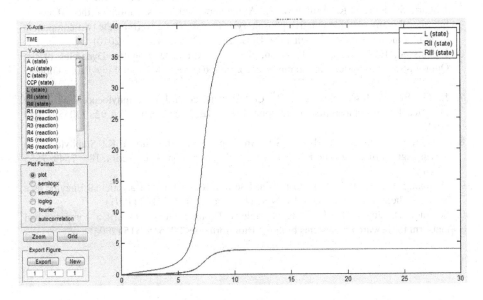

Fig. 6. Level of Lysostaphin RNAII and RNAIII when AIP=2.0 in MatLab GUI

Acknowledgment. This work is partially supported by the China Natural Science Foundation (Grant Number: 30771371, 31271781) , the National High-tech R&D Program of China (863 Program) (Grant Number: 2001AA231091, 2004AA231071), Heilongjiang Province Science Foundation (Grant Number: 2004C0314), Heilongjiang Province key scientific and technological project (Grant Number: WB07C02), HIT Science Foundation (Grant Number: HIT. 2003. 38), National MOST special fund (Grant Number: KCSTE- 2000-JKZX- 021, NCSTE- 2007-JKZX- 022, 2012EG111228).

References

1. Cai, Z., Zhang, T., Wan, X.F.: A computational framework for influenza antigenic cartography. PLoS Computational Biology 6(10), e1000949
2. Cai, Z., Ducatez, M.F., Yang, J., Zhang, T., Long, L.P., Boon, A.C., Webby, R.J., Wan, X.F.: Identifying Antigenicity Associated Sites in Highly Pathogenic H5N1 Influenza Virus Hemagglutinin by Using Sparse Learning. Journal of Molecular Biology 422(1), 145-155 (2012)
3. Edwards, A.M., Massey, R.C.: How does Staphylococcus aureus escape the bloodstream. Trends in Microbiology 19(4), 184–190 (2011)
4. Queck, S.Y., Jameson-Lee, M., Villaruz, A.E., et al.: RNAIII-Independent Target Gene Control by the agr Quorum-Sensing System: Insight into the Evolution of Virulence Regulation in Staphylococcus aureus. Molecular Cell 32(1), 150–158 (2008)
5. Jabbari, S., King, J.R., Williams, P.: A mathematical investigation of the effects of inhibitor therapy on three putative phosphorylation cascades governing the two-component system of theagr operon. Mathematical Biosciences 225(2), 115–131 (2010)
6. Gustafsson, E., Nilsson, P., Karlsson, S., et al.: Characterizing the Dynamics of the Quorum-Sensing System in Staphylococcus aureus. Mol. Microbiology Biotechnology 8(4), 232–242 (2004)
7. Ji, G., Bcavis, R.C., Novick, R.P.: Cell density control of staphylococcal virulence mediated by an octapepitide pheromone. Proc. Natl. Acad. Sci. USA 92, 12055–12059 (1995)
8. Koenig, R.L., Ray, J.L., Maleki, S.J., Smeltzer, M.S., Hurlburt, B.K.: Staphylococcus aureus AgrA binding to the RNAIII- agr regulatory region. J. Bacteriol. 186, 7549–7555 (2004)
9. Oldham, E.R., Daley, M.J.: Lysostaphin: Use of a Recombinant Bactericidal Enzyme as a Mastitis Therapeutic. Journal of Dairy Science 74(12), 4175–4182 (1991)
10. Schmidt, H., Jirstrand, M.: Systems Biology Toolbox for MATLAB: a computational platform for research in systems biology. Bioinformatics 22, 514–515 (2006)

Gene Regulatory Networks from Gene Ontology

Wenting Liu[1,*], Kuiyu Chang[1], Jie Zheng[1,2],
Jain Divya[1], Jung-Jae Kim[1], and Jagath C. Rajapakse[1,3,4]

[1] Bioinformatics Research Center, School of Computer Engineering,
Nanyang Technological University, Singapore
[2] Genome Institute of Singapore, A*STAR
(Agency for Science, Technology, and Research), Biopolis, Singapore
[3] Singapore-MIT Alliance, Singapore
[4] Department of Biological Engineering, Massachusetts Institute of Technology, USA
{wliu7,ZhengJie,jungjae.kim,asjagath}@ntu.edu.sg,
kuiyu.chang@pmail.ntu.edu.sg, divyajain.30@gmail.com

Abstract. Gene Ontology (GO) provides a controlled vocabulary and hierarchy of terms to facilitate the annotation of gene functions and molecular attributes. Given a set of genes, a Gene Ontology Network (GON) can be constructed from the corresponding GO annotations and semantic relations among GO terms. Transitive rules can be applied to GO semantic relations to infer transitive regulations among genes. Using information content as a measure of functional specificity, a shortest regulatory path detection algorithm is developed to identify transitive regulations in GON. Since direct regulations may be overlooked during the detection of gene regulations, gene functional similarities deduced from GO terms are used to detect direct gene regulations. Both direct and transitive GO regulations are then used to construct a Gene Regulatory Network (GRN). The proposed approach is evaluated on seven E.coli sub-networks extracted from an existing known GRN. Our approach was able to detect the GRN with 85.77% precision, 55.7% recall, and 66.26% F1-score averaged across all seven networks.

Keywords: gene ontology, gene regulatory network, transitive gene regulation, semantic similarity, functional similarity.

1 Introduction

Gene regulation denotes the cellular activity that arises when a set of genes interact with one another. Gene regulations can be organized into a gene regulatory network (GRN), which provides insights into complex biological mechanisms. However, ground truths of biological regulatory networks are unknown in most cases. Building a GRN that is accurate and biologically plausible thus remains an open research problem in functional genomics.

Gene Ontology (GO) provides a controlled vocabulary arranged in a hierarchy of terms to facilitate the annotation of gene functions and molecular attributes.

---

* Corresponding author.

Z. Cai et al. (Eds.): ISBRA 2013, LNBI 7875, pp. 87–98, 2013.
© Springer-Verlag Berlin Heidelberg 2013

GO has been widely used for validating functional genomics experiments [1], [2]. In this paper, we present a method to build GRN that captures both direct and transitive regulations based on GO. The semantic relations of these gene annotation terms provide some evidences for gene regulations. Applying the transitive rules of these semantic relations, we can also infer transitive gene regulations.

A transitive regulation is a regulation between two genes via one or more transitive genes in the absence of a direct regulation [3]. For example, suppose gene g_1 regulates gene g_2 directly and gene g_2 have some relation to gene g_3. Then since there is no direct regulation between g_1 and g_3, if we can infer a regulatory relation from g_1 to g_3, we say g_1 transitively regulates g_3 through transitive gene g_2.

The vast majority of previous research has focused on finding direct regulations between genes, using co-expressions [4]. In our approach, we use the information content of GO terms to represent the functional specificity and information flow, thereby determining the most probable transitive regulations between genes. Zhou et al. [3] linked genes of the same biological pathway based on the transitive expression similarity among genes. They determine the transitive co-expression genes by shortest path analysis on large-scale yeast microarray expression data. Instead of finding the shortest path based on distance, we find the shortest path based not only on distance but also on GO regulatory relations, which gives more accurate transitive regulations, as shown in our experiments.

Semantic Similarity of gene annotations can provide some clues for direct gene regulations. Cheng et al. [5] and Kustra et al. [6] incorporate gene similarity score from GO semantic similarity into gene expression data to cluster genes. Franke et al. [4] assume functional interactions among similar genes if they share more GO terms, and incorporate microarray co-expression and protein-protein interaction data to construct a human gene network. Similarly, in order to detect direct regulations, we consider the functional similarity of genes based on the semantic similarity. To better estimate the functional similarity score of gene pairs, we modify the original term probabilities by taking into consideration the chances of a term being annotated to both genes. Since these two GO methods complement each other, we then propose a GO fusion method to combine both direct and transitive regulations to generate the final GRN.

2   Methods

2.1   Gene Ontology Networks

A gene ontology (GO) is a structured controlled biological vocabulary of various gene terminologies and their inter-related functional characteristics. It describes how gene products behave in a cellular context. The ontology covers three domains: biological process (BP), molecular function (MF), and cellular component (CC). BP is a collection of molecular events, MF defines gene functions in the biological process, and CC describes gene locations within a cell. A gene is associated with GO terms that describe the properties of its products (i.e., proteins).

In our approach, only BP and MF terms are used since the cellular component (CC) is not directly related to gene regulation.

There are three defined semantic relations between GO terms: $is - a$ is used when one GO term is a subtype of another GO term, $part - of$ is used to represent part-whole relationship in the GO terms, and $regulate$ is used when the occurrence of one biological process directly affects the manifestation of another process or quality [7]. Let $\mathcal{R} = \{is - a, \ part - of, \ regulate\}$ denote the set of ontological relations. The GO can thus be represented as a hierarchical directed acyclic graph, where each term is related to one or more terms in the same or different domain. The GO has three roots at the topmost level: BP, MF, and CC. Nodes/terms near the root of the directed acyclic graph have broader functions and are hence shared by many genes; leaf nodes/terms on the other hand convey more specific biological functions.

GO annotation (GOA) is the process in which GO terms are annotated to gene products. GOA data can be readily obtained from the GO annotation database [8]. The GO hierarchical structure also allows annotators to assign properties to genes or gene products at different levels, depending on the availability of information about an entity. In general, when inferring information from a gene that is annotated by some hierarchical GO terms, the more specific biological functions at the lower levels should be chosen as an inference base due to its richer information content. As such, we need to come up with a measure to filter the more informative GO terms at the lower levels. A term in the GOA hierarchy that occurs less frequently is considered to be more informative as it has a more specialized function. To capture this frequency-sensitive informativeness of GO terms, the information content of the node was measured using their annotations by [9]. Specifically, the information content $I(t)$ of a GO term $t \in \mathcal{T}$ is given by

$$I(t) = -\log p(t) \tag{1}$$

where $p(t) = N(t)/N(root(t))$ and $root(t) \in \{BP, MF\}$ is the GO root of term t, $N(t)$ is the number of occurrences of term t in the given GOA data. The information content strongly correlates with the hierarchical depth of the term from the GO root. If the GO term is less frequent, it is usually located at a deeper/lower level and therefore has a more specific function.

Consider a set $\mathcal{G} = \{i\}_{i=1}^{n}$ of n genes with gene i associated to GO term-set T_i. The total term-set $\mathcal{T}$ of all genes is given by $\mathcal{T} = \bigcup_{i=1}^{I} T_i$. Let $r(t, t') \in \mathcal{R}$ denote a GO relation between terms $t, t' \in \mathcal{T}$. All GO relations between terms in the term-set $\mathcal{T}$ are represented as $\mathcal{E} = \{r(t, t') : t, t' \in \mathcal{T}, \exists \, r(t, t') \in \mathcal{R}\}$. The pair $(\mathcal{T}, \mathcal{E})$ thus constitute the GO network (GON).

2.2   GRNs from GO Regulatory Paths

Recall that in GON, $regulate$ denotes the occurrence of one biological process that directly affects the manifestation of another process or quality, e.g., process t regulates process t' means that if both processes occur, t always regulates t'. Suppose gene i is annotated by GO term t and gene i' is annotated by GO term t'. If the GO term t has a regulates term t', then we can infer that gene i

might regulate i'. If this inference comes from more specific terms, gene i should high likely regulate i'. In the later GRN inference procedure, we choose the most reliable regulate inference path. There are very few direct *regulate* relations in the current GO database, making it difficult to infer a GRN. As such, we propose to induce transitive *regulate* relations among GO terms, from which we infer a GRN based on both direct and transitive gene regulations derived from the GON.

Consider the transitivity rule:

$$\text{if } t_a \xrightarrow{r_1} t_b \text{ and } t_b \xrightarrow{r_2} t_c, \text{ then } t_a \xrightarrow{r_3} t_c \tag{2}$$

where $t_a, t_b, t_c \in \mathcal{T}$ and $r_1, r_2, r_3 \in \mathcal{R}$. By using the notations of rule deduction, the above transitive relation is written as $r_3 = r_1 \wedge r_2$. According to GO database[1], for any $r \in \mathcal{R} = \{is - a, \; part - of, \; regulates\}$, the following four transitivity relations are valid:

$$r = r \wedge is - a \tag{3}$$
$$r = is - a \wedge r \tag{4}$$
$$regulates = regulates \wedge part - of \tag{5}$$
$$part - of = part - of \wedge part - of \tag{6}$$

Consider a path $(t_j)_{j=0}^{J}$ in GON where t_0 and t_J denote the source and destination terms, respectively; and $r(t_j, t_{j+1})$ is the parent-child relation between parent term t_j and child term t_{j+1}. Using parent-child relations $\{ r(t_j, t_{j+1}) \}_{j=1}^{J-1}$, each term t_j can induce a relation from the source term. Denote path $\pi_J = (t_j)_{j=0}^{J}$, and let $r(\pi_J) = r(t_0, t_J)$ denote the inferred relation along path π_J by applying transitive rules to parent-child relations, we have

$$r(\pi_j) = r(\pi_{j-1}) \wedge r(t_{j-1}, t_j). \tag{7}$$

We then assign a confidence score function $\sigma(\pi)$ for each inferred path π by considering both the number of steps and the information content of the terms along the inferred paths. The confidence score $\sigma(\pi)$ should give preference to paths with fewer inference steps and more informative terms, defined as

$$\sigma(\pi_j) = \sigma(\pi_{j-1}) + \Delta_{r(t_{j-1}, t_j)}(t_{j-1}, t_j) \tag{8}$$

where $\sigma(\pi_j)$ is the score assigned to the inferred path π_j from source t_0 to term $t_j \in \mathcal{T}$, and $\Delta_{r(t, t')}(t, t')$ is the score assigned to relation $r(t, t') \in \mathcal{R}$ between terms $t, t' \in \mathcal{T}$.

The cost for deducing a relation between two terms should facilitate the selection of the most informative inferred path. Semantic similarity of GO terms based on their information content, i.e., Lin's semantic similarity measure [10] and Jiang's semantic distance [11] can be used to define the cost of deducing a relation between two terms. For example, $\Delta_{r(t, t')}(t, t') = 1 - S(t, t')$, where

---

[1] http://www.geneontology.org/GO.ontology.relations.shtml

$S(t, t')$ is the semantic similarity of terms t, t'. Jiang's semantic distance can also be directly used as the cost as $\Delta_{r(t,t')}(t, t')$.

When there exists no relation $r(t, t')$, the cost is assigned as $\Delta_{r(t,t')}(t, t') = \infty$ to avoid inferring empty relations. Thus, the path with the minimum score σ_π is the path inferred collectively using the fewest steps and along the most informative terms, i.e., the most reliable inferred path. Dijkstra's algorithm can be used to find the shortest inferred path. If an inferred path ends with the deduced relation *regulate* at the destination, then there is a regulatory path (RP) between the source and the destination.

We propose an algorithm to detect the most reliable RP between two terms $s, d \in \mathcal{T}$ by using the deduction scores in Dijkstra's shortest path algorithm [12]. For each node $v \in \mathcal{T}$, we use an indicator vector to represent the deduced relations r_v, and $\sigma(r_v, v)$ is a matrix to record the current minimum distance/cost to deduce r_v at term v along path $\pi(r_v, v)$.

Initially, the source term is assigned an $is - a$ relation as it does not change the first transitive relation. Subsequently for each $R \in \mathcal{R}$, S_R denotes the unvisited node set with current relation R, and we iteratively choose $u^* = \arg\min_{u \in S_R}\{\sigma(R, u)\}$ as the starting node of the following iteration. At each inference step, if $\sigma(r_{u^*}, u^*) + \Delta_{r(u^*,v)}(u^*, v) < \sigma(r_v, v)$, we update the current deduced score for the three relations by $r_v = r_{u^*} \wedge r(u^*, v)$, $\sigma(r_v, v) = \sigma(r_{u^*}, u^*) + \Delta_{r(u^*,v)}(u^*, v)$. The iteration stops when all S_R are empty. Upon termination, if $\sigma(regulate, d) < +\infty$, $\pi(regulate, d)$ is the most-reliable RP from term s to d, otherwise no RP from term s to d exists.

Given a source term and target term, Algorithm 1 finds the most reliable RP.

Algorithm 1. Finding the most-reliable Regulatory Path (RP) between two GO terms

---

Step 0. Given source term s, target term d, term set $\mathcal{T}$, and inference cost matrix Δ

Step 1. Set $r_s(is - a) = 1$; $\forall u \in \mathcal{T}, R \in \mathcal{R}$. Set $S_R = \mathcal{T}$, $\pi(R, u) = \{u\}$, $\sigma(R, u) = +\infty$ except $\sigma(is - a, s) = 0$

Step 2. For $R \in \mathcal{R}$, If $S_R \neq \{\}$
  Choose $u^* = \arg\min_{u \in S_R}\{\sigma(R, u)\}$, set $S_R = S_R \backslash \{u^*\}$
  If $\sigma(R, u^*) \neq +\infty$: for each $(u^*, v) \in \mathcal{E}$, if $\sigma(r_{u^*}, u^*) + \Delta_{r(u^*,v)}(u^*, v) < \sigma(r_v, v)$

  Update $r_v = r_{u^*} \wedge r(u^*, v)$, $\sigma(r_v, v) = \sigma(r_{u^*}, u^*) + \Delta_{r(u^*,v)}(u^*, v)$, $\pi(r_v, v) = \{u^*\} \cup \pi(r_v, v)$

Step 3. If $\sigma(regulate, d) < +\infty$, return $\pi(regulate, d)$ as the most-reliable RP

Up till now, we have only determined the most reliable RP between GO terms. For genes i, i' with GO annotations $T_i, T_{i'}$ respectively, if there exists a RP from $t \in T_i$ to $t' \in T_{i'}$, we can infer that gene i regulates gene i'. The confidence score for this inferred path is assigned by the minimum regulatory path score of all RPs (if any): $C(i, i') = \min\{\sigma(\pi(t, t'))|r(\pi(t, t')) = regulate\}$. Then we can construct a GRN with confidence score based on direct and transitive *regulate* relations among the GO terms.

2.3 Complementary GRNs from Functional Similarity

According to the transitive rules, if no *regulate* path exists between two GO terms, then no *regulate* relation can be deduced. As a result, the Regulatory Path method cannot infer a GRN when there is few *regulate* relations among the GO terms in the GOA data.

Recall that Gene Ontology (GO) provides a standard vocabulary of functional terms and allows for coherent annotation of gene products. Gene products are functionally similar if they have comparable molecular functions and are involved in similar biological processes. The more similar genes are, the more likely they belong to the same biological pathway, which involves gene interactions/regulations. We can thus assess the functional similarity of gene products by comparing sets of GO terms, and then recover the direct regulations missing from the GO Regulate Paths method based only on genetic function similarity.

Functional Similarity of Genes Based on Semantic Similarity of GO Annotations. Semantic similarity has been proposed to compare concepts within an ontology. It can evaluate the specificity of a GO term's underlying concept in a given GO annotation. There are three popular semantic similarity measures: Resnik similarity [13] measures the semantic similarity of two terms via the information content of their Lowest Common Ancestors (LCA); Lin's similarity [10] assesses how close the terms are to their LCA. But it does not take into account the level of detail of the LCA; The simRel [14] combines the semantic similarity of Lin's and Resnik. It takes into account how close terms are to their LCA as well as how detailed the LCA is, i.e., it distinguishes between generic and specific terms.

For each term $t \in \mathcal{T}$, let $p(t)$ be the probability of finding t's descendent in the GO annotation database. If t and t' are two terms and $a\,(t, t')$ represents the set of parent terms shared by both t and t', then

$$p(LCA(t, t')) = \min_{t^* \in a(t,t')} p\,(t^*),\qquad(9)$$

The three similarity between two terms is computed as follows

$$Sim_{Resnik}\,(t, t') = -log\,(p(LCA(t, t')))\qquad(10)$$

$$Sim_{Lin}\,(t, t') = \frac{2 \times \log\,(p(LCA(t, t')))}{\log\,(p\,(t)) + \log\,(p\,(t'))}\qquad(11)$$

$$Sim_{Rel}\,(t, t') = \frac{2 \times \log\,(p(LCA(t, t')))}{\log\,(p\,(t)) + \log\,(p\,(t'))}[1 - p(LCA(t, t'))]\qquad(12)$$

In fact, simRel measure reduces to Lin's when $p(LCA(t, t'))$ is very small, i.e., $1 - p(LCA(t, t'))$ approaches to 1. Thus, in our experiments, we consider Lin's and Resnik's measure. Specifically, we consider two genes to be similar if and only if both measures yield high scores.

Gene products annotated with GO terms can be compared on the basis of the aforementioned semantic similarity measures. Let GOscore be the measure of functional similarity between two genes with respect to either their biological process (BPscore) or molecular function (MFscore). Each gene pair receives two similarity values, one for each ontology root. The work in [15] defines the functional similarity between two genes i and i', with annotated GO terms set T_i and $T_{i'}$, respectively, as the *average* inter-set similarity of terms in T_i and $T_{i'}$, as follows

$$GOscore_{avg}(i, i') = \frac{1}{|T_i||T_{i'}|} \sum_{t \in T_i, t' \in T_{i'}} Sim(t, t') \tag{13}$$

The *maximum* similarity measure is also computed as an upper bound, as follows

$$GOscore_{max}(i, i') = \max_{t \in T_i, t' \in T_{i'}} Sim(t, t') \tag{14}$$

Finally, the funSim score is calculated from the BPscore and the MFscore of a pair of gene products as follows

$$funSim(i, i') = \frac{1}{2}[(\frac{BPscore(i, i')}{\max(BPscore)})^2 + (\frac{MFscore(i, i')}{\max(MFscore)})^2] \tag{15}$$

where $\max(BPscore)$ and $\max(MFscore)$ denote the maximum score for biological process and molecular function, respectively.

Modified GO Term Probabilities. GOscore is defined by treating each term equally in the semantic similarity computation. That is, the semantic similarities are defined based on the term probabilities $p(t) = N(t)/N(root(t))$. However, this ignores the hierarchical structure of GO data. Because $N(root(t))$ is in fact the number of genes assigned by $root(t) \in \{BP, MF\}$, i.e., two GO trees of different sizes are involved. $N(t)$ is the number of genes annotated to term t, thus, the definition of $p(t)$ is in fact the distribution of term t conditioned on a specific GOA data instead of all GOA.

Consider the case of two genes in a GO term list where some terms are commonly assigned to two genes, but some are assigned to only one gene. Clearly, the two terms should have different term probabilities. To account for this imbalance, we model the term probability for three different cases as follows: 1) term annotates both genes, 2) term annotates only one gene, and 3) term annotates none of the two genes. Given two genes and a term t, denote by $m = N(root(t))$ and $n = N(t)$, $p(n, m) = p(t) = n/m$, Probability of term t annotated to (i) both genes is $p(n, m, k = 2) = p(n, m) \times \frac{\binom{m}{2}}{\binom{n}{2}}$; (ii) only one gene is $p(n, m, k = 1) = p(n, m) \times \frac{\binom{m}{1}\binom{n-m}{1}}{\binom{n}{2}}$; (iii) none of the two genes has the same defined as the background distribution: $p(n, m, k = 0) = p(n, m) \times \frac{\binom{n-m}{2}}{\binom{n}{2}}$.

Let us consider an example to illustrate the discriminatory power of this modified term probability. Given a term, if it is assigned to both genes, its probability is $p(n, m, k = 2) = \frac{m^2 \times (m-1)}{n^2 \times (n-1)}$; if it is assigned to only one gene, its probability is $p(n, m, k = 1) = \frac{2m^2 \times (n-m)}{n^2 \times (n-1)}$; otherwise, its probability is $p(n, m, k = 0) = \frac{m \times (n-m) \times (n-m-1)}{n^2 \times (n-1)}$. For a specific term, m is always significantly smaller than n, thus, $p(n, m, k = 2) << p(n, m, k = 1) << p(n, m, k = 0)$. Similarly, their information content $-log(p(n, m, k = 2)) >> -log(p(n, m, k = 1)) >> -log(p(n, m, k = 0))$. In other words, the three cases represent three distinct levels of information content.

Note that the definition $p(n, m, k)$ also considers the prior probability, hence, our modified term probabilities is consistent with and improves the original definition. With the modified term probabilities, we can then use the semantic similarity definition and GOscore computation method in Section 2.3 to compute the functional similarity of a gene pair.

Deriving GRN from Gene Functional Similarity. We next propose a method to build a GRN from the computed gene functional similarity scores of all applicable pairs. If Lin's average biological similarity between genes exceeds a threshold θ_1, i.e., $funSim_{Lin,GOavg}(i, i') \geq \theta_1$, and Resnik's maximum biological similarity also exceeds some threshold θ_2, i.e., $funSim_{Resnik,GOmax}(i, i') \geq \theta_2$, then we say that gene i and i' are *functional similar*, and there is a possible regulation between gene i and i'.

The GRN is constructed using the derived gene regulations. In our experiments, we exhaustively evaluated all combinations of semantic measures and GOscore measures to find the combination that gives the best F1-Score for GRN.

Fused GRN from Functional Similarity and RP. Since the GRN from the RP and functional similarity methods are complementary, we propose a method to fuse the two derived regulations into a GRN as: $\forall i, i' \in \mathcal{G}$, (i) If there exists transitive gene regulation from gene i to i' detected by GO RP method, then gene i regulate i'. Or (ii) If gene i, i' are *functional similar*, then there is gene regulation between gene i and i'.

3   Results

We evaluate our GO-inferred GRN against benchmark GRNs from GeneNetWeaver (GNW) [16]. Specifically, we use the E.coli GRN. Since the complete E.coli network from GNW contains many genes that have no corresponding GO annotations, we extract seven sub-networks from it, which are summarized in Table 1. GO terms and relations corresponding to the genes in the networks are obtained from files associated with the GO annotation database[2]. The corresponding informative GO terms related to the target genes are also selected from the gene association files. Only GO terms involved in the molecular function and biological process are considered. To construct a reliable GRN, we choose specific GO terms with information content $I(t) \geq \theta_I$, i.e., above a threshold θ_I. We set

---

[2] http://www.geneontology.org/GO.downloads.annotations.shtml

Table 1. E.coli sub-networks and their GO relations. Each network has 25 genes.

	Net1	Net2	Net3	Net4	Net5	Net6	Net7
No. of edges	18	15	24	11	15	19	29
regulate	2	1	1	1	1	2	4
part-of	3	1	0	2	2	2	2
is-a	26	26	9	28	33	18	36
No. of terms	96	93	44	108	110	82	90

θ_I to $-\log 0.25$ in the experiments. Each GO annotation is classified into one of 5 descending order of quality categories: experimental, computational, author statement, curator statement, and automatic. Annotations derived through direct experiments are deemed higher quality compared to others [17]. We only consider GO terms with the top two quality levels: computational and experimental.

We evaluate the performance of the three GO-inferred methods on the seven networks. The evaluation measurements include accuracy, precision, recall, F1-score, true positive (TP), false positive (FP), true negative (TN), and false negative (FN) numbers. Averaged results over the seven networks are denoted by "Avg". To compare the three GO methods, namely GO Regulatory Path Method (denoted by "RegPath"), GO Functional Similarity Method (denoted by "Fun-Sim"), and GO Fusion Method (denoted by "Fusion"), we list their performance on inferring GRN vs. the seven target networks in Table 2.

From the "RegPath" results in Table 2, we see that our GO RP approach achieved a high averaged precision of 87.43%, with corresponding F1-score 62.93%. In general, very few FP edges were extracted, with two networks (Net4 and Net5) consistently having zero FPs, and the remaining five networks registering less than three FPs. We have discovered seven new gene regulations via our GO RP method: three self-regulations on gene "argP", "fadR" and "flhC"; four gene pairs: "argP" regulate "gyrA"; "argP" regulate "polA"; "dnaA" regulate "dinB"; and "flhC" regulate "flhD". We tried to look up evidences for these seven gene regulation pairs from the MEDLINE database[3], but was unable to find any evidence. We may eventually need experts in the field to confirm or reject these FPs. Moreover, the FPs could have been generated due to (i) human errors in the GO database: incorrect gene annotations or GO relations; (ii) incompleteness of the target network. One limitation of the RP method lies in its poor recall, which averaged only 51.41%. This shows that the GO RP method could not detect enough gene regulations in the target GRN, which can be due to (i) gene's GO annotations are incomplete, and some GO terms are not involved in GON; (ii) there are not enough annotated *regulate* relations among the GO terms; (iii) the other two GO relations, $is-a$ or $part-of$ are incomplete; (iv) the incompleteness of the target ground truth network itself. Clearly, the incomplete GO information will bound the accuracy of our

---

[3] http://www.ncbi.nlm.nih.gov

Table 2. Performance of the three GO methods

		TP	FP	FN	TN	Pre.(%)	Rec.(%)	F1(%)	Acc.(%)
RegPath	Net1	10	3	8	604	76.92	55.56	64.52	98.24
	Net2	6	1	9	609	85.71	40.00	54.55	98.40
	Net3	20	1	4	600	**95.24**	**83.33**	**88.89**	99.20
	Net4	5	0	6	614	**100.00**	45.45	62.50	99.04
	Net5	5	0	10	610	**100.00**	33.33	50.00	98.40
	Net6	5	3	14	603	62.50	26.32	37.04	97.28
	Net7	22	2	7	594	**91.67**	**75.86**	**83.02**	98.56
	Avg	10.43	1.43	8.29	604.86	**87.43**	**51.41**	**62.93**	**98.45**
O:FunSim	Net1	7	51	11	556	12.07	38.89	**18.42**	90.08
	Net2	2	2	13	608	**50.00**	13.33	21.05	97.60
	Net3	10	55	14	546	15.38	41.67	22.47	88.96
	Net4	4	35	7	579	10.26	36.36	16.00	93.28
	Net5	2	4	13	606	33.33	13.33	19.05	97.28
	Net6	16	270	**3**	336	5.59	**84.21**	10.49	56.32
	Net7	19	330	10	266	5.44	**65.52**	10.05	45.60
	Avg	8.57	106.71	10.14	499.57	18.87	**41.90**	16.79	81.30
M:FunSim	Net1	10	3	8	604	76.92	55.56	**64.52**	**98.24**
	Net2	6	1	9	609	85.71	40.00	54.55	**98.4**
	Net3	20	1	4	600	**95.24**	**83.33**	**88.89**	**99.2**
	Net4	5	0	6	614	**100.00**	45.45	62.50	**99.04**
	Net5	5	0	10	610	**100.00**	33.33	50.00	**98.40**
	Net6	5	3	14	603	62.50	26.32	37.04	97.28
	Net7	18	2	11	594	**90.00**	**62.07**	**73.47**	97.92
	Avg	9.86	1.43	8.86	604.86	**87.20**	**49.44**	**61.57**	**98.35**
Fusion	Net1	11	3	7	604	78.57	**61.11**	68.75	**98.40**
	Net2	6	1	9	609	85.71	40.00	54.55	**98.40**
	Net3	20	1	4	600	95.24	**83.33**	**88.89**	99.20
	Net4	6	0	5	614	100.00	54.55	70.59	99.20
	Net5	6	1	9	609	85.71	40.00	54.55	**98.40**
	Net6	6	3	13	603	66.67	31.58	42.86	**97.44**
	Net7	23	3	6	593	88.46	**79.31**	**83.64**	98.56
	Avg	11.14	1.71	7.57	604.57	**85.77**	**55.70**	**66.26**	**98.51**

GO Regulatory Path method, which motivates us to use functional similarity to improve the GRN inference from GO.

The evaluations of GRN predicted from GO Functional Similarity using the original term probabilities (denoted by "O:FunSim") and modified term probabilities (denoted by and "M:FunSim") are shown in Table 2. The "O:FunSim" method resembles existing works which extract gene interactions based on gene similarity from GO Semantic Similarity [5,4,6]. It can be seen that "M:FunSim" outperforms "O:FunSim" notably on the averaged precision, recall, F1-score, and accuracy. Clearly, the modified term probabilities are more effective in capturing the functional similarity of genes.

Due to the complementary strengths of last two methods, they can be fused to capture more information from the GO databases. From the Fusion method in

Table 2, we see that the GO Fusion approach gave the best performance in terms of F1 measure and accuracy. The overall low recall of the GO Fusion method is due to the incompleteness of the GO database and the target network. The extremely high TNs (reflected in the accuracy rates of 98% or higher) suggest that our GO method is able to filter the vast majority of negative edges. Our GO method also yields very low false positive rates. As a result, high precision rates averaging 85.77% are achieved as shown in Table 2. This conservative behaviour is desirable because gene regulation is hard to validate in general and a GRN should have as high a precision as possible.

4   Conclusion

We proposed a method to detect both direct and transitive regulations between genes by using the corresponding GO annotations and their inter-relations. By developing a novel shortest path detection algorithm, we detected the most likely regulatory paths from GON. Experimental results show that transitive regulations play an important role in GRN and their detection improves the accuracy of the generated GRN. We show that GO can be used effectively to detect transitive regulations.

Due to the incomplete information of the source GO database, the GRN from the GO Regulate Path method may overlook some important direct regulations. Inspired by the fact that gene regulations occur between functionally similar genes, we propose the GO FunSim method to detect direct regulations. Gene function similarity scores are computed from the semantic similarities of their corresponding GO terms, using their occurrence probabilities. We then modified the term occurrence probabilities to account for GO term imbalance, e.g., the likelihood of a term being assigned to each, both, or neither of the two terms. Experimental results show that our GO FunSim method based on the modified term probabilities are extremely adept at capturing pairwise gene function similarities.

Lastly, we proposed a simple fusion method to combine the results of the proposed FunSim and Regulate Path methods to generate a fused GRN. Experiments show that our GO Fusion method yielded the best GRN in terms of F1-score.

The errors on predicting the networks may arise from (i) the incompleteness of the target networks; (ii) the incompleteness of the GO database due to lack of updates; (iii) the erroneous annotations of GO database due to human error; (iv) GO allows us to annotate genes and their products with a limited set of attributes, its scope is limited to the three domains, which is not comprehensive. Hence, the GRN we extracted from GO are in fact based upon partial evidence provided by the current GO. The false negatives of the networks could be further reduced by fusing the GO generated GRN with additional data sources such as wet-lab data. One extension to this work is to identify ontology terms that are specific to the pathways under consideration, e.g., terms related to cell-cycle functions in our experiments. The GRN developed by our method could be useful for validation of networks built by other experimental or computational approaches.

Acknowledgement. This research was supported in part by Singapore's Ministry of Education Academic Research Tier 2 fund MOE2010-T2-1-056 and MOE AcRF Tier 2 grant of ARC9/11.

References

1. Steuer, R., Humburg, P., Selbig, J.: Validation and functional annotation of expression-based clusters based on gene ontology. BMC Bioinformatics (2006)
2. Mundra, P.A., Rajapakse, J.C.: SVM-RFE with MRMR filter for gene selection. IEEE Transactions on Nanobiosciences 9 (2010)
3. Zhou, X.H., Kao, M.J., Wong, W.H.: Transitive functional annotation by shortest-path analysis of gene expression data. In: PNAS, vol. 99, pp. 12783–12788 (2002)
4. Franke, L., van Bakel, H., Fokkens, L., de Jong, E.D., Egmont-Petersen, M., Wijmenga, C.: Reconstruction of a functional human gene network, with an application for prioritizing positional candidate genes. American Journal of Human Genetics 78, 1011–1025 (2006)
5. Cheng, J., Cline, M., Martin, J., Finkelstein, D., Awad, T., Kulp, D., Siani-Rose, M.: A knowledge-based clustering algorithm driven by gene ontology. Journal of Biopharmaceutical Statistics 14, 687–700 (2004)
6. Kustra, R., Zagdański, A.: Data-fusion in clustering microarray data: Balancing discovery and interpretability. TCBB 7, 59–63 (2010)
7. Ashburner, M., Ball, C.A., Blake, J.A.: Gene ontology: tool for the unification of biology. Nature Genetics 25, 25–29 (2000)
8. Barrell, D., Dimmer, E., Huntley, R.P.: The goa database in 2009 - an integrated gene ontology annotation resource. NAR 37, 396–403 (2009)
9. Alterovitz, G., Xiang, M., Mohan, M.: Go pad: the gene ontology partition database. NAR 35, 322–327 (2007)
10. Lin, D.: An information theoretic definition in similarity. In: ICML, pp. 266–304 (1998)
11. Jiang, J., Conrath, D.: Semantic similarity based on corpus statistics and lexical taxonomy. In: ROCLING (1997)
12. Johnson, D.B.: A note on dijkstra's shortest path algorithm. JACM (1973)
13. Resnik, P.: Semantic similarity in a taxonomy: An information-based measure and its application to problems of ambiguity in natural language. In: JAIR, vol. 11, pp. 95–130 (1999)
14. Schlicker, A., Domingues, F.S., Rahnenfuhrer, J., Lengauer, T.: A new measure for functional similarity of gene products based on gene ontology. BMC Bioinformatics (2006)
15. Lord, P., Stevens, R., Brass, A., Goble, C.A.: Investigating semantic similarity measures across the gene ontology: the relationship between sequence and annotation. Bioinformatics 19, 1275–1283 (2003)
16. Schaffter, T., Marbach, D., Floreano, D.: GeneNetWeaver: In silico benchmark generation and performance profiling of network inference methods. Bioinformatics 27, 2263–2270 (2011)
17. Rhee, S.Y., Wood, V., Dolinski, K.: Use and misuse of the gene ontology annotations. Nature Reviews Genetics 9, 509–515 (2008)

Partitioning Biological Networks
into Highly Connected Clusters
with Maximum Edge Coverage

Falk Hüffner[1], Christian Komusiewicz[1], Adrian Liebtrau[2],
and Rolf Niedermeier[1]

[1] Institut für Softwaretechnik und Theoretische Informatik, TU Berlin, Germany
{falk.hueffner,christian.komusiewicz,rolf.niedermeier}@tu-berlin.de
[2] Institut für Informatik, Friedrich-Schiller-Universität Jena, Germany

Abstract. We introduce the combinatorial optimization problem
HIGHLY CONNECTED DELETION, which asks for removing as few edges
as possible from a graph such that the resulting graph consists of highly
connected components. We show that HIGHLY CONNECTED DELETION
is NP-hard and provide a fixed-parameter algorithm and a kernelization.
We propose exact and heuristic solution strategies, based on polynomial-
time data reduction rules and integer linear programming with column
generation. The data reduction typically identifies 85 % of the edges
that need to be deleted for an optimal solution; the column generation
method can then optimally solve protein interaction networks with up
to 5 000 vertices and 12 000 edges.

1 Introduction

A key idea of graph-based data clustering is to identify densely connected subgraphs (clusters) that have many interactions within themselves and few with
the rest of the graph. Hartuv and Shamir [8] proposed a clustering algorithm producing so-called *highly connected* clusters. Their method has been successfully
used to cluster cDNA fingerprints [9], to find complexes in protein–protein interaction (PPI) data [10], and to find families of regulatory RNA structures [15].
Hartuv and Shamir [8] formalized the connectivity demand for a cluster as follows: the *edge connectivity* $\lambda(G)$ of a graph G is the minimum number of edges
whose deletion results in a disconnected graph, and a graph G with n vertices is
called *highly connected* if $\lambda(G) > n/2$. An equivalent characterization is that a
graph is highly connected if each vertex has degree at least $\lfloor n/2 \rfloor + 1$ [5]. Thus,
highly connected graphs are very similar to *0.5-quasi-complete graphs* [11], that
is, graphs where every vertex has degree at least $(n-1)/2$. Further, being highly
connected also ensures that the diameter of a cluster is at most two [8].

The algorithm by Hartuv and Shamir [8] partitions the vertex set of the
given graph such that each partition set is highly connected, thus guaranteeing
good intra-cluster density (including maximum cluster diameter two and the
presence of more than half of all possible edges). Moreover, the algorithm needs

Z. Cai et al. (Eds.): ISBRA 2013, LNBI 7875, pp. 99–111, 2013.
© Springer-Verlag Berlin Heidelberg 2013

no prespecified parameters (such as the number of clusters) and it naturally extends to hierarchical clustering. Essentially, Hartuv and Shamir's algorithm iteratively deletes the edges of a minimum cut in a connected component that is not yet highly connected. While Hartuv and Shamir's algorithm guarantees to output a partitioning into *highly connected* subgraphs, it does not guarantee to achieve this by minimizing inter-cluster connectivity. Thus, we propose a formally defined *combinatorial optimization problem* that additionally specifies the goal to minimize the number of edge deletions.

HIGHLY CONNECTED DELETION
Instance: An undirected graph $G = (V, E)$.
Task: Find a minimum subset of edges $E' \subseteq E$ such that in $G' = (V, E \setminus E')$ all connected components are highly connected.

Note that, by definition, isolated edges are *not* highly connected. Hence, the smallest clusters are triangles; we consider all singletons as *unclustered*. The problem formulation resembles the CLUSTER DELETION problem [17], which asks for a minimum number of edge deletions to make each connected component a clique; thus, CLUSTER DELETION has a much stronger demand on intra-cluster connectivity. Also related is the 2-CLUB DELETION problem [13], which asks for a minimum number of edge deletions to make each connected component have a diameter of at most two. Since highly connected clusters also have diameter at most two [8], 2-CLUB DELETION poses a looser demand on intra-cluster connectivity.

It could be expected that the algorithm by Hartuv and Shamir [8] yields a good approximation for the optimization goal of HIGHLY CONNECTED DELETION. However, we can observe that in the worst case, its result can have size $\Omega(k^2)$, where $k := |E'|$ is the size of an optimal solution. For this, consider two cliques with vertex sets $u_1, \ldots, u_n$ and $v_1, \ldots, v_n$, respectively, and the additional edges $\{u_i, v_i\}$ for $2 \leq i \leq n$. Then these additional edges form a solution set of size $n - 1$; however, Hartuv and Shamir's algorithm will (with unlucky choice of minimum cuts) transform one of the two cliques into an independent set by repeatedly cutting off one vertex, thereby deleting $n(n+1)/2 - 1$ edges. This also illustrates the tendency of the algorithm to cut off size-1 clusters, which Hartuv and Shamir counteract with postprocessing [8]. This tendency might introduce systematic bias [12]. Hence, exact algorithms for solving HIGHLY CONNECTED DELETION are desirable.

Preliminaries. We consider only undirected and simple graphs $G = (V, E)$. We use n and m to denote the number of vertices and edges in the input graph, respectively, and k for the minimum size of an edge set whose deletion makes all components highly connected. The *order* of a graph G is the number of vertices in G. We use $G[S]$ to denote the *subgraph induced* by $S \subseteq V$. Let $N(v) := \{u \mid \{u, v\} \in E\}$ denote the *(open) neighborhood* of v and $N[v] := N(v) \cup \{v\}$. A *minimum cut* of a graph G is a smallest edge set E' such that deleting E' increases the number of connected components of G. For the notions of fixed-parameter tractability and kernelization, see e. g. [14]. Due to the lack of space, we defer some proofs and details to the full version of this paper.[1]

2   Computational Complexity

We can prove the hardness of HIGHLY CONNECTED DELETION by a reduction from PARTITION INTO TRIANGLES on 4-*regular neighborhood-restricted graphs* [19].

Theorem 1. HIGHLY CONNECTED DELETION *on 4-regular graphs is NP-hard and cannot be solved in* $2^{o(k)} \cdot n^{O(1)}$, $2^{o(n)} \cdot n^{O(1)}$, *or* $2^{o(m)} \cdot n^{O(1)}$ *time unless the exponential-time hypothesis (ETH) is false.*

Problem Kernel. We now present four data reduction rules that preserve optimal solvability and whose exhaustive application results in an instance with at most $10 \cdot k^{1.5}$ vertices. The first data reduction rule is obvious.

Rule 1. Remove all connected components from G that are highly connected.

The following lemma can be proved by a simple counting argument.

Lemma 1. *Let G be a highly connected graph and let u, v be two vertices in G. If u and v are connected by an edge, then they have at least one common neighbor; otherwise, they have at least three common neighbors.*

A simple data reduction rule follows directly from Lemma 1.

Rule 2. If there are two vertices u and v with $\{u, v\} \in E$ that have no common neighbors, then delete $\{u, v\}$ and decrease k by one.

Interestingly, Rules 1 and 2 yield a linear-time algorithm for HIGHLY CONNECTED DELETION on graphs of maximum degree three, which together with Theorem 1 shows a complexity dichotomy with respect to the maximum degree.

Theorem 2. HIGHLY CONNECTED DELETION *can be solved in linear time when the input graph has degree at most three.*

The next two data reduction rules are concerned with finding vertex sets that have a small edge cut. For $S \subseteq V$, we use $D(S) := \{\{u, v\} \in E \mid u \in S \text{ and } v \in V \setminus S\}$ to denote the set of edges outgoing from S, that is, the edge cut of S.

The idea behind the next reduction rule is to find vertex sets that cannot be separated by at most k edge deletions. We call two vertices u and v *inseparable* if the minimum edge cut between u and v is larger than k. Analogously, a vertex set S is inseparable if all vertices in S are pairwise inseparable.

Rule 3. If G contains a maximal inseparable vertex set S of size at least $2k$, then do the following. If $G[S]$ is not highly connected, then there is no solution of size at most k. Otherwise, remove S from G and set $k := k - |D(S)|$.

---

[1] http://fpt.akt.tu-berlin.de/publications/hcd.pdf

Lemma 2. *Rule 3 preserves optimal solvability and can be exhaustively applied in $O(n^2 \cdot mk \log n)$ time.*

Note that a highly connected graph of size at least $2k$ is an inseparable vertex set. Hence, after exhaustive application of Rule 3, every cluster has bounded size. While Rule 3 identifies clusters that are large with respect to k, Rule 4 identifies clusters that are large compared to their neighborhood.

Rule 4. If G contains a vertex set S such that $|S| \geq 4$, $G[S]$ is highly connected, and $|D(S)| \leq 0.3 \cdot \sqrt{|S|}$, then remove S from G and set $k := k - |D(S)|$.

Lemma 3. *Rule 4 preserves optimal solvability and can be exhaustively applied in $O(n^2 \cdot mk \log n)$ time.*

Proof. We show that there is an optimal solution in which S is a cluster. To this end, suppose that there is an optimal solution which produces some clusters $C_1, \ldots, C_q$ that contain vertices from S and vertices from $V \setminus S$. We show how to transform this solution into one that has S as a cluster and needs at most as many edge deletions. First, we bound the overall size of the C_i's. Note that deleting all edges between S and $\{C_i \setminus S \mid 1 \leq i \leq q\}$ cuts each C_i. By the condition of the rule, such a cut has at most $0.3\sqrt{|S|}$ edges. Since each $G[C_i]$ is highly connected, this implies that $\sum_{1 \leq i \leq q} |C_i| < 0.6\sqrt{|S|}$.

Now, transform the solution at hand into another solution as follows. Make S a cluster, that is, undo all edge deletions within S and delete all edges in $D(S)$, and for each C_i, delete all edges in $G[C_i \setminus S]$. This is indeed a valid solution since $G[S]$ is highly connected, and all other vertices that are in "new" clusters are now in singleton clusters.

We now compare the number of edge modifications for both edge deletion sets and show that the new solution needs less edge modifications. To this end, we consider each vertex $u \in S$ that is contained in some C_i. On the one hand, since $G[S]$ is highly connected, and since there is at least some $v \in S$ that is not contained in any C_i we undo at at least $|S|/2$ edge deletions between vertices of S. On the other hand, an additional number of up to $0.3\sqrt{|S|} + \binom{\lfloor 0.6\sqrt{|S|} \rfloor}{2}$ edge deletions may be necessary to cut all the C_i's from S and to delete all edges in each $G[C_i \setminus S]$. By the preconditions of the rule we have $\sqrt{|S|} \leq |S|/2$ and thus the overall number of saved edge modifications for u is at least

$$|S|/2 - 0.3\sqrt{|S|} - \binom{\lfloor 0.6\sqrt{|S|} \rfloor}{2} > |S|/2 - 0.6|S|/2 - 0.36|S|/2 > 0. \quad (1)$$

Hence, the number of undone edge modifications is larger than the number of new edge modifications. Consequently, S is a cluster in every optimal solution. The running time can be bounded analogously to the running time of Rule 3. □

Theorem 3. HIGHLY CONNECTED DELETION *can be reduced in $O(n^2 \cdot mk \log n)$ time to an equivalent instance, called problem kernel, with at most $10 \cdot k^{1.5}$ vertices.*

Proof. Let $I = (G, k)$ be an instance that is reduced with respect to Rules 1, 3 and 4. We show that every yes-instance has at most $10 \cdot k^{1.5}$ vertices. Hence, we can answer no for all larger instances.

Assume that I is a yes-instance and let $C_1, \ldots, C_q$ denote the clusters of a solution. Since I is reduced with respect to Rule 3, we have $|C_i| \leq 2k$ for each C_i. Furthermore, for every C_i we have $D(C_i) \geq 0.3\sqrt{|C_i|}$ since I is reduced with respect to Rules 1 and 4. In other words, every cluster C_i "needs" at least $0.3\sqrt{|C_i|}$ edge deletions. Hence, the overall instance size is at most

$$\max_{(c_1, \ldots, c_q) \in \mathbb{N}^q} \sum_{i=1}^{q} c_i \text{ s.t. } \forall i \in \{1, \ldots, q\} : c_i \leq 2k, \sum_{1 \leq i \leq q} 0.3 \cdot \sqrt{c_i} \leq 2k.$$

A simple calculation shows that there is an assignment to the c_i's maximizing the sum such that at most one c_i is smaller than $2k$. Hence, the sum is maximized when a maximum number of c_i's have value $2k$. Each of the corresponding clusters is incident with at least $0.3\sqrt{2k}$ edge deletions. Hence, there are at most $2k/0.3\sqrt{2k} = 10\sqrt{2k}/3$ such clusters. The overall instance size follows. □

Fixed-Parameter Algorithm. We sketch a fixed-parameter algorithm for HIGHLY CONNECTED DELETION. Since any highly connected graph has diameter at most two, if there is a connected component with diameter three or more, we can find a shortest path $uvwx$ between two vertices u and x, and then branch into three cases according to which edge of this path gets deleted. At the leaves of this search tree, we have a graph where every connected component has diameter at most two. Using Rule 3, we can ensure that each component has at most $4k$ vertices. We can solve an arbitrary HIGHLY CONNECTED DELETION instance by dynamic programming in $O(3^n \cdot m)$ time; applying this to each component yields the following theorem.

Theorem 4. HIGHLY CONNECTED DELETION *can be solved in* $O(3^{4k} \cdot k^2 + n^2 mk \cdot \log n)$ *time.*

3 Further Data Reduction and ILP Formulation

The fixed-parameter tractability results for HIGHLY CONNECTED DELETION (Theorem 3) are currently mostly of theoretical nature. Hence, we follow an algorithmic approach that consists of two main steps: First, apply a set of *data reduction rules* that exploit the structure of biological networks and yield a new instance that is significantly smaller than the original one. Second, solve the new, smaller instance by devising an *integer linear programming* (ILP) formulation.

Further Data Reduction. As we demonstrate in the computational experiments presented in Section 4, Rule 2 tremendously simplifies many real-world input instances. In particular, as shown by Theorem 2, it is useful to reduce vertices of small degree, as found in protein interaction networks. However, Rules 3 and 4 that produce a kernel have the downside of requiring relatively large substructures. To improve performance in practice, we use the following two rules.

We try to identify triangles uvw that must form highly connected clusters. For a triangle edge $\{x, y\}$, let $N_{xy} := (N(x) \cup N(y)) \setminus \{u, v, w\}$ be the common neighbors of the edge outside the triangle. Let the value of an edge e be 3 if $N_e \neq \emptyset$ and 0 otherwise. Let the value of a vertex x be the size of the largest connected component in $G[N(x) \setminus \{u, v, w\}]$, or 0 if this size is 1.

Rule 5. Assume that for a triangle uvw the following conditions hold:

- for no two triangle edges $\{x, y\}, \{x, z\}$ ($\{x, y, z\} = \{u, v, w\}$) there is an edge in G between some vertex in N_{xy} and some vertex in N_{xz};
- for no triangle edge e is there an edge in $G[N_e]$;
- for any $\{x, y, z\} = \{u, v, w\}$, the value of $\{x, y\}$ plus the value of z is at most 3;
- the sum of the values of u, v, and w is at most three.

Then isolate the triangle by deleting all edges incident on u, v, and w except the triangle edges.

Proof (preservation of optimality). By case distinction: if the triangle is not a solution cluster, then it must be part of a larger cluster, or the vertices are divided into two or three clusters. The conditions ensure that none of these situations yield a better solution than isolating the triangle.   □

The following rule reduces some low-degree vertices.

Rule 6. Let u be a vertex and $N_2(u)$ be the neighbors of u that have degree 2. If $G[N_2(u)]$ contains an edge, then isolate all vertices of degree 0 in $G[N_2(u)]$. Otherwise, if there is a vertex v that is in G a neighbor of a vertex w in $N_2(u)$ and has degree 3 in G, then delete the edge from v to the neighbor that is not u or w.

Proof (preservation of optimality). The vertex u can be contained in at most one triangle. Each of the deleted edges could only be part of a triangle with u, and for each such triangle there is another triangle which destroys fewer opportunities of using vertices for other clusters.   □

Integer Linear Programming with Column Generation. We now consider integer linear programming (ILP) based approaches. With these, we can utilize the decades of engineering that went into commercial solvers like CPLEX or Gurobi to be able to tackle large instances. Our main approach is somewhat involved due to the use of column generation. We additionally tried a more straightforward approach based on a CLIQUE PARTITIONING formulation and row generation. Our experiments show that the extra complexity pays off and the column generation approach can solve larger instances exactly.

We describe an ILP formulation of HIGHLY CONNECTED DELETION, which in its basic scheme is similar to that of Aloise et al. [1] for modularity maximization;

however, we need a new approach for solving the column generation subproblem. Let $\mathcal{T}$ be the set of all vertex sets that induce a highly connected subgraph. We use binary variables z_T to indicate that the cluster $T \in \mathcal{T}$ is part of the solution. Then the model is

$$\text{maximize} \sum_{T \in \mathcal{T}} c_T z_T, \tag{2}$$

$$\text{s.t.} \sum_{\{T \in \mathcal{T} \mid u \in T\}} z_T = 1 \quad \forall u \in V, \tag{3}$$

$$z_T \in \{0, 1\} \quad \forall T \in \mathcal{T}, \tag{4}$$

where c_T is the number of edges in the subgraph induced by t. The objective (2) maximizes the number of edges within clusters, which equivalent to minimizing the number of inter-cluster edges (deletions). The constraints of type (3) ensure that each vertex is contained in exactly one cluster.

Due to the large number of variables, this model cannot be solved directly except for tiny instances. Thus, the idea is to only consider "relevant" variables. More precisely, we start with an initial set of z_T variables that yields a feasible solution (e.g., all singleton clusters). Then we successively add variables ("columns") that improve the objective, until this is no longer possible. Due to the structure of real-world instance, typically only a small subset of possible variables needs to be added.

Now the improvement of adding a column for cluster T is c_T minus the contribution of the vertices in T to the objective function. This contribution for some vertex u can be calculated as the value of the dual variable λ_u for the corresponding constraint of type (3) in the continuous relaxation of the problem (2)–(4) (see e.g. Aloise et al. [1] for details). The values of the dual variables can be easily calculated by a linear programming solver. Thus, we need to find a cluster T that maximizes $c_T - \sum_{u \in T} \lambda_u$. In other words, we need to find a highly connected cluster that maximizes the number of edges minus vertex weights. For this, we again use an ILP formulation, using binary edge variables e_{uv} and binary vertex variables v_u to describe the cluster selected, and a positive integral variable d to describe the cluster size:

$$\text{maximize} \sum_{\{u,v\} \in E} e_{uv} - \sum_{u \in t} \lambda_u v_u, \tag{5}$$

$$\text{s.t.} \; d = \sum_{u \in V} v_u, \tag{6}$$

$$e_{uv} \leq v_u, e_{vu} \leq v_v \quad \forall \{u, v\} \in E, \tag{7}$$

$$\text{if } v_u \text{ then} \sum_{v \in N(u)} e_{uv} > d/2 \quad \forall u \in V, \tag{8}$$

where the constraint (8) can be linearized using the big-M method (that is, by adding $M(1 - v_u)$ on the left-hand side with a sufficiently large constant M); in our implementation, we instead use indicator constraints as supported by CPLEX.

Table 1. Instance properties and data reduction results. Here, K is the number of connected components, n' and m' are the number of vertices and edges in the largest connected component, respectively, Δk is the number of edges deleted during data reduction, K' is the number of connected components after data reduction, and n'' and m'' are the number of vertices and edges in the largest connected component after data reduction, respectively.

	n	m	K	n'	m'	Δk	Δk [%]	K'	n''	m''
CE phys.	157	153	39	23	24	100	92.6	137	11	38
CE all	3613	6828	73	3434	6721	5204	80.1	3202	373	1562
MM phys.	4146	7097	114	3844	6907	5659	85.3	3656	426	1339
MM all	5252	9640	135	4890	9407	7609	84.8	4566	595	1893
AT phys.	1872	2828	82	1625	2635	2057	83.1	1605	187	619
AT all	5704	12627	128	5393	12429	8797	79.5	4579	866	3323
SP all	2698	16089	17	2661	16065	2936	$\geq$ 18.2	1299	1372	13111

We can make use of the fact that it is not necessary to find a maximally improving column. Therefore, we can solve the column generation problem heuristically, and only solve it optimally using the ILP when no improving solution was found. As heuristic, we use a simple greedy method that starting from each vertex repeatedly adds the vertex that maximizes the value of the cluster, and records the best cluster that was highly connected. Further, we abort solving the column generation ILP as soon as an improving solution is found.

4   Experimental Evaluation

We implemented the data reduction in OCaml and the ILPs in C++ using the CPLEX 12.4 ILP solver. For the minimum cut subroutine of the algorithm of Hartuv and Shamir [8] (called *min-cut method* below), a highly optimized implementation in C was used [6]. Our source code and sample instances are available at http://www.user.tu-berlin.de/hueffner/hcd/. The test machine is a 3.6 GHz Intel Xeon E5-1620 with 10 MB L3 cache and 64 GB main memory, running under Debian GNU/Linux 7.0. Only a single thread was used.

We used protein interaction networks available at the BIOGRID repository [18]. The three species for which we illustrate our results are *A. thaliana*, *C. elegans*, and *M. musculus*. For each species, we extracted one network with physical interactions only, and one with all interactions. In Fig. 2, we also consider the network of all interactions of *S. pombe*. Table 1 shows some basic properties of these networks. For the computation of the enrichment of annotation terms, we used the GO:TermFinder tool [3] with *A. thaliana* annotation data from the TAIR database [2]. The computed p-values are corrected for multiple hypothesis testing. We used a significance threshold of $p \leq 0.01$.

Running Time Evaluation. Table 1 shows the effect of data reduction. Knowing the optimal k (see Table 2) allows us to state that typically 85 % of the edges that need to be deleted are identified. Since connected components can be treated separately,

Table 2. Results for the instances of Table 1. Here, k is the number of edges deleted, s and K are the number of singleton and nonsingleton clusters, respectively, n and m are the number of vertices and edges in the largest cluster, respectively, and t is the running time in seconds.

	min-cut without DR						min-cut with DR						column generation with DR					
	k	s	K	n	m	t	k	s	K	n	m	t	k	s	K	n	m	t
CE-p	111	136	5	9	30	0.01	108	133	6	9	30	0.01	108	133	6	9	30	0.06
CE-a	6714	3589	2	17	94	86.46	6630	3521	22	17	94	6.36	6499	3436	45	19	113	2088.35
MM-p	7004	4116	5	12	57	126.30	6882	4003	41	12	57	7.42	6638	3845	80	11	41	898.13
MM-a	9563	5227	5	13	65	267.63	9336	5044	61	13	65	17.84	8978	4812	120	13	65	3858.62
AT-p	2671	1796	19	14	76	5.82	2567	1723	39	14	76	0.68	2476	1675	49	14	76	60.34
AT-a	12096	5559	23	23	190	434.52	11590	5213	122	23	190	32.09	11069	4944	180	23	190	34121.23

the most important time factor is the size of the largest connected component. Here, the number of edges is reduced to typically 23 %. This demonstrates the effectivity of the data reduction, which preserves exact solvability, and suggests it should be applied regardless of the actual solution method.

Table 2 shows the clustering results and running times. Doing data reduction before running the min-cut method actually improves the running time, since it reduces the number of costly min-cut calls. The column generation method is able to solve all six test instances, although the hardest one takes more than 9 hours. However, it is not able to solve e. g. the network of all interactions of *S. pombe* with 1541 vertices and 3036 edges; this is probably because this is a denser network, making data reduction less effective.

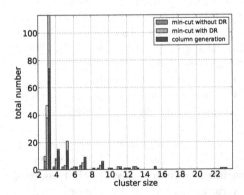

Fig. 1. Clusters in the *A. thaliana* network with all interactions. The brighter part of each bar shows the fraction of clusters without significant enrichment of biological process annotation terms.

Biological Evaluation. For the biological evaluation, we studied the *A. thaliana* network with all interactions in more detail since it was the largest instance for which the exact algorithm finished. Our findings are summarized in Figure 1. Solving HIGHLY CONNECTED DELETION exactly produces more clusters than using the min-cut algorithm with data reduction which in turn produces more clusters than the min-cut algorithm without data reduction. This behavior can be observed for small and for larger clusters.

To assess the biological relevance of these clusters, we determined for each cluster whether the corresponding protein set has a statistically significant enrichment of annotations describing processes in which the protein take part. As shown in Fig. 1, for all three

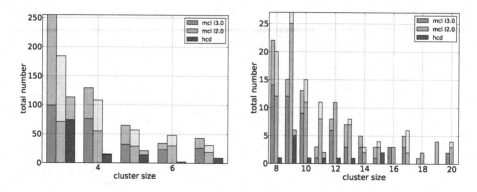

Fig. 2. Clusters in the *A. thaliana* networks produced by the MCL algorithm and our algorithm (HCD) for small clusters (left) and medium-size clusters (right).

methods a large portion of clusters shows such an enrichment. The min-cut algorithm with data reduction clearly outperforms the min-cut algorithm without data reduction: it produces more clusters without producing a larger fraction of nonenriched clusters. For the exact algorithm the results are less clear: it produces even more clusters, but a larger fraction is nonenriched. This behavior is particularly pronounced for small clusters of size at most three, but also for some larger cluster sizes.

Comparison with Markov Clustering. Next, we compare our clustering algorithm with a popular clustering algorithm for protein interaction networks. As comparison, we choose the so-called Markov Clustering Algorithm (MCL). For details concerning MCL refer to [7]; in the experiments, we used the MCL-implementation available at http://micans.org. One parameter that can be set when using MCL is the "inflation" I. We performed experiments with the default value of $I = 2.0$ and with $I = 3.0$ which produces a more fine-grained clustering (as does our algorithm). Unless stated otherwise, we use MCL to refer to the algorithm with default setting.

When comparing the two algorithms, our exact approach (in the following referred to as HCD) and the MCL algorithm, there are some clear advantages of the MCL algorithm: MCL finishes within less than a second, MCL assigns almost all proteins to nonsingleton clusters, and MCL produces more clusters than HCD. MCL also produces larger clusters than HCD. For instance, it finds 30 clusters of size more than 20, and the largest cluster has size 280. As shown in Figure 2, the number of produced clusters is higher across all cluster sizes. The fraction of clusters whose proteins share a significantly enriched GO annotation term, however, is for small and medium-size clusters much lower in the clustering produced by MCL than in the clustering produced by HCD. For large clusters (not shown), 85% of the clusters produced by MCL show a significant enrichment of some annotation term.

To provide a more systematic analysis of the similarity of annotation terms for the clusters, we computed for each protein pair in the same cluster the semantic similarity score for the GO annotations proposed by Wang et al. [20]. The computed scores lie in $[0, 1]$; a higher score indicates higher similarity between the two considered proteins. The average semantic similarity score for a protein pair in the same cluster is 0.410 for HCD and 0.192 for MCL. This pure numeric score, however, could be skewed in favor of HCD. We therefore further examined the effect of the cluster size on the average semantic score for protein pairs in the same cluster. We found that across all cluster sizes, the clusters produced by HCD show better similarity than those produced by MCL. Summarizing, our results for the *A. thaliana* network indicate that HCD outperforms MCL in terms of quality of the reported clusters while MCL shows better coverage and a better running time.

Variants & Extensions. The comparison of HCD with the MCL clustering algorithm showed that two drawbacks of HCD are the running time explosion and the fact that a large fraction of proteins remains unclustered in the optimal HCD solution. We discuss here two strategies to lessen both drawbacks. First, the exact column generation approach is not able to solve the hardest instances. Therefore, we consider a heuristic variant, where we stop the column generation process after a time limit is exceeded. Comparisons with the min-cut algorithm show that with a time limit of one hour, this heuristic variant can find 120 additional clusters compared to the min-cut algorithm.

Another intrinsic problem of demanding highly-connected clusters is the fact that biological networks contain many low-degree vertices: these vertices cannot be contained in any highly connected cluster and HCD computes a clustering of the dense core of the network. Similar to a post-processing suggested by Hartuv and Shamir [8], we used the following simple post-processing to "readd" the proteins not included in any cluster returned by HCD: add each unclustered protein to some cluster if its interactions are exclusively with proteins of this cluster. A first examination of the enrichment statistics indicates that this version of HCD produces better clusters than MCL concerning cluster quality while clustering a significantly larger number of proteins than the pure HCD approach.

5   Outlook

We conclude with a few promising directions for future work. We plan to perform further evaluation of the quality of the clusters found by our approach. First, we plan to evaluate the column-generation-based heuristic on larger standard protein interaction networks such as *S. cerevisiae* and perform comparisons with further clustering algorithms, for example the RN algorithm [16]. Second, a main feature of HIGHLY CONNECTED DELETION is that the cluster definition is easy to interpret. This makes it easy to modify the produced clustering as shown in Section 4. There, the presented post-processing is just a first step, more sophisticated approaches are conceivable and should be explored to further increase clustering quality. Finally, it seems useful to consider edge-weighted HIGHLY CONNECTED DELETION, that is, to maximize the sum of edge weights in the clustering. This could be useful to model different degrees of reliability in the data [4]. Our ILP can be adapted to solve this problem as well.

Acknowledgments. We are indebted to Nadja Betzler and Johannes Uhlmann for their early contributions in the theoretical part of this research.

References

[1] Aloise, D., Cafieri, S., Caporossi, G., Hansen, P., Perron, S., Liberti, L.: Column generation algorithms for exact modularity maximization in networks. Physical Review E 82, 046112 (2010)

[2] Berardini, T.Z., Mundodi, S., Reiser, R., Huala, E., Garcia-Hernandez, M.: et al. Functional annotation of the Arabidopsis genome using controlled vocabularies. Plant Physiology 135(2), 1–11 (2004)

[3] Boyle, E.I., Weng, S., Gollub, J., Jin, H., Botstein, D., Cherry, J.M., Sherlock, G.: GO:TermFinder–open source software for accessing gene ontology information and finding significantly enriched gene ontology terms associated with a list of genes. Bioinformatics 20(18), 3710–3715 (2004)

[4] Chang, W.-C., Vakati, S., Krause, R., Eulenstein, O.: Exploring biological interaction networks with tailored weighted quasi-bicliques. BMC Bioinformatics 13(S-10), S16 (2012)

[5] Chartrand, G.: A graph-theoretic approach to a communications problem. SIAM Journal on Applied Mathematics 14(4), 778–781 (1966)

[6] Chekuri, C., Goldberg, A.V., Karger, D.R., Levine, M.S., Stein, C.: Experimental study of minimum cut algorithms. In: Proc. 8th SODA, pp. 324–333 (1997)

[7] van Dongen, S.: Graph Clustering by Flow Simulation. PhD thesis, University of Utrecht (2000)

[8] Hartuv, E., Shamir, R.: A clustering algorithm based on graph connectivity. Information Processing Letters 76(4-6), 175–181 (2000)

[9] Hartuv, E., Schmitt, A.O., Lange, J., Meier-Ewert, S., Lehrach, H., Shamir, R.: An algorithm for clustering cDNA fingerprints. Genomics 66(3), 249–256 (2000)

[10] Hayes, W., Sun, K., Pržulj, N.: Graphlet-based measures are suitable for biological network comparison. Bioinformatics (to appear, 2013)

[11] Jiang, D., Pei, J.: Mining frequent cross-graph quasi-cliques. ACM Transactions on Knowledge Discovery from Data 2(4), 16:1–16:42 (2009)

[12] Koyutürk, M., Szpankowski, W., Grama, A.: Assessing significance of connectivity and conservation in protein interaction networks. Journal of Computational Biology 14(6), 747–764 (2007)

[13] Liu, H., Zhang, P., Zhu, D.: On editing graphs into 2-club clusters. In: Snoeyink, J., Lu, P., Su, K., Wang, L. (eds.) AAIM 2012 and FAW 2012. LNCS, vol. 7285, pp. 235–246. Springer, Heidelberg (2012)

[14] Niedermeier, R.: Invitation to Fixed-Parameter Algorithms. OUP (2006)

[15] Parker, B.J., Moltke, I., Roth, A., Washietl, S., Wen, J., Kellis, M., Breaker, R., Pedersen, J.S.: New families of human regulatory RNA structures identified by comparative analysis of vertebrate genomes. Genome Research 21(11), 1929–1943 (2011)

[16] Ronhovde, P., Nussinov, Z.: Local resolution-limit-free Potts model for community detection. Physical Review E 81(4), 046114 (2010)

[17] Shamir, R., Sharan, R., Tsur, D.: Cluster graph modification problems. Discrete Applied Mathematics 144(1-2), 173–182 (2004)

[18] Stark, C., Breitkreutz, B.-J., Chatr-aryamontri, A., Boucher, L., Oughtred, R., et al.: The BioGRID interaction database: 2011 update. Nucleic Acids Research 39, 698–704 (2011)

[19] van Rooij, J.M.M., van Kooten Niekerk, M.E., Bodlaender, H.L.: Partition into triangles on bounded degree graphs. Theory of Computing Systems (to appear, 2013)

[20] Wang, J.Z., Du, Z., Payattakool, R., Yu, P.S., Chen, C.-F.: A new method to measure the semantic similarity of GO terms. Bioinformatics 23(10), 1274–1281 (2007)

Reconstructing k-Reticulated Phylogenetic Network from a Set of Gene Trees

Hoa Vu[1], Francis Chin[1], W.K. Hon[2], Henry Leung[1], K. Sadakane[3],
Ken W.K. Sung[4], and Siu-Ming Yiu[1]

[1] Department of Computer Science, The University of Hong Kong
[2] Department of Computer Science, National Tsinghua University, Taiwan
[3] Informatics Research Division, National Institute of Informatics, Japan
[4] Department of Computer Science, National University of Singapore, Singapore

Abstract. The time complexity of existing algorithms for reconstructing a level-x phylogenetic network increases exponentially in x. In this paper, we propose a new classification of phylogenetic networks called *k-reticulated network*. A k-reticulated network can model all level-k networks and some level-x networks with $x > k$. We design algorithms for reconstructing k-reticulated network ($k = 1$ or 2) with minimum number of hybrid nodes from a set of m binary trees, each with n leaves in $O(mn^2)$ time. The implication is that some level-x networks with $x > k$ can now be reconstructed in a faster way. We implemented our algorithm (ARTNET) and compared it with CMPT. We show that ARTNET outperforms CMPT in terms of running time and accuracy. We also consider the case when there does not exist a 2-reticulated network for the input trees. We present an algorithm computing a maximum subset of the species set so that a new set of subtrees can be combined into a 2-reticulated network.

1 Introduction

The study of evolutionary history of a species plays a crucial role in biomedical research. For example, understanding the evolutionary history of a virus (e.g. SARS) may help us deduce the natural reservoirs of the virus, thus identifying the source of the virus. The details of how the virus evolves may help to uncover clues to treat or vaccinate the virus and understand how it evolves resistance to existing drugs. A traditional representation of evolutionary history is phylogenetic tree (a rooted, unordered tree with distinctly labeled leaves, each represents a species or a strain of the species). To construct a phylogenetic tree, a common practice is to select a group of genes, which are believed to be critical for evolution, to represent the species. However, selecting a different set of genes may end up with a different phylogenetic tree (called a gene tree). To deal with this issue, researchers may try to extract the subtrees which are common in all trees (known as the maximum agreement subtree problems, see [1-3] for examples) and ignore the other non-common parts. This may result in a small tree. Also, information not in the common subtree will be lost.

It is now well-known that the differences in the gene trees are not due to errors. There exist evolutionary events (known reticulation events), such as hybridization,

Z. Cai et al. (Eds.): ISBRA 2013, LNBI 7875, pp. 112–124, 2013.
© Springer-Verlag Berlin Heidelberg 2013

horizontal gene transfer, and recombination, that may cause the genes to evolve differently and a phylogenetic tree is not powerful enough to model the resulting evolutionary history [4]. To model the evolutionary history better, phylogenetic network is proposed. Phylogenetic network is a generalization of phylogenetic tree (note that in this paper, we focus on rooted bifurcating (each node has at most 2 descends) phylogenetic tree/network). Phylogenetic network is defined as a rooted, directed acyclic graph in which (1) exactly one node has indegree 0 (the root), and all other nodes have indegree 1 or 2; (2) all nodes with indegree 2 (hybrid nodes or reticulation nodes) have outdegree 1, and all other nodes have outdegree 0 or 2; and (3) all nodes with outdegree 0 (leaves) are distinctly labeled. For a hybrid node h in a phylogenetic network, every ancestor s of h such that h can be reached using two disjoint directed paths starting from the children of s is called a split node of h (and h is called a hybrid node of s). The edges attached to a hybrid node is called hybrid edges. Figure 1 shows an example. Typically, a split node is used to represent a speciation event (two different species are evolved) while a hybrid node is used to represent the reticulation event between the two descendants of the split node.

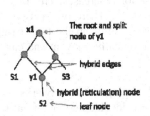

Fig. 1. An example phylogenetic network

Fig. 2. A level-4 network but is an 2-reticulated network for a set of 15 HIV-1 sequences resulting from 9 gene tr

We say that a phylogenetic network *N* is compatible with (*induces* or *displays*) a set of gene trees if each tree can be obtained from *N* by deleting one of the hybrid edges of each hybrid node and contracting all nodes with outdegree and indegree equal 1 (see Figure 3 for an example). If there is no restriction, for any given set of trees, we can always have a phylogenetic network that induces the trees. However, reticulation events are hard to occur, so a more biological meaningful question is to ask for such a phylogenetic network with the minimum number of hybrid nodes.

Fig. 3. Network N is compatible with T_1, T_2. T_3

A common classification of phylogenetic network is the *level-x network* [11 – 12]. A level-x network is one in which each biconnected component (also known as *blob* [5]) of the network contains at most x hybrid nodes. A level-0 network is a phylogenetic tree, a level-1 network is also known as a galled tree [5] or a galled network [6]. There are algorithms that reconstruct a level-x network, however, the time complexity increases exponentially in x even if we only consider some restricted cases. Thus, in practice, if $x > 2$, the algorithm is not fast enough. On the other hand, the evolutionary history of quite many viruses can only be modeled by high level networks (with $x > 2$). For example, to capture all known reticulation events of HIV [7], we need to use a level-4 network (Figure 2 shows the network).

In this paper, we propose to consider a new classification of networks by restricting the maximum number of hybrid nodes each split node may have, namely a *k-reticulated network* is one in which each split node can correspond to at most k hybrid nodes. This new classification is also supported by evidence in real life cases. Several studies of recombination in bacteria have shown that recombination rates decrease as sequence divergence increases [8-9]. These studies imply that the number of recombination events of a split node will be limited as the descendants from the same split node will diverge more as the number of generations increases. This observation is also supported by a computer simulation study [10]. Therefore, networks with limited reticulation events for each split node while no limit on the total number of reticulation events in each blob seem to be more biologically relevant and can model the recombination events in nature more appropriately.

Our Contributions. This new classification of phylogenetic networks is more powerful than level-x networks. By definition, every level-x network is also an *x-reticulated* network. And some level-x network can be modeled by a *k-reticulated* network with $k < x$ (see Figure 2 for an example of a level-4 network which is also a 2-reticulated network). So, even solving the problem for *k-reticulated* network with k as small as 2, some of the meaningful high level networks can be constructed efficiently. We show that given a set of binary gene trees, one can reconstruct an 1-reticulated or 2-reticulated network (if one exists) with minimum number of hybrid nodes compatible with all trees in $O(mn^2)$ time where m is the number of trees and n is the number of leaves. We also consider the problem that when a compatible 2-reticulated network does not exist, compute a subset of species with maximum size so that a 2-reticulated network exists. This problem is believed to be NP-hard and we provide an $O(2^{mm}n^{3m})$ algorithm to solve it. We implement the 2-reticulated network reconstruction algorithm (ARTNET) and compare it with the program CMPT [13] that reports a phylogenetic network with the smallest number of hybrid nodes. We only consider the case when a 2-reticulated network exists for the input set of trees. The experiments show that ARTNET is more efficient than CMPT. When the number of hybrid nodes increases, the running time of ARTNET only increases slightly while that of CMPT increases rapidly. Regarding accuracy evaluation, ARTNET also outperforms CMPT.

Related Work. Several methods of constructing phylogenetic networks have been proposed. Nakhleh *et al.* [6] have developed an algorithm for constructing a level-1 phylogenetic network from two phylogenetic trees running in polynomial time.

However, Nakhleh *et al.*'s algorithm can handle two trees only. Huynh *et al.* [12] have succeeded in providing a $O(|T|^2 n^2)$ algorithm reconstructing a galled network from a set T of multiple phylogenetic trees of arbitrary degree. Huson and Klopper [14] gave an $O(n^k)$ algorithm constructing restricted level-*k* network from a set of trees. A rooted phylogenetic tree can be uniquely represented by the set of triplets obtained by taking all combinations of three leaves in the tree [12]. It takes $O(n^3)$ running time to construct a galled network in the algorithm designed by Jansson, Nguyen and Sung [15]. Extending to level-2 network, Van Iersel *et al.* [11] developed an $O(n^8)$ time algorithm. Habib and To [16] have solved the general problem of constructing level-*k* network from a dense triplet set T in exponential running-time $O(|T|^{k+1} n^{\lfloor \frac{4k}{3} \rfloor + 1})$. Gambette *et al.* [17] have shown that we can decide in optimal $O(n^4)$ time whether there exists a simple unrooted level-1 network for a set of all quartets.

Notations. Let u is a node in a tree T, $T[u]$ = the subtree of T rooted at u, and $L(T)$ = the leaf label set of T. If u is a node in network N, a subnetwork $N[u]$ is obtained from N by only retaining all nodes and their incident edges which are reachable from u, and $L(N)$ is the set of leaf labels of N. Given a subtree t of T, $T\backslash t$ is a subtree obtained by removing t from T. Similarly, with a subnetwork N' of N, $N\backslash N'$ is a network obtained by removing N' from N. Given a tree T with the leaf set L, and $L' \subseteq L$. $T|L'$ denotes a subtree obtained by first deleting all nodes which are not on any directed path from the root to a leaf in L' along with their incident edges, and then, for every node with outdegree 1 and indegree less than 2, contracting its outgoing edge.

2   Algorithms for Reconstructing *k*-Reticulated Network (*k* = 1, 2)

Denote $P(N, T_i, L)$ the procedure to reconstruct a *k*-reticulated network N compatible with $T_1, T_2, ..., T_m$, where $k = 1$ or 2. We employ the divide-and-conquer technique.

2.1   Reconstructing 1-Reticulated Network

Base Case: if each input tree is a single node with the same label, return a network which is that single node of the same label; otherwise consider the following cases:

Case I: *Bipartition*
$\{T_1, ..., T_m\}$ admit a ***leaf-set-bipartition*** (L_1, L_2) if for every tree T_i with root r_i and its children r_{i1} and r_{i2}, $L(T_i[r_{i1}]) = L_1$ and $L(T_i[r_{i2}]) = L_2$, then find $P(N_1, T_i[r_{i1}], L_1)$ and $P(N_2, T_i[r_{i2}], L_2)$. If N_1 and N_2 exist, network N is obtained by creating a new node r becoming the parent of the roots of N_1 and N_2. (Fig. 4)

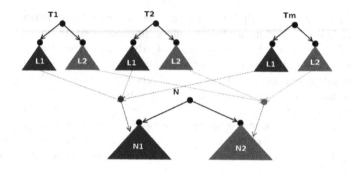

Fig. 4. The tree set admit a leaf set bipartition

Case II: *Tripartition*

$\{T_1,...,T_m\}$ admit a ***leaf-set-tripartition (L_1, L_h, L_2)*** if for every tree T_i with root r_i and its children r_{i1} and r_{i2}, there exists a node $h_i \neq r_i$ such that $L(T_i[h_i]) = L_h$; and if $h_i \neq r_{i1}$ and $h_i \neq r_{i2}$ for every $i = 1...m$; $\{T_i' = T_i \backslash T_i[h_i]\}$ admit a leaf-set-bipartition (L_1, L_2). Otherwise, $L_1 = L(T_i')$ and $L_2 = \emptyset$.

If $h_i \neq r_{i1}$ and $h_i \neq r_{i2}$ for $i = 1...m$, the problem can be divided into 3 subproblems: $\mathbf{P}(N_1, T_i'[r_{i1}], L_1)$; $\mathbf{P}(N_2, T_i'[r_{i2}], L_2)$; and $\mathbf{P}(N_h, T_i[h_i], L_h)$. Network N can be combined from N_1, N_2 and N_h by first creating a new node r to be the parent of the roots of N_1 and N_2. Find node u_1 in N_1 and u_2 in N_2 such that for $i = 1...m$, either u_1 or u_2 corresponds to h_i's sibling s_i. Let v_1 and v_2 be the parent of u_1 and u_2 respectively, create nodes p_1 and p_2 on edges (v_1, u_1) and (v_2, u_2) respectively. A new hybrid node h is created, and let h be a child of p_1 and p_2, and h be the parent of N_h's root (Fig. 5).

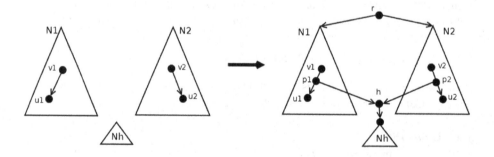

Fig. 5. Combining N_1, N_2 and N_h to get network N

Given network N compatible with tree T, *a node u in N is said to correspond to a node s in T* if T can be converted from N by a series of cuts in which any edge contraction related to node u will create a new node that is labeled u, then u becomes s.

If there is a tree T_i in which h_i is a child of the root r_i, the network constructed in this case is skew (i.e., there is a split node such that the path from the split node to its hybrid node is 1). The problem can be divided into 2 sub-problems: $\mathbf{P}(N', T_i', L_1)$; and $\mathbf{P}(N_h, T_i[h_i], L_h)$. If N' and N_h can be constructed, N can be obtained by first creating a node r and making r become the parent of the root of N'. Find a node u in N' such that for every tree T_i in which h_i is not a child of the root r_i, u corresponds to s_i in T_i', which is the sibling of h_i before removing $T_i[h_i]$. Let v be the parent of u, and a new node p on edge (v, u), create a hybrid node h that is the child of p and r, and h is the parent of the root of N_h (Fig. 6).

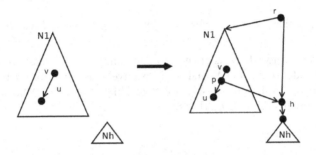

Fig. 6. Combining N_1 and N_h to get a skew network N

2.2   Reconstructing 2-Reticulated Network

To solve this problem, we also consider the base case, Case I and Case II as in the above. In addition, we need to consider Case III – Quadripartition as follows.

Case III: *Quadripartition*
The tree set $\{T_1, T_2, ..., T_m\}$ is said to admit a ***leaf-set-quadripartition (L_1, L_{h1}, L_{h2} L_2)*** if for every tree T_i with root r_i and its children r_{i1} and r_{i2}, there exists a node $h_{i2} \notin \{r_i, r_{i1}, r_{i2}\}$ such that $L(T_i[h_{i2}]) = L_{h2}$; and $\{T_i' = T_i \backslash T_i[h_{i2}], i = 1, 2, ..., m\}$ admit a leaf-set-tripartition (L_1, L_{h1}, L_2). If there exist a 2-reticulated network N' compatible with $\{T_1', ..., T_m'\}$; and a 2-reticulated network N_{h2} compatible with $\{T_i[h_{i2}], i = 1, ..., m\}$.

If N' is a non-skew network, N' is created by combining three 2-reticulated networks N_1, N_2 and N_{h1} (as case II). Find two nodes a and b in two distinct networks out of three networks N_1, N_2 and N_{h1} such that either a or b corresponds to node s_i in T_i', which is the sibling of h_{i2} in T_i, for $i = 1, 2, ..., m$. Attaching N_{h2} to N' is done similarly to case II by creating a new hybrid node h_2 (Fig. 7).

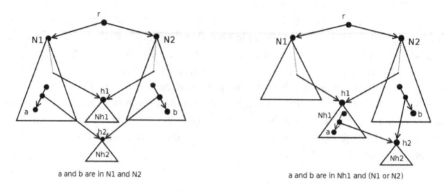

Fig. 7. Combining non-skew network N' and N_{h2} to get a 2-reticulated network N

If N' is a skew-network, N' is created by combining two 2-reticulated networks N_1 and N_{h1}. Find nodes a and b in N_1 and N_{h1} respectively such that for $i = 1...m$, either a or b corresponds to node s_i in T_i', which is the sibling of h_{i2} in T_i. Attaching N_{h2} to N' by creating a new hybrid node h_2. (Fig.8)

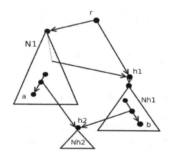

Fig. 8. Combining skew network N' and N_{h2} to get a 2-reticulated network N

2.3   Algorithm Correctness

Lemma 1. Given a network N compatible with tree T and a node v in N, if all nodes in a subnetwork $N[v]$ cannot be reached from any other nodes outside $N[v]$ without passing through node v, then there exists a node u in T such that its subtree $T[u]$ and $N[v]$ have exactly the same leaf label set, and $N[v]$ is compatible with $T[u]$.

Theorem 1. The algorithm described in section 3.1 and 3.2 can construct a 2-reticulated network compatible N with a given set of trees if and only if N exists.

Proof. Assume there is a 2-articulated compatible network N for $\{T_1, T_2,..., T_m\}$. Consider the root r of N:

Case 1: r is the only node in N. The theorem is obviously correct.
Case 2: r does not correspond to any hybrid node

Let r_1 and r_2 be two children of r, and L_1 and L_2 be the leaf set of $N[r_1]$ and $N[r_2]$ respectively. As r does not correspond to any hybrid node, any node outside $N[r_1]$ (resp. $N[r_2]$) has to pass through r_1 before reaching any node inside $N[r_1]$ (resp. $N[r_2]$). From Lemma 1, there are nodes u_1 and u_2 in every tree T_i such that $L(T[u_1]) = L_1$, and $L(T[u_2]) = L_2$, and $N[r_1]$ and $N[r_2]$ are compatible with $T[u_1]$ and $T[u_2]$ respectively. We have $L_1 \cap L_2 = \emptyset$ and $L_1 \cup L_2 = L$, so in N and every tree T_i, their root is the only common ancestor of any node in L_1 and any node in L_2. This means the input tree set admit a leaf set bipartition, corresponding to case I in the algorithm.

Case 3: r corresponds to one hybrid node h
All nodes in $N[h]$ cannot be reached by any other node not in $N[h]$ without passing through node h, otherwise, there must exist another hybrid node of root r in $N[h]$, contradicting to the fact that r corresponds to exact one hybrid node. By **Lemma 1**, there exists a node h_i in every tree T_i such that $N[h]$ is compatible with $T_i[h_i]$, and $L(N[h]) = L(T_i[h_i])$; hence, $N\!\backslash\!N[h]$ is compatible with $T_i\backslash T_i[h_i]$. As the root of $N\!\backslash\!N[h]$ does not correspond to any hybrid node, the argument can be turn back to *Case 2*. This implies that the tree set admit a leaf set tripartition, corresponding to case 2 of the algorithm.

Case 4: r corresponds to two hybrid nodes h_1 and h_2
Let p_1 and q_1 be the parents of h_1, and p_2 and q_2 be the parents of h_2, then either h_1 lies on one of the merge paths from the root r to h_2, or none of the merge paths from r to h_2 (resp. h_1) go through h_1 (resp. h_2) (figure 7).

In both cases, all nodes in the subnetwork $N[h_2]$ cannot be reached by any other node outside $N[h_2]$ without passing through node h_2; otherwise, r would correspond to another hybrid node in $N[h_2]$.

From Lemma 1, there exists a node h_{i2} in every tree T_i such that $N[h_2]$ is compatible with $T_i[h_{i2}]$, and $L(N[h_2]) = L(T_i[h_{i2}])$. Plus, $N\!\backslash\!N(h_2)$ is a compatible 2-articulated network of $T_i\backslash T_i[h_{i2}]$. As the root of $N\!\backslash\!N(h_2)$ corresponds to exact 1 hybrid node h_1, the argument can turn back to Case 3 above. This implies the tree set admit a leaf set quadripartition, corresponding to Case 3 of the algorithm.   □

2.4   Time Complexity

Lemma 2. Determining whether $\{T_1, T_2, \ldots, T_m\}$ admit a leaf set bipartition or tripartition or quadripartition and partitioning every tree can be done in $O(mn)$.

Proof. Denote $LCA(X)$ the lowest common ancestor of all nodes in set X.
For every tree T_i, $i = 1, 2, \ldots, m$, denote r_i the root of T_i, and r_{i1} and r_{i2} are two children of r_i. Define a subset $L^* \subset L$:

$L^* = \emptyset$ if the tree set admit the leaf set bipartition.
$L^* = L_h$ if the tree set admit a leaf set tripartition (L_1, L_h, L_2).
$L^* = L_{h1} \cup L_{h2}$ if the tree set admit a leaf set quadripartition $(L_1, L_{h1}, L_{h2}, L_2)$.

Determine L^*: Let $L_c = L(T_1[r_{11}])$; $L_d = L(T_1[r_{12}])$. It takes $O(mn)$ to divide L_c into two disjoint subsets L_{c1} and L_{c2}, and L_d into two disjoint subsets L_{d1} and L_{d2} (Fig. 9). Pick one leaf node v in L_c, then

For every leaf node $u \in L_c$: If $LCA(v, u)$ is not the root of every tree T_i; u is put in L_{c1}; else, u is put in L_{c2}.

For every leaf node $w \in L_d$: If $LCA(v, w)$ is the root of every tree T_i; w is put in L_{d1}; else w is put in L_{d2}.

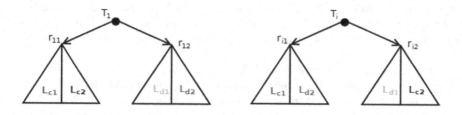

Fig. 9. Partition the leaf set

Claim 1: The tree set admits a leaf set bipartition iff $L_{c2} = L_{d2} = \emptyset$; otherwise check tripartition property.

Claim 2: The tree set admits a leaf set tripartition iff either $L^* = L_{c2} \cup L_{d2}$ or $L^* = L_{c1} \cup L_{d1}$, and there exists a node h_i in every tree T_i such that $L(T_i[h_i]) = L^*$, and $T_i \backslash T_i[h_i]$ admit a leaf set bipartition if h_i is not a child of r_i for every $i = 1...m$, taking $O(mn)$ time; otherwise, check quadripartition property.

Claim 3: If the tree set admits a leaf set quadripartition $(L_1, L_{h1}, L_{h2}, L_2)$, one of two sets $L_{c1} \cup L_{d1}$ or $L_{c2} \cup L_{d2}$ can be either (1) L_{h1}, or (2) L_{h2}, or (3) $L_{h1} \cup L_{h2}$.
Pick any tree, say T_1, to find the $p_1 = LCA(L_{c1} \cup L_{d1})$ and $p_2 = LCA(L_{c2} \cup L_{d2})$.

1. One of two nodes p_1 or p_2 is the root r_1 and the other is not. Assume $p_1 = r_1$, and p_2 is a proper descendant of r_1, then $L_{c2} \cup L_{d2} = L_{h1}$ or $L_{c2} \cup L_{d2} = L_{h2}$.
2. If both p_1 and p_2 are r_1, find $j = 1$ or 2 such that $LCA(L_{cj})$ and $LCA(L_{dj})$ are the children of the root r_1. If j does not exist, return *"null"*; otherwise, assume $j = 1$, then $L_{h1} \cup L_{h2} = L_{c2} \cup L_{d2}$.

It takes $O(n)$ time to determine L' which is either L_{h1}, or L_{h2} or $L_{h1} \cup L_{h2}$.

- If L' is L_{h1} or $L_{h2} \rightarrow L^* = L'$
  Check if there is a node h_i in every tree T_i such that $L(T_i[h_i]) = L'$ in $O(mn)$. If yes, check if $\{T_i \backslash T_i[h_i], i = 1,..., m\}$ admit a leaf set tripartition (L_1, L_h, L_2). If yes, $L_{h2} = L'$ and $L_h = L_{h1}$; else return *"null"*.
  If there is a tree T_j in which there does not exist any node w such that $L(T_j[w]) = L'$ (*Fig.*10). If h_j is $LCA(L')$, then $L' = L_{h1} \cup L_{h2}$, which is examined as case 2 below.

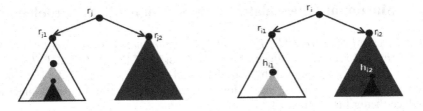

Fig. 10. With a leaf set L_{h1}, find a node w in T_j such that $L(T_j[w]) = L_{h1} \cup L_{h2}$

- If $L' = L_{h1} \cup L_{h2}$.
  Let T_k be a tree in which there is no node satisfying $L' = L(T_k[w])$. If there are exact two nodes h_{k1} and h_{k2} in T_k such that $L(T_k[h_{k1}]) \cup L(T_k[h_{k2}]) = L'$, then either $L(T_k[h_{k1}])$ or $L(T_k[h_{k2}])$ is L_{h2}; otherwise, return *"null"*. Finding h_{k1} and h_{k2} takes $O(n)$. It then takes $O(mn)$ to determine which one, $L(T_k[h_{k1}])$ or $L(T_k[h_{k2}])$, is L_{h2}, and partition every tree T_i into T_i' and $T_i[h_{i2}]$, for $i = 1, 2, \ldots, m$.

In total, checking whether the input trees admit a leaf set bipartition or tripartition or quadripartition, and partition every tree into proper subtrees takes $O(mn)$.   □

Lemma 3. Given a network N compatible with m trees $\{T_1, T_2, \ldots, T_m\}$ with the same leaf label set L of size n, and a node s_i in T_i, for $i = 1, 2, \ldots, m$, then finding whether there is a node u in N corresponding to $s_1, s_2, \ldots, s_m$ can be done in $O(n)$.

Proof. For $i = 1, \ldots, m$, let u_i be a node in N having the lowest height in N such that $L(T_i[s_i]) \subseteq L(N[u_i])$.

Claim: Node u exits iff u_i is u or a descendant of u such that all nodes on the path from u to u_i are either hybrid node of a skew split node or non-hybrid nodes whose siblings are hybrid nodes. It takes $O(mn)$ to find the set $\{u_1, u_2, \ldots, u_m\}$ from N (note that u_x can be u_y), and $O(n)$ time to check (*i*) all nodes $\{u_1, u_2, \ldots, u_m\}$ lie on the same directed path; and (*ii*) The siblings of u_i, $i = 1, 2, \ldots, m$, are all hybrid nodes. If two conditions are satisfied, the node u will be the starting node x of the path created by $\{u_1, u_2, \ldots, u_m\}$ or x's sibling if x's sibling is the hybrid node of a skew split node ; otherwise, return *"null"*.   □

Theorem 2. Constructing a k-reticulated network ($k = 1$ or 2) from a set of m binary trees with the same leaf label set L of size n can be done in $O(mn^2)$.

Proof. From *Lemma 2* and *Lemma 3*, dividing and conquering take $O(mn)$ time complexity. There are $O(n)$ nodes in a tree with n leaf nodes. Hence the time complexity of our algorithm is $O(mn^2)$.

3   Maximum 2-Reticulated Network Compatibility Problem

Given a set of binary trees $\{T_1, T_2, ... T_m\}$, compute the maximum leaf set L^* such that there exists a 2-reticulated network N compatible with $\{T_1|L^*, T_2|L^*, ..., T_m|L^*\}$.

Using brute-force approach by considering all possible subsets of the leaf set, the problem can be done in $O(2^n mn^2)$. However, when $m \ll n$, the following algorithm produces better time complexity.

MCN(T_i) denotes the Maximum compatible 2-reticulated network for $T_1, T_2, ..., T_m$. **MCLS(T_i)** denotes the **MCN(T_i)**'s Leaf Set

Let v be a node in a tree T, **child$_1$(v)** and **child$_2$(v)** denote two children of node v.

Let **sib_hyb(l, h)** be an array of pointers in which $sib\_hyb(l, h)[i]$ is pointing to a node w_i in tree T_i; where l is a positive integer number, and h receives a value of 1 or 2. The following rules are applied in the process of removing nodes and edges when computing MCLS(T_i):

- If w_i is one of the end node of the edge that is contracted, $sib\_hyb(l, h)[i]$ will point to a new node created after doing contraction.
- If a whole subtree rooted at w_i is deleted from T_i, $sib\_hyb(l, h)[i]$ points to the sibling of w_i.
- If a whole subtree rooted at p_i, which is an ancestor of w_i, is deleted, $sib\_hyb(l, h)[i]$ points to the sibling of p_i.

Theorem 3. Given m trees $T_1, ..., T_m$ rooted at $r_1,..., r_m$ respectively. l_1 and l_2 are global variables initialized = 0; *Base case*: there is a tree that is a single node, $MCLS(T_i, i = 1, ..., m) = \bigcap_{i=1}^{m} L(T_i)$.

$MCLS(T_i)$ is the set having the maximum size of the following terms:

1. $\max\{MCLS(T_{ia}[child_1(r_{ia})], T_{ib}), MCLS(T_{ia}[child_2(r_{ia})], T_{ib})$; with $i_1 \geq 1$ and $i_2 \geq 1$ $\{T_{ia}, a = 1 ... i_1\} \cup \{T_{ib}, b = 1 ... i_1\} = \{T_1, ..., T_m\}$ and $\{T_{ia}, a = 1 ... i_1\} \cap \{T_{ib}, b = 1 ... i_1\} = \emptyset\}$;
2. $MCLS1(T_i) = \max\{ MCLS(T_i[child_{c_i}(r_i)]) + MCLS(T_i[child_{d_i}(r_i)])$; where $(c_i, d_i) \in \{(1, 2), (2, 1)\}$, for $i = 1, ..., m\}$;
3. $MCLS2(T_i) = \max \{MCLS1(T_i \backslash T_i[v_i]) + MCLS(T_i[v_i])$; where v_i is some node in T_i, The sibling of v_i is pointed by $sib\_hyb(++l_1, 1)[i]$, for $i = 1...m\}$;
4. $\max \{MCLS2(T_i \backslash T_i[w_i]) + MCLS(T_i[w_i])$, where w_1 is a node in T_i. The sibling of w_i is pointed by $sib\_hyb(++l_2, 2)[i]$, for $i = 1...m\}$;

Before computing $MCLS(t_i)$, if there is a tree t_p whose root is pointed by any $sib\_hyb(l_1, 1)[p]$ (resp. $sib\_hyb(l_2, 2)[i])$ with a specific value l_1 (resp. l_2), then for every other tree t_j containing a node s_j that is pointed by $sib\_hyb(l_1, 1)[j]$ (resp. $sib\_hyb(l_2, 2)[j])$, replace $t_j = t_j[s_j]$ in computing $MCLS(t_i)$.

Time Complexity: By applying dynamic programming, and backtracking on the recursive equations, the problem can be computed in $O(2^m mn^{3m})$.   □

4   Experiments

We evaluate and compare the performance of our method, namely ARTNET, with the program CMPT [13] which constructs a network with the smallest number in reticulation from a set of binary trees. NETGET [10] is used to generate random networks. For every 2-reticulated network simulated, we produce a certain number of induced binary trees which are the input of both programs ARTNET and CMPT. We use *n* (number of leaf node) = 40. Figure 11 shows that we run faster than CMPT. Following [11], we use split-based false negative (FN) and false positive (FP) rates to measure the error rates of the methods. Figure 12 shows that ARTNET produces fewer false positives than CMPT. On the other hand, CMPT and ARTNET have similar performance in false negative rates.

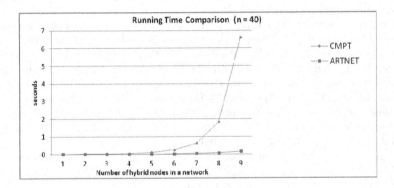

Fig. 11. Time comparison between ARTNET and CMPT

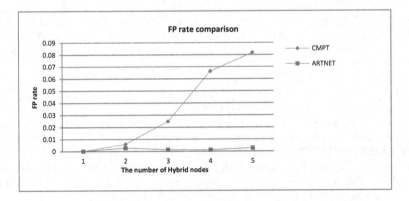

Fig. 12. Comparing the false positive rate between ARTNET and CMPT

Acknowledgement. The project is partially supported by in part by the General Research Fund (GRF) of the Hong Kong Government (HKU 719709E) and Kunihiko Sadakane is supported in part by KAKENHI 23240002.

References

[1] Lam, T.W., Sung, W.K., Ting, H.F.: Computing the unrooted maximum agreement subtree in sub-quadratic time. Nordic Journal of Computing 3(4), 295–322 (1996)

[2] Farach, M., Thorup, M.: Sparse dynamic programming for evolutionary-tree comparison. SIAM Journal on Computing 26(1), 210–230 (1997)

[3] Steel, M., Warnow, T.J.: Kaikoura tree theorems: Computing the maximum agreement subtree. Information Processing Letters 48, 77–82 (1993)

[4] Ford Doolittle, W.: Phylogenetic classification and the universal tree. Science 284(5423), 2124–2128 (1999)

[5] Gusfield, D., Bansal, V.: A fundamental decomposition theory for phylogenetic networks and incompatible characters. In: Miyano, S., Mesirov, J., Kasif, S., Istrail, S., Pevzner, P.A., Waterman, M. (eds.) RECOMB 2005. LNCS (LNBI), vol. 3500, pp. 217–232. Springer, Heidelberg (2005)

[6] Nakhleh, L., Warnow, T., Linder, C.R.: Reconstructing reticulate evolution in species – theory and practice. In: Proceedings of the 8th Annual International Conference on Research in Computational Molecular Biology (RECOMB 2004), pp. 337–346 (2004)

[7] Lee, W.-H., Sung, W.-K.: RB-finder: An improved distance-based sliding window method to detect recombination breakpoints. In: Speed, T., Huang, H. (eds.) RECOMB 2007. LNCS (LNBI), vol. 4453, pp. 518–532. Springer, Heidelberg (2007)

[8] Falush, D., Torpdahl, M., Didelot, X., Conrad, D.F., Wilson, D.J., Achtman, M.: Mismatch induced speciation in salmonella: model and data. Philos. Trans. R Soc. Lond. B Biol. Sci. 361(1475), 2045–2053 (2006)

[9] Majewski, J.: Sexual isolation in bacteria. FEMS Microbiol. Lett. 199(2), 161–169 (2001)

[10] Fraser, C., Hanage, W.P., Spratt, B.G.: Recombination and the nature of bacterial speciation. Science 315(5811), 476–480 (2007)

[11] van Iersel, L., Keijsper, J., Kelk, S., Stougie, L., Hagen, F., Boekhout, T.: Constructing level-2 phylogenetic networks from triplets. In: Vingron, M., Wong, L. (eds.) RECOMB 2008. LNCS (LNBI), vol. 4955, pp. 450–462. Springer, Heidelberg (2008)

[12] Huynh, T.N.D., Jansson, J., Nguyen, N.B., Sung, W.-K.: Constructing a smallest refining galled phylogenetic network. In: Miyano, S., Mesirov, J., Kasif, S., Istrail, S., Pevzner, P.A., Waterman, M. (eds.) RECOMB 2005. LNCS (LNBI), vol. 3500, pp. 265–280. Springer, Heidelberg (2005)

[13] Zhi-Zhong, C., Lusheng, W.: Algorithms for Reticulate Networks of Multiple Phylogenetic Trees. IEEE/ACM Transactions on Computational Biology and Bioinformatics (TCBB) 9(2), 372–384 (2012)

[14] Huson, D.H., Klöpper, T.H.: Beyond galled trees - decomposition and computation of galled networks. In: Speed, T., Huang, H. (eds.) RECOMB 2007. LNCS (LNBI), vol. 4453, pp. 211–225. Springer, Heidelberg (2007)

[15] Jansson, J., Nguyen, N.B., Sung, W.-K.: Algorithms for combining rooted triplets into a galled phylogenetic network. In: Proceedings of the 16th Annual ACM-SIAM Symposium on Discrete Algorithms (SODA), pp. 349–358 (2005)

[16] Habib, M., To, T.-H.: Constructing a minimum phylogenetic network from a dense triplet set. J. Bioinformatics and Computational Biology 10(05) (2012)

[17] Gambette, P., Berry, V., Paul, C.: Quartets and Unrooted Phylogenetic Networks. Journal of Bioinformatics and Computational Biology (2011)

LCR_Finder: A de Novo Low Copy Repeat Finder for Human Genome

Xuan Liu, David Wai-lok Cheung, Hing-Fung Ting,
Tak-Wah Lam[*], and Siu-Ming Yiu[*]

Department of Computer Science, The University of Hong Kong, Hong Kong
{xliu,dcheung,hfting,twlam,smyiu}@cs.hku.hk

Abstract. Low copy repeats (LCRs) are reported to trigger and mediate genomic rearrangements and may result in genetic diseases. The detection of LCRs provides help to interrogate the mechanism of genetic diseases. The complex structures of LCRs render existing genomic structural variation (SV) detection and segmental duplication (SD) tools hard to predict LCR copies in full length especially those LCRs with complex SVs involved or in large scale. We developed a de novo computational tool LCR_Finder that can predict large scale (>100Kb) complex LCRs in a human genome. Technical speaking, by exploiting fast read alignment tools, LCR_Finder first generates overlapping reads from the given genome, aligns reads back to the genome to identify potential repeat regions based on multiple mapping locations. By clustering and extending these regions, we predict potential complex LCRs. We evaluated LCR_Finder on human chromosomes, we are able to identify 4 known disease related LCRs, and predict a few more possible novel LCRs. We also showed that existing tools designed for finding repeats in a genome, such RepeatScout and WindowMasker are not able to identify LCRs and tools designed for detecting SDs also cannot report large scale full length complex LCRs.

1   Introduction

Complex low copy repeats (LCRs, also termed as segmental duplications), which are composed of multiple repeat elements either in direct or inverse directions, provide the structural basis for diverse genomic variations and combinations of variations (Zhang et al. 2009). Around 5% of sequenced portion of human genome is composed of LCRs. LCRs usually range from 10Kb to several hundred Kb. Among the copies, there are common regions (with sequence similarity as high as 95% (Babcock et al. 2007). However, there may also be big gaps between common regions and not all common regions exist in all copies. This makes the detection process very difficult. Complex LCRs are found to trigger and mediate genomic rearrangement including deletion, duplication etc., by altering gene dosage, and result in human genetic disorders (Stankiewicz et al. 2002; Zhang et al. 2009). Several human genetic disorders caused by genomic recombination are reported to be triggered and mediated by LCRs,

---

[*] Joint corresponding authors.

Z. Cai et al. (Eds.): ISBRA 2013, LNBI 7875, pp. 125–136, 2013.
© Springer-Verlag Berlin Heidelberg 2013

such as Familial Juvenile Nephronophthisis at chromosome band 2q13 caused by 2.9Mb deletion (Saunier et al. 2000; OMIM 607100), William-Beuren syndrome (WBS) at chromosome band 7q11.23 caused by 1.5Mb to 1.8Mb deletion (Valero, M. C. et al. 2000; OMIM 194050), Sotos syndrome (Sos) by 5q35 deletion (Kurotaki, N. et al. 2005; OMIM 117550) and Smith-Magenis syndrome (SMS) by deletion on 17p11.2 (Claudia M.B. Carvalho et al. 2008 ; OMIM 182290). Genes inside the regions of LCRs have a high chance to develop into diseases. Identifying the locations of LCRs thus is important and the pattern of LCRs can help to unveil the complicated mechanism of genetic diseases.

Several computational methods have been proposed to identify copy number repeats and duplications such as RepeatScout (Price et al. 2005) and WindowMasker (Morgulis et al. 2006). Although detecting repeats on genome sequence is relatively well studied, there are only a limited number of tools for LCRs/SDs such as DupMasker (Jiang et al. 2007). The outcome of existing software for detecting LCRs, especially complex LCRs is not satisfactory.

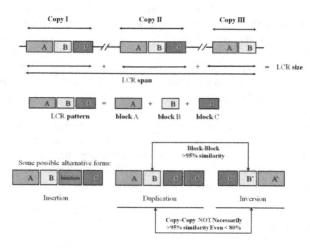

Fig. 1. Illustration of LCR structure. An LCR is composed of several copies, and each copy is an alternative form of pattern of this LCR. LCR pattern is a combination of several tandem/disconnected blocks. LCR span is defined as the minimum interval that all copies are included. LCR size is length sum of all copies. There is a high similarity between the blocks with the same label in different copies. Insertion, deletion, inversion, translocation, duplication etc. could be introduced to blocks or gaps between blocks to make an alternative form.

As mentioned in the above, LCR structure is complicated (Fig 1). Unlike repeats, complex LCRs are higher level combination of different but associated repeats transposed to specific genomic regions, creating duplication blocks with mosaic architecture of juxtaposed duplicated segments (Jiang et al. 2007). Each copy of LCR pattern is an alternative form of block patterns and not necessarily the same as others and may even quite different in structure. The similarity between block patterns in different LCR copies varies and can range from 30% to 95%. Due to the gaps between blocks, the similarity between the whole LCR copies is even lower. Note that not all block patterns appear in all copies of the LCR and duplication, translocation, and inversion of patterns are not uncommon. Copy number of LCR is usually below 5, while the repeats it contains may either highly or lowly repetitive.

Despite the fact that LCRs could be considered as repeats combination with SVs introduced, Repeat/SV tools are not able to detect them. SV-targeting tools reveal variance at the same loci between two input sequences, instead of different loci of one single sequence. There are Repeat-targeting methods designed to identify repetitive sequences in a single input sequence, such as RepeatMasker (Smit et al. 2011), RepeatScout (Price et al. 2005) and WindowMasker (Morgulis et al. 2006). However, most of these tools were not designed for large repeat patterns, thus many small-scale highly repetitive patterns are reported which cannot be easily grouped to locate the LCRs. For example, we ran RepeatMasker on human chromosome 17 (GRCh37.p9), 165,382 repetitive sequences, covering 37,757,301/81,195,210 bp of chromosome 17 with an average length of repeat as 228.30 are reported. These 165,382 repeats were mapped to only 1268 repeat patterns (55 repeat families, including SINE, LINE, small RNAs, low complex sequences and so on). It is not trivial how to "glue" these short patterns together to locate LCRs.

On the other hand, there are database dependent tools that limit the search scope to existing human segmental duplication libraries and are not applicable to other species. DupMasker reported 47, 62 and 82 duplicons for each SMS-LCRs copy respectively instead of 3 full length LCR copies. Also, DupMasker uses the result of RepeatMasker as part of the input, it is very time consuming.

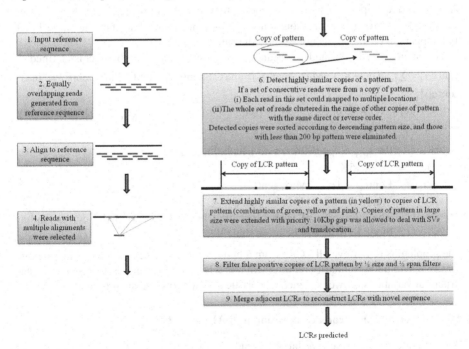

Fig. 2. Workflow of LCR_Finder

In this paper, we introduce a novel de novo LCR detection tool, called LCR_Finder, which does not rely on known information on segmental duplications and can large scale LCRs in complex structures, on a given genomic sequence (e.g. chromosome level). Technical speaking, we exploit existing read alignment tools to solve the problem as follows. We (1) computationally generate single end reads from the given genome, (2) align them back to the genome, (3) locate large repetitive patterns using reads with multiple aligned positions, (4) extend copies with large gap allowed and (5) filter out false positives and report potential LCRs (Fig 2). We evaluate LCR_Finder on human chromosomes and show that it can identify four known diseases related LCR loci and report a few more potential novel LCRs. We also compare the results reported by RepeatScout and WindowMasker and found that their results are not as good as given by LCR_Finder.

2   Methods

2.1   Problem Definition

We define a block as a DNA pattern with size > 5K. A LCR is defined as a set of blocks $B:=\{b_1,b_2,...,b_m\}$, $m\geq1$. A LCR(C) consists of 2 to 5 repeated regions represented as $C:=\{c_1,c_2,...,c_n\}$, $2\leq n\leq5$. Each copy is a collection of blocks (with mutations) in set B, $c_i :=\{ b_{i_1},b_{i_2}, ... , b_{i_t}\}$. Note that the same block can appear more than once in each copy (probably with different mutations).

For each i_k, $1\leq i_k\leq i_t$, b_{i_k} is represented by a pair of genomic locations indicating the starting and end positions of that block on reference, $bi=(s_i,e_i)$, $s_i<e_i$. $b_{i_1},b_{i_2}, ... , b_{i_t}$ are sorted according to increasing starting position order. Note that the same block that appears in each copy is not exactly the same.

The span of c_i is defined as $Span(c_i) =(e_{i_t} - s_{i_1})$, and length of c_i is defined as $Len(c_i) = \sum_{j=1}^{i_t} \left(e_{i_j} - s_{i_j}\right)$. $Max\_span(C):=max\{Span(c_i)\}$ and $Max\_len(C):= max\{ Len(c_i)\}$.

Each copy $c_i := \{ b_{i_1},b_{i_2}, ... , b_{i_t}\}=\{(e_{i_1} - s_{i_1}),(e_{i_2} - s_{i_2}), ... , (e_{i_t} - s_{i_t})\}$ satisfies the following properties:

i) $0< s_{i_{j+1}} - e_{i_j}<100K$ for all j such that $i_t\leq j\leq i_j$;

ii) $ei_j -si_j>5K$;

iii) $Len(c_i) \geq 1/2 \times Max\_len(C)$;

iv) $Span(c_i) \geq 1/2 \times Max\_span(C)$;

v)   The similarity of each block that appears in c_i and the corresponding block in B should be more than 30%.

The problem is to retrieve all LCRs from a human genome. The problem is in high complexity due to a low similarity requirement and a large number of possible mutations and structural variations, so we designed a heuristic solution to solve the problem.

2.2   Overlapping Reads Generation and Alignments

We generate consecutive overlapping reads from the input genomic sequence. The locations of where the reads come from are marked. For highly similar repeats, a set of consecutive reads should form a similar sorted overlapping pattern (either in direct or reverse order) when they were mapped to these repeat regions. Read length, overlapping length and mismatches in alignment can be adjusted (default values: 100 bp read, 20 overlapping length, with 4 mismatches in alignments) according to similarity requirement. Longer read length and fewer mismatches go with higher similarity requirement, otherwise lower.

2.3   Small-Size Highly Similar Sequences Detection

We mapped those overlapping reads back to reference sequence using Soap3[1](Liu et al. 2012), and selected those reads with multiple alignments. To identify reads from repetitive sequences, we clustered the reads as follows. Let us start with read x. For every position read x could be mapped to, we considered the next consecutive read (x+1). If read (x+1) could be mapped to at least two corresponding locations with ± 10bp, it was chained to x, and we went on with (x+2), otherwise stop and report the chain starting from x. x is the chain head.

We filtered out chains with less than 50 supporting reads, and sorted the rest in descending order. Each chain was supposed to correspond to a repeat pattern. So far, efforts have been put on how to detect repeats using stringent criteria (95%-98% similarity requirement). In LCR structure, those repeats were acted as skeletons, and in the next step, we consider each set of repeats as skeleton of a LCR and extend them one at a time. If a set of repeats were covered by others during extension, this set of repeats would be eliminated.

2.4   Basic Extension

Given copies of a repeat pattern, we chose the copy with smallest starting coordinates as model and regarded others as candidates. We applied an extension procedure to model repeat on two directions, one at a time (Fig 3).

Boundary of model was extended by L = 5 Kb region. All valid alignments of reads within this region consecutive to model boundary were clustered. Initially, each alignment was considered as a cluster. Iteratively merge two clusters if there was at least one alignment in each of them, had a distance below 200 bp (10 times read overlapping length), until no more clusters could be merged. The minimum interval on reference sequence that covered all alignments in a cluster was calculated for each cluster. Only intervals larger than of L/2were kept (Fig 3B).

In order to deal with large SVs – insertion, deletion, inversion and translocation with large translocated distance, we applied large extending gap G (default 100Kb) to cover SVs smaller than G. For each candidate, if there was at least one interval whose gap to this candidate on either side was less than G, then update the corresponding boundary to include the nearest interval. If at least one candidate was updated, then this extension step was considered as successful. Once three consecutive unsuccessful extensions were found, extension to this direction was stopped. Meanwhile, the corresponding boundary of model scrolled 3 steps back to exclude the last three failed extensions at the end (Fig 3C).

---

[1] Other alignment tools can be used.

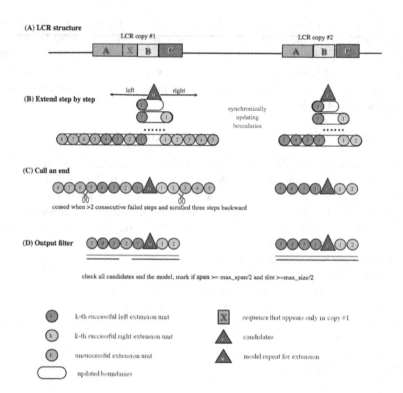

Fig. 3. Illustration of basic extension. (A) Example of LCR structure. (B) Extension process. Candidate regions marked as triangles were input for extension. Each circle was an extension unit consecutively to current boundaries. If reads from this unit could cluster within certain distance within boundaries of candidates, this extension unit was successful. (C) Call and end and cut the failed extension tails. One side extension was called to an end when three consecutive failed units were witnessed at the boundary. These three failed units were cut and extension to this direction ceased. When both directions reached an end, the whole extension process stopped. (D) Output filter. A filter was applied to eliminate false positives.

When both directions reached to an end, size and span were calculated for model and candidates to eliminate false positives. Size was defined as sum of model/candidate initial size and all extended region for this model/candidate. Span was defined as the minimum interval that covered the initial model/candidate and all other extended regions. For each copy of LCR, calculate the maximum span and size. If more than one copy passed ½ span and ½ size filters, those eligible copies were reported (Fig 3D). For example, suppose one LCR pattern was 50Kb, and there was a 4Kb deletion at the end of one candidate X but the deletion region appeared 80Kb away from this candidate. During extension, this region was incorrectly included as well as 80Kb non-LCR sequence by X (size = 50Kb; span = 130Kb). However, due to ½ span filter, X and other copies were abandoned.

2.5   Merge Adjacent LCRs to Deal with Large Size Novel Sequences

We merged adjacent LCRs to deal with large gaps (2.5 L <gap< M) between LCR blocks. If there was novel sequence in model (sequence that didn't appear in other copies) with >2.5 L length made model copy failed to go across the gap during extension. However, we were able to connect LCRs on both sides of the gap and reconstruct the original LCRs (Fig 4). If two LCRs 1 and 2 had the same copy number and for each copy in LCR1 there was a copy in LCR2 that they were either overlapping or within distance < M (default 200Kb), the minimum interval that covered the pair of copies was reported as a reconstructed copy. All reconstructed copies were filtered by ½- span filter to reduce false positives.

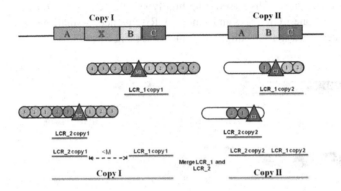

Fig. 4. Merge adjacent LCRs to handle novel insertions. Two LCRs were found around novel insertion in Copy I. LCR_1 and LCR_2 were merged to form full-length LCR copies.

3   Results

In order to evaluate the performance of LCR_Finder on real data, we tested our tool on 4 human chromosomes, and to each of them there is a known disease-related LCR. We also compared the results of LCR_Finder with RepeatScout, windowMasker and RepeatMasker (chr17 only) on human chromosomes 2, 5, 7, 17.

3.1   Performance of LCR_Finder

Human chromosomes (GRCh37.p9) were downloaded from NCBI (http://www.ncbi.nlm.nih.gov). Considering there is at least 95% sequence similarity between repeats, we simulate 100 bp single end reads, and each read overlaps with the next one by 20bp. When they are mapped to reference sequence using Soap3, 4 mismatches are allowed and both orientations are valid.

Our experiments were implemented on Linux86 64 system with 8G memory using Perl. We were able to predict 59, 34, 86, 44 LCRs for chromosome 2, 5, 7, 17 respectively. 15, 9, 16, 9 out of 59, 34, 86, 44 LCRs had LCR copies larger than 100Kb including the 4 known disease-related LCRs. Comparing our results with 4 known disease-related LCRs (one for each chromosome), we successfully identified all 4 known LCRs.

3.2   Supporting Evidence on Novel LCRs

We further investigated the 45 novel LCRs (>100Kb copy size). We were able to see some high similarity blocks (>95%) tandem or interspersed arranged in one LCR copy when it was aligned to other LCR copies using BLASTN (http://blast.ncbi.nlm.nih.gov). The overall similarity between LCR copies was not necessarily to be high, but highly similar blocks should be observed. When we used one copy of LCR as query and Blast other copies (subject) to it, subject sequence were divided into several tandem/dispersed long highly similar sequences to query. In contrast to the case of one LCR copy and a non-LCR sequence, a small number of short subject sequences were sparsely aligned to query (see Fig 5).

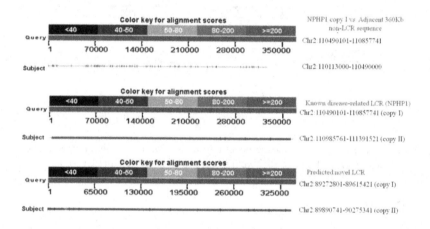

Fig. 5. Blast results of non-LCR sequences, known related disease LCR sequences, and predicted LCR sequences. We blast NPHP1 copy I to similar size adjacent sequence (top), two copies of NPHP1 LCR (middle) and two copies of predicted LCR (bottom).

3.3   LCR_Finder Limitations

Limitations of LCR_Finder are discussed in this section. Results of LCR_Finder are dependent on how parameters are set, including extending unit length L and gap tolerance G. We compared the results under L = 2Kbp, 5Kbp and 10Kbp and G =50Kbp, 100Kbp, 200Kbp separately. On each (L ,G) pair, we ran LCR_Finder on chromosome 17 and show visualized results on region chr17:15Mb-21Mb using Circos (Krzywinski et al., 2009) (Fig 6). (i) LCR copy size reported by LCR_Finder is related to extending unit length L. The higher L is, the less the boundary resolution is. Because the extending unit at the boundary is limited to at least L /2, there are at most L /2 non-LCR region being counted as part of LCR copy, which means at most L /2 non-LCR region was incorrectly reported or at most L /2 LCR region was missed. For example, in SMS proximal copy, it could be seen that boundaries found under parameters (10K,100K) were less precise than (5K,100K). (ii) LCR_Finder were not able to identify LCRs with novel insertion in model copy with length larger than 2.5L (2L in some cases depending on extending start position) during extension, and deletion in model copy with deletion size larger than M. For example, LCR_Finder failed to report full length SMS-REPs when (L ,G) =(5K, 50K), because there was ~10Kb region between block A to block C in model copy (distal copy) deleted in proximal and middle copies. Although 1Kb block B existed for all copies, it was too small for minimum successful extension requirement (L /2=2.5Kb). Three consecutive unsuccessful extensions were found after block A, so LCR_Finder reported only block A for all copies. However when (L ,G) =(10K, 50K), ~10K gap was smaller than 2L (20K), thus full-length copies were reported. Besides when (L ,G) =(2K, 50K), block B made an extension successful since it was larger than L /2 (1Kb) and followed by 2 failed extensions, the next extension was successful, so full length SMS-REPs were reported. (iii) Insufficient gap tolerance G caused loss of LCR copy. LCR_Finder could not handle insertion, deletion and inversion larger than G and translocation with

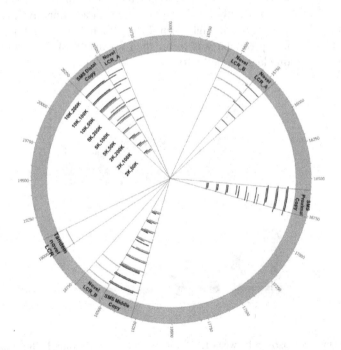

Fig. 6. Results of LCR_Finder on various extending parameters. Only >10Kb results were presented. The outermost circle (in blue) with ticks represents region chr17:15Mb-21Mb on the given sequence. Ticks are in Kb scale. Results of LCR_Finder on different parameters (L ,G) were arranged as inner circles. Copies of the same LCR were marked with the same color. SMS-REPs reported by LCR_Finder were colored in red. 2 novel LCRs were predicted.

transcending distance larger than G. Successful extension happened only when extension unit was found within G of candidate boundaries (Fig 3). Considering the complex structures of LCRs, we recommend users to adopt a set of various parameters to capture LCRs with a wide range of sizes and structures.

3.4   Tools Comparison

To compare LCR_Finder to RepeatScout (RS) and WindowMasker (WM), we ran RS and WM on human chromosomes 2, 5, 7 and 17, one chromosome at a time. Numbers of detected LCRs are listed in Table 1.

RS and WM had lower time cost than LCR_Finder and predicted a larger number of CNVs. But most of CNVs they predicted were small-scale, and none of them captured LCRs listed in Table A1. The output of RS was CNV patterns while WM reported intervals covered all copies of a pattern. Thus we calculated the total length (sum of intervals) and average length (average of intervals) of LCR_Finder and WM. Besides, average patten length (average of patterns reported by RS and average copy length by LCR_Finder) comparison was conducted for LCR_Finder and RS. We can observe from Fig 7 that WM captured larger total length on chromosome 2 and 5, but the average length was significantly lower than out tool. In addition, average pattern length of LCR_Finder was much high than RS. Although the total number of CNVs/LCRs LCR_Finder predicted was lower than RS and WM, LCR_Finder was more efficient in detecting large-scale LCRs.

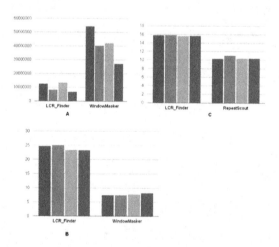

Fig. 7. Performance comparison between LCR_Finder and other software. LCR_Finder, WindowMasker (WM) and RepeatScout (RS) were tested using 4 human chromosomes. We calculated   (A) Total Length (bp) (B) Average Length (Log2 bp) for LCR_Finder and WindowMasker; and (C)Average Pattern Length (Log2 bp) for LCR_Finder and RepeatScout. 3/4 chromosomes were reported with larger span covered by WM, but the average span and average pattern length of LCR_Finder reported were significantly larger than WM and RS.

Table 1. Time cost of RS, WM and LCR_Finder running on chromosome 2, 5, 7 and 17

Software	Running time (min)		Number of LCRs/CNVs detected
RS	Chr2	59	1,146
	Chr5	44	1,331
	Chr7	35	1,656
	Chr17	18	1,477
WM	Chr2	55	323,086
	Chr5	41	242,639
	Chr7	34	207,383
	Chr17	17	94,783
LCR_Finder	Chr2	302	59
	Chr5	205	34
	Chr7	180	86
	Chr17	106	44

4   Conclusions and Discussions

Although various genomic structural variation detection tools have been developed using the next-generation sequencing data, due to the difficulty in capturing the characteristics of complex low copy repeats, existing methods are not yet satisfactory. In this paper, we presented a novel tool focusing on complex low copy repeats. Besides basic repeats discovery, our tool is capable of combine different sets of repeats according to their genomic locations and report large-scale complex low copy repeats coordinates despite their complex structures. Our tool helps to interrogate genomic coordinates and understand mechanisms of genetic diseases.

Several issues are remained to be better understood and investigated in the future. A more precise formulation and definition of LCR, a more systematical parameter setting up in less ad hoc manner and a more comprehensive evaluation method such as validating putative LCRs with existing LCR database will facilitate the detection of complex low copy repeats.

References

1. Smit, A.F.A., Hubley, R., Green, P.: unpublished data (2011); Current Version: open-3.3.0 (RMLib: 20110920)
2. Babcock, M., et al.: Hominoid lineage specific amplification of low-copy repeats on 22q11.2 (LCR22s) associated with velo-cardio-facial/digeorge syndrome. Hum. Mol. Genet. 16, 2560–2571 (2007)
3. Bailey, J.A., et al.: Recent segmental duplications in the human genome. Science 297, 1003–1007 (2002)
4. Cheung, V.G., et al.: Integration of cytogenetic landmarks into the draft sequence of the human genome. Nature 409, 953–958 (2001)

5. Jiang, Z., et al.: Ancestral reconstruction of segmental duplications reveals punctuated cores of human genome evolution. Nat. Genet. 39, 1361–1368 (2007)

6. Kurotaki, N., et al.: Sotos syndrome common deletion is mediated by directly oriented subunits within inverted Sos-REP low-copy repeats. Hum. Molec. Genet. 14, 535–542 (2005)

7. Krzywinski, M., et al.: Circos: an Information Aesthetic for Comparative Genomics. Genome Res. 19, 1639–1645 (2009)

8. Li, H., et al.: Mapping short DNA sequencing reads and calling variants using mapping quality scores. Genome Res. 18, 1851–1858 (2008)

9. Liu, C.M., et al.: SOAP3: Ultra-fast GPU-based parallel alignment tool for short reads. Bioinformatics 28, 878–879 (2012)

10. Morgulis, A., et al.: WindowMasker: window-based masker for sequenced genomes. Bioinformatics 22, 134–141 (2006)

11. Park, S.S., et al.: Structure and Evolution of the Smith-Magenis Syndrome Repeat Gene Clusters, SMS-REPs. Genome Res. 12, 729–738 (2002)

12. Price, A.L., et al.: De novo identification of repeat families in large genomes. Bioinformatics 21, i351–i358 (2005)

13. Saunier, S., et al.: Characterization of the NPHP1 locus: mutational mechanism involved in deletions in familial juvenile nephronophthisis. Am. J. Hum. Genet. 66, 778–789 (2000)

14. Stankiewicz, P., Lupski, J.R.: Genome architecture, rearrangements and genomic disorders. Trends Genet. 18, 74–82 (2002)

15. Valero, M.C., et al.: Fine-scale comparative mapping of the human 7q11.23 region and the orthologous region on mouse chromosome 5G: the low-copy repeats that flank the Williams-Beuren syndrome arose at breakpoint sites of an evolutionary inversion(s). Genomics 69, 1–13 (2000)

16. Zhang, F., et al.: Copy number variation in human health, disease, and evolution. Annu. Rev. Genomics Hum. Genet. 10, 451–481 (2009)

Heuristic Algorithms
for the Protein Model Assignment Problem

Jörg Hauser[1], Kassian Kobert[1], Fernando Izquierdo Carrasco[1],
Karen Meusemann[3], Bernhard Misof[3], Michael Gertz[2],
and Alexandros Stamatakis[1]

[1] Heidelberg Institute for Theoretical Studies, Heidelberg, Germany
[2] Heidelberg University, Institute of Computer Science, Heidelberg, Germany
[3] Zentrum für molekulare Biodiversitätsforschung, Zoologisches Forschungsmuseum
Alexander Koenig, Bonn, Germany

Abstract. Assigning an optimal combination of empirical amino acid
substitution models (e.g., WAG, LG, MTART) to partitioned multi-gene
datasets when branch lengths across partitions are linked, is suspected
to be an NP-hard problem. Given p partitions and the approximately 20
empirical protein models that are available, one needs to compute the
log likelihood score of 20^p possible model-to-partition assignments for
obtaining the optimal assignment.

Initially, we show that protein model assignment (PMA) matters for
empirical datasets in the sense that different (optimal versus suboptimal) PMAs can yield distinct final tree topologies when tree searches
are conducted using RAxML.

In addition, we introduce and test several heuristics for finding near-
optimal PMAs and present generally applicable techniques for reducing
the execution times of these heuristics. We show that our heuristics can
find PMAs with better log likelihood scores on a fixed, reasonable tree
topology than the naïve approach to the PMA, which ignores the fact
that branch lengths are linked across partitions. By re-analyzing a large
empirical dataset, we show that phylogenies inferred under a PMA calculated by our heuristics have a different topology than trees inferred
under a naïvely calculated PMA; these differences also induce distinct
biological conclusions. The heuristics have been implemented and are
available in a proof-of-concept version of RAxML.

Keywords: phylogenetic inference, maximum likelihood, model assignment, protein data.

1 Introduction

An important task in phylogenetics consists in computing the (maximum) likelihood score on a given tree topology. Typically, the logarithm of the likelihood
is computed for numerical reasons. Throughout the paper, we use likelihood and
log likelihood as synonyms. The likelihood score represents the probability of
observing the data (a set of aligned molecular sequences), given a strictly bifurcating unrooted tree. A statistical model of evolution is required to specify how

Z. Cai et al. (Eds.): ISBRA 2013, LNBI 7875, pp. 137–148, 2013.
© Springer-Verlag Berlin Heidelberg 2013

the observed data (e.g., an alignment of amino acid sequences) was generated by the given topology, that is, the model provides transition rates between possible states (e.g., amino acid characters).

For DNA data, a general time reversible substitution model [1] is typically being used, which requires a direct maximum likelihood estimate of the transition rates. For amino acid data, this is mostly not considered, because it may result in over-parametrizing the model (DNA has 5 rates, protein data has 189 transition rates). Therefore, a plethora of empirical protein substitution models such as MTART [2], WAG [3], and LG [4], have been derived from large collections of real-world protein alignments. Some of these models are intended for general use (e.g., WAG and LG) and some have been optimized for specific organisms (e.g., the MTART model for *Arthropoda*).

Selecting an appropriate empirical protein substitution model for the data at hand represents an important and generally non-trivial task. This is because using an inappropriate model that does not fit the data well, can lead to erroneous phylogenetic estimates (see, e.g., [5] or [6]).

Here, we consider the case of protein model assignment for partitioned (different sets of sites evolve under distinct evolutionary models) multi-gene amino acid sequence alignments. Note that, determining an appropriate partitioning scheme is also a non-trivial problem (e.g., [7]) but outside the scope of this paper. Therefore, we assume that an appropriate partitioning scheme is given. We denote this task as *protein model assignment (PMA) problem*. Given a fixed, reasonable (i.e., non-random) tree we want to assign the best-fit empirical protein substitution model to each partition such that the overall likelihood is maximized. Note that, using the optimal (with respect to the likelihood score) PMA does not increase the number of parameters in the model. Hence, over-fitting the data is not an issue and we can directly obtain the optimal PMA by finding the assignment that maximizes the likelihood. However, finding the optimal PMA is challenging if we assume that branch lengths are shared across partitions, that is, partitions are linked via a joint branch length estimate.

Using a joint branch length estimate across partitions is important because it drastically reduces the number of free parameters in the model. The number of inner branches in a strictly binary unrooted tree is $2n - 3$, where n is the number of taxa. Thus, each set of independent branch lengths that is estimated increases the number of model parameters by $2n - 3$. Therefore, joint branch length estimates can be deployed to avoid over-parametrizing the model.

Simply calculating the maximum likelihood score for all possible PMAs on a fixed, reasonable (i.e., non-random) tree, for p partitions and the approximately 20 available protein substitution models, is computationally prohibitive because of the exponential number (20^p) of possible assignments. We have already developed a proof (preprint available at http://www.exelixis-lab.org/Exelixis-RRDR-2012-9.pdf) that shows that the PMA problem is NP-hard. Here, we introduce and evaluate three heuristic strategies for computing 'good' PMAs for partitioned protein alignments under joint branch length estimates.

For small problem instances with $p := 3$ partitions (extracted from publicly available real datasets [8] and [9]) we observed substantial differences in final RAxML-based tree topologies inferred under the optimal PMA obtained from the exhaustive algorithm and suboptimal PMAs obtained via a naïve approach that is currently being used for determining the PMA. On simulated datasets, which generally tend to exhibit stronger signal (see, e.g., [10]), we did not observe that the PMA has an impact on final tree topologies, presumably because simulated data tend to be 'too perfect'. As we show here, finding a 'good' PMA is important for empirical analyses of real biological data because it changes the results, that is, the final tree topologies. Our heuristic PMA search strategies consistently find better PMAs, with respect to the likelihood score (without increasing the number of parameters in the model!) than the commonly used naïve heuristics that disregard the fact that partitions are linked via the branch lengths.

The remainder of this paper is organized as follows: In Section 2 we briefly review related work on the general problem of protein model selection. In Section 3 we introduce our heuristics and computational shortcuts for reducing the computational burden of computing likelihood scores for candidate PMAs. In Section 4 we discuss the experimental setup and provide experimental results. We conclude in Section 5 and discuss directions of future work.

2   Related Work

To the best of our knowledge, this paper and the paper addressing the NP-hardness proof [11] are the first to identify and address the PMA problem.

Hence, we will briefly review work on the protein model selection methods in phylogenetics. There exists an extensive literature on methods for selecting models of nucleotide or amino acid substitution (see [12] for an in-depth review).

Initially, model testing pipelines applied likelihood ratio tests for selecting the best fit model. However, these tests require the models to be nested, which is not always the case. Therefore, tests relying on the Akaike Information Criterion (AIC) and the Bayesian Information Criterion (BIC), that do not require the models to be nested, have recently gained momentum.

One of the most widely used tools for selecting protein models is ProtTest [13]. Another, fairly similar tool, for protein model selection is Aminosan [14].

As stated above, none of the existing pipelines address the PMA problem. Keep in mind that, PMA is essentially not a model selection problem, but an optimization problem because the number of model parameters is constant for all 20^p possible PMAs. As such, computing a 'good' PMA (finding the optimal PMA is NP-hard!) for partitions that are linked via a joint branch length estimate forms part of the general model selection process that is implemented by the above tools.

3   Heuristics

For all heuristics described here, we assume that a reasonable (i.e., non-random) tree is given. Such a tree can be obtained by executing a neighbor joining or parsimony tree search. It is broadly accepted that using a fixed, parsimony or neighbor joining tree for estimating model parameters is sufficient to obtain accurate parameter estimates [15]. Hence, given such a reasonable fixed tree *and* a data partitioning scheme with p data partitions, our goal is to find the PMA that maximizes the likelihood. This PMA can then be used for a subsequent maximum likelihood (ML) tree search using, for instance, RAxML.

Initially, we briefly describe the *naïve heuristics* that represent a simple and straight-forward approach to obtain a somewhat reasonable PMA. The naïve heuristics simply ignore the fact that partitions are linked via branch lengths and determine the best-scoring protein substitution model independently for each partition (by looping over the protein models) using a per-partition branch length estimate. The PMA obtained by this naïve approach can be used as initial seed for the search algorithms presented in Sections 3.3 and 3.4 to accelerate convergence.

If the number of partitions is small (e.g., $p := 3$) one can also perform an *exhaustive search* by computing the maximum likelihood scores for all possible $20^3 = 8000$ PMAs to obtain the global maximum, that is, the exact solution.

In our heuristics, we want to explore an as large as possible fraction of the search space by evaluating as many candidate PMAs as possible. However, computing the likelihood on candidate PMAs is expensive, because model parameters such as the α shape parameter of the Γ model of rate heterogeneity [16] and the joint branch lengths need to be re-optimized for each PMA. Therefore, we initially discuss some general computational shortcuts to reduce the computational cost of calculating likelihood scores for candidate PMAs.

3.1   Accelerating the Evaluation of Candidate PMAs

In the course of the searches we need to compute the maximum likelihood score for a large number of candidate PMAs. This entails fully re-optimizing all model parameters such as the branch lengths and the α shape parameter for each new PMA from scratch, that is, from some initial default values for α and the branch lengths. These parameters are optimized via standard numerical optimization procedures such as Brent's algorithm (α) and the Newton-Raphson procedure (branch lengths). Instead of re-optimizing all parameters from scratch, we can, re-use the parameter values of the current PMA i as initial values for optimizing the parameters and scoring a new PMA $i + 1$. This will generally be faster, because the parameter estimates (especially the α parameter) for assignment i will not differ substantially from those of assignment $i + 1$. The differences in model parameter estimates between PMAs i and $i + 1$ are also small because in the heuristics presented below, we only change the protein model of one partition at a time to obtain PMA $i + 1$ from PMA i. Hence, the numerical optimization routines will require less iterations to converge because the initial parameter values are 'good'. In our tests, this modification only yielded minimal deviations in likelihood scores (less than 0.5 log likelihood units) while improving execution times by a factor of 2.8 on average (see [17] for details, available at http://www.exelixis-lab.org/pubs/JoergHauserMasterThesis.pdf).

The second approach to reducing execution times of candidate PMAs strives to avoid evaluating candidate PMAs that are not promising. In other words, given a PMA that needs to be scored by computing its maximum likelihood score, we deploy an inexpensive pre-scoring criterion to determine whether or not it is worth to evaluate this PMA. To pre-score PMAs we use the per-partition likelihood scores for each partition and each protein substitution model that can be computed using the naïve approach outlined above. These scores, albeit obtained under a per-partition branch length estimate, can be used to pre-score candidate PMAs because of a strong correlation between the overall (across all partitions) likelihood scores under a joint branch length estimate and the likelihood scores under a per-partition branch length estimate. In Figure 1 we depict the full (left y-axis) and approximate (right y-axis) likelihood scores for 100 random PMAs on a dataset with 50 partitions and 50 taxa that was sub-sampled from the real biological dataset [8] used in Section 4.

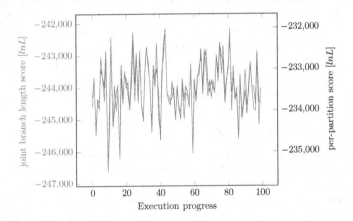

Fig. 1. Full likelihoods and approximate likelihoods for 100 random PMAs on a real biological dataset

Because of this strong correlation, the per-partition likelihood scores as obtained under a per-partition branch length estimate can be used to omit the evaluation of candidate PMAs that do not appear to be promising. For details on computing the threshold for deciding which candidate PMA evaluations to skip, please refer to [17]. By using this technique we were able to accelerate the heuristics by a factor of 1.5 to 2.

3.2   Greedy Partition Addition Strategy

The greedy partition addition heuristics represent a constructive approach that gradually extends the alignment by adding one partition (and model) at a time. We start with the first partition and determine and fix the best protein substitution model for this partition. Then, we add the second partition and compute the likelihood scores for all 20 possible protein model assignments to this second partition while keeping the model for the first partition fixed. Once we have determined the best protein model for the second partition, we fix the model for the second partition as well. Thereafter, we add the third partition and compute the likelihood scores for all possible 20 model assignments to this third partition while keeping the models for the first and second partition fixed. Note that, the per-partition likelihood scores are re-computed for *all* partitions each time a new model is assigned to the new partition that is being added because the joint branch lengths are re-estimated for the entire alignment.

We continue extending the alignment (and PMA) in this way until all partitions have been added to the alignment. For this algorithm, we need to evaluate $p * 20$ candidate PMAs, where p is the number of partitions. Note that, the final PMA obtained by applying this strategy can be different depending on the order by which we add partitions. Therefore, we have implemented a fixed partition addition order by sorting the partitions by their length in terms of number of sites and adding them in descending order (longest partition first). We chose to optimize the model for the longest partition first because the longest partition typically contributes most to the overall likelihood score of the full alignment. However, this had no notable effect on performance of the heuristics with respect to the final likelihood scores of the best PMA that was found [17].

3.3   Steepest Ascent Strategy

The steepest ascent approach implements a classic neighborhood-based hill climbing strategy. Given some initial PMA, which can either be a random assignment, the result of the naïve heuristics, or the assignment computed by the greedy addition strategy (see above), we proceed as follows: We evaluate the likelihood scores of all PMAs that differ by one model-to-partition assignment from the current assignment. In other words, we explore a neighborhood of size 1. We need to calculate the likelihood scores of $(20 - 1) * p$ PMAs to explore the size 1 neighborhood of the current assignment (when not using the pre-scoring approach). Once all $19 * p$ scores have been calculated, we select the PMA that yields the largest likelihood improvement. We then explore the neighborhood of this new assignment. If there does not exist a PMA in the size 1 neighborhood that further improves the likelihood, we have reached a local optimum and the algorithm terminates.

3.4   Simulated Annealing Strategy

We also implemented a simulated annealing algorithm because of its ability to navigate out of local maxima [18].

We can initialize the PMA for the simulated annealing strategy either at random or with the result of the naïve heuristics. As for the steepest ascent algorithm, we explore the size 1 neighborhood of the current PMA. There are nonetheless some fundamental differences. We iteratively evaluate the neighboring assignments of the current PMA and compute their corresponding likelihood scores.

For each neighboring assignment that is evaluated, we carry out an acceptance/rejection step. Thus, if the likelihood of the candidate PMA is better than that of the current PMA, we accept it immediately and use it as current assignment (in analogy to a greedy hill climbing strategy). If the likelihood of the candidate PMA is worse than that of the current PMA we need to decide whether to accept a backward step or not. We accept a PMA that decreases the likelihood if $r < e^{-\frac{l-l'}{T_k}}$, where r is a uniform random number in $[0; 1]$, l is the likelihood of the current assignment, and l' the likelihood of the candidate PMA. Finally, T_k is the temperature of the annealing process at iteration k (evaluation of the kth PMA). This procedure is also known as *Metropolis criterion* (see [19]). We implemented a standard cooling schedule $T_k = \lfloor T_0 \beta^k \rfloor$, where T_0 is the starting temperature and $\beta \in [0; 1]$ represents a parameter that needs to be tuned. We empirically determined a setting of $\beta := 0.992$ (see [17] for details). The simulated annealing process terminates at iteration n when $T_n = 0$ *and* when a PMA is generated that has a worse likelihood score than the currently best one.

4 Performance Assessment

The modified RAxML code, the test datasets, as well as the wrapper scripts (greedy algorithm) are available at `http://exelixis-lab.org/joerg/pma.tar.gz`.

4.1 Experimental Setup

We implemented the three search strategies outlined in Sections 3.2 through 3.4 as well as the naïve and exhaustive search algorithms in the standard RAxML version [20] and via wrapper scripts. We also used RAxML to compute Robinson-Foulds distances [21] between trees. Computational experiments were performed on our institutional cluster, which is equipped with 50 48-core AMD Magny-Cours nodes (equipped with 128GB RAM each) and connected via Infiniband.

4.2 Test-Datasets

We used real (empirical) *and* simulated datasets to test (i) whether a good PMA 'matters' with respect to the final tree topology and (ii) to evaluate our heuristic search strategies. We used three partitioned real-world data sets from two studies [8,22] that encompass data from all three domains of life. The properties of the datasets are summarized in Table 1. We simulated datasets using INDELible [23] on random 'true' tree shapes with 40 taxa that were generated via a R script provided by David Posada (included in the on-line data archive) and 2, 4, 8, 16, 32, 64, and 128 partitions, respectively. Partition lengths were randomly generated and ranged between 300 and 500 sites. Protein substitution models to simulate the data along the tree for each partition were also assigned at random.

Table 1. Properties of the empirical datasets

Domain	# taxa	# partitions	length	reference
Eukaryotes	117	129	37,476	[8]
Bacteria	992	56	20,609	[22]
Archaea	86	68	17,639	[22]

4.3   Results

Does the PMA Matter? Initially, we address the question whether obtaining a good PMA actually matters, that is, if it alters the final tree topology when applying a standard RAxML maximum likelihood tree search. For this purpose, we randomly sub-sampled 50 datasets containing three partitions and 50 taxa from each of the three real-world datasets listed in Table 1. We thereby generated a total of 150 small real-world test datasets. For each sub-sampled alignment we then computed a PMA using the naïve algorithm and the exhaustive algorithm to obtain the globally optimal PMA. Note that, running the exhaustive algorithm on more than 3 ($20^3 = 8000$ distinct possible PMAs) partitions was computationally not feasible. Model assignments differed for 86 out of the 150 alignments. Thus, the naïve approach yields suboptimal PMAs for more than half of the datasets. For those 86 datasets where the PMAs differed we executed 10 standard RAxML tree searches (staring from distinct randomized addition order parsimony trees) per dataset to obtain the best-known ML tree under the naïve and optimal PMA. We only obtained topologically identical ML trees for 14% of the 86 datasets. The average topological RF-distance between the trees inferred under the naïve and the optimal assignment was 9%. As expected the naïve PMA never yielded a final tree with a better likelihood than the optimal PMA. Hence, on real data, investing computational effort to finding a 'good' PMA is important, because it has a noticable impact on the structure of the final tree topology.

We then also calculated the PMAs using the steepest ascent heuristics on these 150 datasets. The inferred PMAs differed from the optimal PMA obtained by the exhaustive algorithm for only 10 out of 150 datasets (7%). Furthermore, the best-known ML trees inferred on those 10 datasets showed an average RF-distance of only 3%. We conclude that, (i) the steepest ascent heuristics are able to infer the optimal PMA in the majority of the cases and (ii) when the heuristics yield a suboptimal PMA, the inferred PMA nonetheless induces a substantially smaller topological error than the naïve PMA.

For the simulated datasets, we inferred ML trees using a random PMA, the naïve assignment, and the known, true PMA under which the data was generated. Thereafter, we calculated the RF distances between the ML trees inferred using the random PMA and the true PMA as well as the RF-distance between the trees inferred under the naïve PMA and the true, known PMA. We found differences in RF distance to be negligible in both cases (random and naïve PMAs) on simulated data. We suspect that this is due to the fact that simulated data tends to be more perfect than real data [10].

The important finding is that determining a PMA that fits the data well has a substantial impact on real word data analyses.

Performance of Heuristics. To assess the relative performance and quality of the three heuristic strategies we propose, we sub-sampled 15 datasets containing 50 taxa and 50 partitions from each of the three real-world datasets. This was done to reduce the computational burden of these analyses.

We intend to determine which strategy performs best with respect to execution times and result quality which we quantify as the maximum likelihood score of the respective PMAs. Note that, the number of free model parameters is identical for all candidate PMAs, hence a likelihood-based comparison of PMAs is meaningful. As a reference, we used the likelihood score and the execution time required by the naïve heuristics. The simulated annealing and steepest ascent algorithms were seeded with the PMAs obtained from the naïve heuristics. These two search strategies were also seeded with a random seed, but performed worse (results not shown).

We summarize the results in Figure 2. The figure contains average execution times in seconds and average score improvements in *log* likelihood units over the 15 test datasets for the three PMA heuristics we propose. The execution times displayed for the simulated annealing and steepest ascent strategy include the execution time of the naïve algorithm whose assignment is used as a seed. For *all* 15 test datasets, we were able to find a PMA with a better likelihood than obtained via the naïve algorithm on the same, fixed, reasonable tree topology. Overall, the steepest ascent algorithm performs best with respect to execution times *and* result quality.

Re-analysis of a Biological Dataset. We inferred ML trees and bootstrap support values on the main empirical dataset used in [8] with (i) the PMA as used in the original study (WAG assigned to all partitions; denoted as allWAG) and (ii) the PMA as obtained from the steepest ascent heuristics (denoted as optimized). The relative RF distance between the resulting best-known ML trees was 8%. Hence, an optimized PMA can change the shape of final tree topologies as well as the biological conclusions which we discuss in the following.

The most conspicuous difference between the two trees is the position of the bristle tail (*Lepismachilis ysignata*, a wingless insect of the *Archaeognatha* insect order), which belongs to the primarily wingless *hexapods*. Insects in the *Archaeognatha* order are typically assumed to be a sister group (neighboring subtree) of the so-called *Dicondylia* that include all winged insects (the so-called *Pterygota*). Therefore, the phylogenetic position of the bristle tail within the winged insects in the allWAG analysis is rather implausible, since it also shows

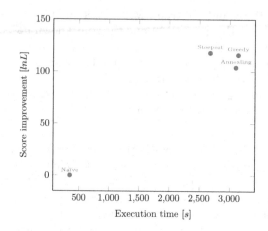

Fig. 2. Execution times in seconds of the three strategies and average improvement in terms of *log* likelihood units over the PMA obtained from the naïve approach

low bootstrap support. Moreover, its position in the `optimized` phylogeny received strong bootstrap support. Its phylogenetic position as a sister group of the winged insects as obtained from the `optimized` analysis has also been observed in prior studies based on molecular and morphological data [8,24,25].

Another notable difference is the placement of *Locusta migratoria* from the order *Orthoptera*. *Orthoptera* (grasshoppers, crickets, weta, and locusts) are commonly assumed to form a monophyletic clade (be located in a single, distinct subtree). Hence, the placement of *Locusta migratoria* is more plausible in the `allWAG` analysis in which *Orthoptera* are monophyletic. However, its position in the `optimized` tree only received moderate bootstrap support, such that it is difficult to draw conclusions regarding its placement based on the dataset at hand. Note that, the phylogenetic position of *Locusta migratoria* is generally considered difficult and hard-to-resolve [8]. The placement of *Locusta migratoria* highly depends on the dataset being used [26]. There is some evidence that *Locusta migratoria* might be a so-called rogue taxon [27].

Overall, from a biological perspective, the tree obtained via the `optimized` tree inference has to be favored. Furthermore, our re-analysis shows that biologically meaningful differences can be observed when inferring trees under an appropriately optimized PMA.

5   Conclusion and Future Work

We addressed the problem of assigning empirical protein substitution models to partitioned datasets that are analyzed under a joint branch length estimate across partitions. This paper is the first paper addressing this problem empirically. We show that obtaining a 'good' PMA (with respect to the likelihood score) matters on empirical datasets, because tree inferences under a naïve PMA can yield a topologically and biologically different phylogeny with worse likelihood scores than inferences under the optimal PMA. We specifically use the term 'good' PMA because finding the optimal PMA is NP-hard. While we can compute the globally optimal assignment for datasets with three partitions via an exhaustive search, finding a 'good' PMA on datasets with more partitions requires heuristic search strategies. We introduce, make available, and test three 'classic' search strategies for combinatorial optimization problems and adapt them to the problem at hand. We show that all three strategies can produce PMAs with better likelihood scores than the naïve search on all test data sets. Moreover, we presented two techniques for reducing the computational cost of our heuristics.

On a large biological dataset [8], we demonstrate that investing computational effort to optimize the PMA is important because it has an impact on the final tree topology as inferred with RAxML and on the biological interpretation of the tree.

We are currently integrating the steepest ascent strategy that performed best in our experiments into the standard RAxML version. Moreover, we also intend to parallelize the heuristic strategies by using a hybrid MPI/PThreads approach. In this setting, the evaluation of candidate PMAs can be distributed among MPI processes that conduct the likelihood calculations in parallel using the fine-grain PThreads parallelization of the phylogenetic likelihood function in RAxML.

References

1. Tavaré, S.: Some probabilistic and statistical problems in the analysis of DNA sequences. Some Mathematical Questions in Biology-DNA Sequence Analysis 17, 57–86 (1986)
2. Abascal, F., Posada, D., Zardoya, R.: Mtart: a new model of amino acid replacement for arthropoda. Mol. Biol. Evol. 24(1), 1–5 (2007)
3. Whelan, S., Goldman, N.: A general empirical model of protein evolution derived from multiple protein families using a maximum-likelihood approach. Mol. Biol. Evol. 18(5), 691–699 (2001)
4. Le, S., Gascuel, O.: An improved general amino acid replacement matrix. Mol. Biol. Evol. 25(7), 1307–1320 (2008)
5. Sullivan, J., Swofford, D.: Are guinea pigs rodents? The importance of adequate models in molecular phylogenetics. J. Mamm. Evol. 4(2), 77–86 (1997)
6. Keane, T., Creevey, C., Pentony, M., Naughton, T., McInerney, J.: Assessment of methods for amino acid matrix selection and their use on empirical data shows that ad hoc assumptions for choice of matrix are not justified. BMC Evol. Biol. 6(1), 29 (2006)
7. Lanfear, R., Calcott, B., Ho, S., Guindon, S.: Partitionfinder: combined selection of partitioning schemes and substitution models for phylogenetic analyses. Mol. Biol. Evol. 29(6), 1695–1701 (2012)
8. Meusemann, K., von Reumont, B., Simon, S., Roeding, F., Strauss, S., Kück, P., Ebersberger, I., Walzl, M., Pass, G., Breuers, S., et al.: A phylogenomic approach to resolve the arthropod tree of life. Mol. Biology Evol. 27(11), 2451–2464 (2010)
9. Yutin, N., Puigbò, P., Koonin, E., Wolf, Y.: Phylogenomics of Prokaryotic Ribosomal Proteins. PloS ONE 7(5) (2012)

10. Stamatakis, A., Ludwig, T., Meier, H.: RAxML-III: A Fast Program for Maximum Likelihood-based Inference of Large Phylogenetic Trees. Bioinformatics 21(4), 456–463 (2005)
11. Kobert, K., Hauser, J., Stamatakis, A.: Is the Protein Model Assignment Problem NP-hard?; Exelixis-RRDR-2012-9; Technical report, Heidelberg Institute for Theoretical Studies (October 2012),
    http://sco.h-its.org/exelixis/pubs/Exelixis-RRDR-2012-9.pdf
12. Posada, D.: In: Selection of Phylogenetic Models of Molecular Evolution. John Wiley & Sons, Ltd. (2001)
13. Abascal, F., Zardoya, R., Posada, D.: Prottest: selection of best-fit models of protein evolution. Bioinformatics 21(9), 2104–2105 (2005)
14. Tanabe, A.: Kakusan4 and aminosan: two programs for comparing nonpartitioned, proportional and separate models for combined molecular phylogenetic analyses of multilocus sequence data. Mol. Ecol. Resources 11(5), 914–921 (2011)
15. Yang, Z.: Among-site rate variation and its impact on phylogenetic analyses. Trends Ecol. & Evol. 11(9), 367–372 (1996)
16. Yang, Z.: Maximum likelihood phylogenetic estimation from DNA sequences with variable rates over sites. J. Mol. Evol. 39, 306–314 (1994)
17. Hauser, J.: Algorithms for Model Assignment in Multi-Gene Phylogenetics. Master's thesis, Ruprecht-Karls University Heidelberg (2012)
18. Kirkpatrick, S., Gelatt, C., Vecchi, M.: Optimization by simulated annealing. Science 220(4598), 671 (1983)
19. Aarts, E., Laarhoven, P.: Simulated annealing: an introduction. Stat. Neerland. 43(1), 31–52 (1989)
20. Stamatakis, A.: RAxML-VI-HPC: maximum likelihood-based phylogenetic analyses with thousands of taxa and mixed models. Bioinformatics 22(21), 2688–2690 (2006)
21. Robinson, D., Foulds, L.: Comparison of phylogenetic trees. Math. Biosci. 53(1-2), 131–147 (1981)
22. Yutin, N., Puigbò, P., Koonin, E., Wolf, Y.: Phylogenomics of Prokaryotic Ribosomal Proteins. PloS ONE 7(5), e36972 (2012)
23. Fletcher, W., Yang, Z.: Indelible: a flexible simulator of biological sequence evolution. Mol. Biol. Evol. 26(8), 1879–1888 (2009)
24. Grimaldi, D.: 400 million years on six legs: On the origin and early evolution of Hexapoda. Arthropod Struct. & Dev. 39(2), 191–203 (2010)
25. Trautwein, M., Wiegmann, B., Beutel, R., Kjer, K., Yeates, D.: Advances in insect phylogeny at the dawn of the postgenomic era. Ann. R. Entomol. 57, 449–468 (2012)
26. Letsch, H., Meusemann, K., Wipfler, B., Schütte, K., Beutel, R., Misof, B.: Insect phylogenomics: results, problems and the impact of matrix composition. Proc. Royal Soc. B 279(1741), 3282–3290 (2012)
27. von Reumont, B., Jenner, R., Wills, M., Dell'Ampio, E., Pass, G., Ebersberger, I., Meyer, B., Koenemann, S., Iliffe, T., Stamatakis, A., et al.: Pancrustacean phylogeny in the light of new phylogenomic data: support for Remipedia as the possible sister group of Hexapoda. Mol. Biol. Evol. 29(3), 1031–1045 (2012)

Alignment of DNA Mass-Spectral Profiles Using Network Flows*

Pavel Skums[1], Olga Glebova[2], Alex Zelikovsky[2], Zoya Dimitrova[1],
David Stiven Campo Rendon[1], Lilia Ganova-Raeva[1], and Yury Khudyakov[1]

[1] Laboratory of Molecular Epidemiology and Bioinformatics,
Division of Viral Hepatitis, Centers for Disease Control and Prevention,
1600 Clifton Road NE, 30333 Atlanta, GA, USA
[2] Department of Computer Science, Georgia State University, 34 Peachtree Str.,
30303, Atlanta, GA, USA

Abstract. Mass spectrometry (MS) of DNA fragments generated by base-specific cleavage of PCR products emerges as a cost-effective and robust alternative to DNA sequencing. MS has been successfully applied to SNP discovery using reference sequences, genotyping and detection of viral transmissions. Although MS is yet to be adapted for reconstruction of genetic composition of complex intra-host viral populations on the scale comparable to the next-generation DNA sequencing technologies, the MS profiles are rich sources of data reflecting the structure of viral populations and completely suitable for accurate assessment of genetic relatedness among viral strains. However, owing to a data structure, which is significantly different from sequences, application of MS profiles to genetic analyses remains a challenging task. Here, we develop a novel approach to aligning DNA MS profiles and assessment of genetic relatedness among DNA species using spectral alignments (MSA). MSA was formulated and solved as a network flow problem. It enables an accurate comparison of MS profiles and provides a direct evaluation of genetic distances between DNA molecules without invoking sequences. MSA may serve as accurately as sequence alignments to facilitate phylogenetic analysis and, as such, has numerous applications in basic research, clinical and public health settings.

1 Introduction

Mass spectrometry (MS) of DNA fragments generated by base-specific cleavage of PCR products is a cost-effective and robust alternative to DNA sequencing. MS is cheaper and less labor-intensive than most of the next-generation sequencing technologies [4][8], and also is not prone to the errors characteristic for these technologies. MS has been successfully applied to the reference-guided single nucleotide polymorphism (SNP) discovery [1][17][13], genotyping [7][10],

---

* The rights of this work are transferred to the extent transferable according to title 17 U.S.C. 105.

Z. Cai et al. (Eds.): ISBRA 2013, LNBI 7875, pp. 149–160, 2013.
© Springer-Verlag Berlin Heidelberg 2013

viral transmission detection [8], identification of pathogens and disease suscep-
tibility genes [15][19], DNA sequence analysis [9], analysis of DNA methylation
[18], simultaneous detection of bacteria [14] and viruses [16][20].

MS technology is based on matrix-assisted laser desorption/ ionization time-
of-flight (MALDI-TOF) analysis of complete base-specific cleavage reactions of a
target RNA obtained from PCR fragments [10][17]. RNA transcripts generated
from both strands of PCR fragment are cleaved by RNaseA at either U or C, thus
querying for every of the 4 nucleotides (A, C, U and G) in separate reactions.
Cleavage at any one nucleotide; e.g. U, generates a number of short fragments
corresponding to the number of U's in the transcript. The mass and size of the
fragments differ based on the number of A, C and G nucleotides residing between
the U's that flank each short fragment. The fragments are resolved by MALDI-
TOF-MS, resulting in mass spectral profiles, where each peak defines a specific
mass measured in Daltons and has intensity that corresponds to the number of
molecules of identical masses.

It should be noted that in MALDI-TOF-MS technology all molecules are
equally singly charged, so the actual molecular weights could be obtained simply
by subtracting the mass of a single hydrogen from every mass from MS profile.
Therefore, in the paper, we assume that MS profiles reflect molecular weights of
the corresponding DNA molecules.

Unlike sequencing, MS is not readily applicable to reconstruction of the genetic composition of DNA/RNA populations. Algorithms for reconstruction of
sequences from MS data were proposed [2]; but, owing to technological and computational limitations, none is widely used.

MS may serve as a rich source of information about the population structure
and the genetic relations among populations without sequences reconstruction.
One of the most important applications of sequences is to phylogenetic reconstructions. However, construction of phylogenetic trees requires knowledge of
genetic distances among species rather than sequences, with sequences being
merely used to estimate the distances. Comparison of MS profiles may also accurately approximate genetic distances. The problem of calculating the distance
between two MS samples is known as spectral alignment problem [3][11]. It is
usually formulated as follows: match the masses from two MS profiles in such a
way that some predefined objective function is maximized or minimized. We discuss the most common objective functions and methods for solving the spectral
alignment problem in section 2.

Spectral alignment is crucial for the most applications of MS based on the
matching of the sample and reference spectra, with the reference MS spectrum
generated *in silico*. Spectral alignments are also used for MS data of proteins
[12], but the protein technology and, therefore, the problem formulation and
algorithm for its solution are completely different.

In this paper we propose a new formulation of the problem of aligning of
the base-specific cleavage MS profiles (MS-Al) and present a method for its
finding. The method is based on the reduction of the problem to the network
flow problem. MS-Al allows *de novo* comparison of sampled populations and may

be used for phylogenetic analysis and viral transmission detection. For conserved genomes (such as human genome) it allows accurate estimation of actual genetic distance between DNA sequences.

2   Problem Formulation

MS profile $P = \{p_1, ..., p_n\}$ consists of n peaks, where each peak $p_i = (m(p_i), f(p_i))$ is represented by a mass $m(p_i)$ and intensity $f(p_i)$. Further without loss of generality we assume that $f(p_i)$ is an integer proportional to the number of occurrences of the mass $m(p_i)$ in the sample. In the simplest version, the spectral alignment problem could be formulated as follows [3]:

Problem 1
   Input: Two MS profiles $P^1 = \{p_1^1, ..., p_{n_1}^1\}$ and $P^2 = \{p_1^2, ..., p_{n_2}^2\}$
   Find: Two subsets $P_*^1 \subseteq P^1$ and $P_*^2 \subseteq P^2$ of matched peaks and a bijection $\pi : P_*^1 \to P_*^2$ such that the following objective function is maximized:

$$score(P_*^1, P_*^2, \pi) - \sum_{p_i^1 \in P^1 \backslash P_*^1} pen(p_i^1) - \sum_{p_i^2 \in P^2 \backslash P_*^2} pen(p_i^2) \qquad (1)$$

Here $score$ is a matching score function and pen is a mismatch penalty function. Usually it is assumed [3] that the function score is additive, which means that matches between different peaks are independent:

$$score(P_*^1, P_*^2, \pi) = \sum_{p_i^1 \in P_*^1} score(p_i^1, \pi(p_i^1)) \qquad (2)$$

Most of known score functions are based on matches of peaks with close masses. In the simplest case we can put $pen \equiv 0$ and

$$score(p_i^1, p_j^2) = \begin{cases} 1, & |m_i^1 - m_j^2| < \epsilon; \\ 0, & \text{otherwise.} \end{cases} \qquad (3)$$

Using these functions and a greedy algorithm for solving Problem 1, authors of [4][8] accurately identified HCV transmission clusters.

In general, Problem 1 with a score function (2) could be efficiently solved using dynamic programming [3][11]. However, it assumes that matches between different peaks are independent. In some cases this is not true, and taking into account dependencies between peak matches may significantly improve the quality of an alignment. One such case is MS based on a complete base-specific cleavage. Further we formulate spectral alignment problem in that case.

Let $\Sigma = \{\sigma_1, ..., \sigma_4\} = \{C, A, G, T\}$ be an alphabet, and let Σ^* be the set of strings over Σ. We assume that Σ^* contains the empty string o. Let $s = (s_1, ..., s_n) \in \Sigma^*$ and let $\Sigma_k = \Sigma \backslash \{\sigma_k\}$, $k = 1, ..., 4$. For each $\sigma_k \in \Sigma$ define $s(\sigma_k) = s(k)$ as

$$s(k) = \begin{cases} \{s\}, & s_i \neq \sigma_k \text{ for every } i = 1, ..., n; \\ \{x \in \Sigma_k^* : s \in \{x\sigma_k y, z\sigma_k x, z\sigma_k x\sigma_k y\} \text{ for some } y, z \in \Sigma^*\}, & \text{otherwise.} \end{cases} \qquad (4)$$

(see [2]). In other words, $s(k)$ is the set of all maximal substrings of s, which does not contain σ_k. For $s^1, s^2 \in \Sigma^*$ denote by $r_{s^1}(s^2)$ the number of substrings of s^1 equal to s^2.

Let $m(\sigma_k)$, $k = 1, ..., 4$ be the mass of the nucleotide σ_k and $m(s) = \sum_{i=1}^{n} m(s_i)$ be the mass of molecule represented by a sequence s.

Suppose that $S = \{s^1, ... s^m\}$, $s^j \in \Sigma^*$, is a sample tested using MS with base-specific cleavage. Let $S(k) = \bigcup_{j=1}^{m} s^j(k)$. MS profile P of S is partitioned into four subprofiles: $P = P(A) \cup P(G) \cup P(C) \cup P(T)$, where

$$P(\sigma_k) = \{p_i^{\sigma_k} = (m, f) : m \in \{m(s) : s \in S(k)\}, f = \sum_{\substack{s \in S(k): \\ m(s) = m}} \sum_{j=1}^{m} r_{s^j}(s)\} \quad (5)$$

Example 1. Let $S = \{s\}$ and $R = \{r\}$ be two samples each containing one sequence, $s = $AAGCTAGTTCA, $r = $AAGCTCGTTCA. Then

$$s(C) = \{\text{AAG,TAGTT,A}\}, s(A) = \{\text{GCT,GTTC}\},$$

$$s(G) = \{\text{AA,CTA,TTCA}\}, s(T) = \{\text{AAGC,AG,CA}\}$$

$$r(C) = \{\text{AAG,T,GTT,A}\}, r(A) = \{\text{GCTCGTTC}\},$$

$$r(G) = \{\text{AA,CTC,TTCA}\}, r(T) = \{\text{AAGC,CG,CA}\}$$

If $P_S = P_S(C) \cup P_S(A) \cup P_S(G) \cup P_S(T)$ and $Q_R = Q_R(C) \cup Q_R(A) \cup Q_R(G) \cup Q_R(T)$ are MS profiles of S and R, respectively, then they have the following form:

$P_S(C)$	$Q_R(C)$
$p_1^C = (2m(A)+m(G),1)$	$q_1^C = (2m(A)+m(G),1)$
$p_2^C = (3m(T)+m(A)+m(G),1)$	$q_2^C = (m(T),1)$
$p_3^C = (m(A),1)$	$q_3^C = (2m(T)+m(G),1)$
	$q_4^C = (m(A),1)$
$P_S(A)$	$Q_R(A)$
$p_1^A = (m(G)+m(C)+m(T),1)$	$q_1^A = (3m(T)+3m(C)+2m(G),1)$
$p_2^A = (2m(T)+m(G)+m(C),1)$	
$P_S(G)$	$Q_R(G)$
$p_1^G = (2m(A),1)$	$q_1^G = (2m(A),1)$
$p_2^G = (m(C)+m(T)+m(A),1)$	$q_2^G = (2m(C)+m(T),1)$
$p_3^G = (2m(T)+m(C)+m(A),1)$	$q_3^G = (2m(T)+m(C)+m(A),1)$
$P_S(T)$	$Q_R(T)$
$p_1^T = (2m(A)+m(G)+m(C),1)$	$q_1^T = (2m(A)+m(G)+m(C),1)$
$p_2^T = (m(A)+m(G),1)$	$q_2^T = (m(C)+m(G),1)$
$p_3^T = (m(C)+m(A),1)$	$q_3^T = (m(C)+m(A),1)$

6 of 11 peaks from P_S could be matched by the equal masses and the cleavage base with peaks from Q_R (p_1^C and q_1^C, p_3^C and q_4^C, p_1^G and q_1^G, p_3^G and q_3^G, p_1^T and q_1^T, p_3^T and q_3^T). However, it is easy to see that a single A-C SNP at position 6 between s and r causes the following relations between masses of remaining peaks:

$$m(p_2^C) = m(q_2^C) + m(q_3^C) + m(A) \tag{6}$$

$$m(p_1^A) + m(p_2^A) + m(C) = m(q_1^A) \tag{7}$$

$$m(p_2^G) - m(A) = m(q_2^G) - m(C) \tag{8}$$

$$m(p_2^T) - m(A) = m(q_2^T) - m(C) \tag{9}$$

If peaks and pairs of peaks are matched according to the relations (6)-(9) (p_2^C and (q_2^C, q_3^C), (p_1^A, p_2^A) and q_1^A, p_2^G and q_2^G, p_2^T and q_2^T), then all peaks from P_S and Q_R will be matched. Moreover, masses of single nucleotides and subprofiles involved in (6)-(9) allow to guess the corresponding SNP between s and r and in some cases the number of such type of matches allows to estimate the number of SNP's (in this example 1 SNP).

In general, the relations analogous to (6)-(9) have the following form:

$$m(p_i^{\sigma_{k_1}}) = m(q_{i_1}^{\sigma_{k_1}}) + m(q_{i_2}^{\sigma_{k_1}}) + m(\sigma_{k_2}) \tag{10}$$

$$m(p_{j_1}^{\sigma_{k_2}}) + m(p_{j_2}^{\sigma_{k_2}}) + m(\sigma_{k_1}) = m(q_j^{\sigma_{k_2}}) \tag{11}$$

$$m(p_{h_1}^{\sigma_{k_3}}) - m(\sigma_{k_2}) = m(q_{h_2}^{\sigma_{k_3}}) - m(\sigma_{k_1}) \tag{12}$$

$$m(p_{l_1}^{\sigma_{k_4}}) - m(\sigma_{k_2}) = m(q_{l_2}^{\sigma_{k_4}}) - m(\sigma_{k_1}) \tag{13}$$

Usually there are many possible alternative matches between peaks according to (10)-(13). The goal is to choose the optimal assignments such that the alignment score is maximized. Therefore the problem could be formulated as follows. Let $P_{(2)}$ be a set of all 2-element subsets of a set P. For $p \in P$ denote by $P_{(2)}(p)$ the set of all 2-subsets containing p. If P is a MS-profile, add to P an auxiliary empty peak $p_\epsilon = (0, \infty)$ with 0 mass and unbounded intensity. We will call such profile an extended MS profile. We assume without loss of generality that all other peaks have intensity 1 (otherwise, if peak p_i has intensity $f(p_i) > 1$ replace it with $f(p_i)$ peaks of intensity 1). Further, extend an alphabet Σ by addition of an auxiliary empty symbol ϵ with $m(\epsilon) = 0$. Those additional objects are needed to include insertions, deletions and mutations in homopolymers (i.e. sequences of identical nucleotides) in the model.

Problem 2

Input: Two extended MS profiles $P^1 = \{p_1^1, ..., p_{n_1}^1\} = P^1(C) \cup P^1(A) \cup P^1(G) \cup P^1(T) \cup \{p_\epsilon\}$ and $P^2 = \{p_1^2, ..., p_{n_2}^2\} = P^2(C) \cup P^2(A) \cup P^2(G) \cup P^2(T) \cup \{p_\epsilon\}$

Find: Two subsets $P_*^1 \subseteq P^1 \cup P_{(2)}^1$ and $P_*^2 \subseteq P^2 \cup P_{(2)}^2$ of matched peaks and pairs of peaks and a bijection $\pi : P_*^1 \to P_*^2$ such that the following conditions hold:

(i) $|P_*^j \cap (P_{(2)}(p_l^j) \cup \{p_l^j\})| \le 1$ for every $p_l^j \in P^j \setminus \{p_\epsilon^j\}$, $j = 1, 2$ (every peak is matched at most once either as a singleton or as a member of a pair)

(ii) $\pi(\{p_i^1, p_j^1\}) \in P^2$ for every pair $\{p_i^1, p_j^1\} \in P_{(2)}^1$ (pair of peaks should be matched to a single peak);

(iii) there exists a bijection $\psi : P_*^1 \cap P_{(2)}^1 \to P_*^2 \cap P_{(2)}^2$ (matchings of pairs of peaks go in pairs)

and the objective function (1) is maximized. The objective function should be defined in such a way that

a) a pair of peaks is matched to a peak and vise versa only if (10) and (11) holds for them; the bijection ψ maps pairs which are conjugate by (10) and (11);

b) the number of matches involving pairs is as small as possible. Each such match potentially corresponds to an insertion, deletion or replacement and we are trying to align MS profiles with the smallest number of involved mismatches as possible - analogously to alignment of sequences using edit distance.

In the next section we show how to define such a function and present an algorithm for its calculation. This is a new approach, which, as Example 1 shows, is more accurate than the approaches based on the direct peak matching, and, moreover, in many cases allows to estimate the actual number and types of SNPs.

Note that (10)-(13) holds for a certain SNP, if it is isolated, which means that substrings between it and the closest SNPs contain all four nucleotides. For the conserved genomes this is a reasonable assumption: it was shown in [1] that the overwhelming majority of SNPs in human genome are isolated (for the data analyzed in [1] the average and minimal distance between two neighbor SNPs is 231bp and 14bp, respectively). Therefore for such genomes a solution of Problem 2 provides a reliable estimation for the number and types of SNPs. If two mutations happen in close proximity, then the relation between peaks caused by them is more complex than (10)-(13). Moreover, if sample contains more than one unknown sequence, it is usually impossible to assign peaks to each sequence. Therefore for a highly mutable genomes, such as viral genomes, solution of Problem 2 provides a distance, which specifies and generalizes the most commonly used distance with the score function (3), instead of direct estimation of the number of mismatches.

3 Network Flow Method for Spectral Alignment

For a directed graph (or network) N with a vertex set V, an arcs set A, pair of source and sink $s, t \in V$, arcs capacities cap and possibly arc costs $cost$ a network flow is a mapping $f : A \to \mathbb{R}_+$ such that $f(a) \leq cap(a)$ for every $a \in A$ (capacity constraints) and $\sum_{uv \in A} f(uv) - \sum_{vw \in A} f(vw) = 0$ for every $v \in V \setminus \{s, t\}$ (flow conservation constraints). The value of flow is $|f| = \sum_{sv \in A} f(sv)$. The classical network flow problem either searches for a flow of maximum value (Maximum Flow Problem) or for a flow with a given value of a minimum cost (Minimum-cost Flow Problem).

It is well-known that in discrete optimization many matching-related problems (such as Maximum Bipartite Matching Problem, Assignment problem, Minimum Cost Bipartite Perfect Matching Problem, Linear Assignment Problem, etc.) could be solved using either network flows or shortest path - based algorithms. It suggests that a similar approach could be used for Problem 2. However, the formulation of Problem 2 is more complex that of the above-mentioned problems, so the reduction of Problem 2 to the network flow-based problem appeared to be rather complex. Below we present that reduction.

Let $P^1 = \{p_1^1, ..., p_{n_1}^1\} = P^1(C) \cup P^1(A) \cup P^1(G) \cup P^1(T) \cup \{p_\epsilon\}$ and $P^2 = \{p_1^2, ..., p_{n_2}^2\} = P^2(C) \cup P^2(A) \cup P^2(G) \cup P^2(T) \cup \{p_\epsilon\}$ be extended MS profiles. Let also $\delta \in \mathbb{R}_+$ be the mass precision, $g \in \mathbb{R}_+$ be the mismatch penalty and $p, q \in \mathbb{R}_+$ be the mutation (i.e. replacement, insertion, deletion) penalties corresponding to pairs of relations (10),(11) and (12),(13), respectively. Construct the network

$$N = (V, A, l, m, cost, cap) \tag{14}$$

where $l : V \to \Sigma^*$ is a vertices labels function, $m : V \to \mathbb{R}_+$ is vertices weights function, $cost : A \to \mathbb{R}_+$ and $cap : A \to \mathbb{R}_+$ are cost and capacity functions of arcs, respectively. Vertex set

$$V = \{s, t\} \cup V_1 \cup V_2 \cup V_{p_1} \cup V_{p_2} \cup V_{a_1} \cup V_{a_2} \cup V_{d_1} \cup V_{d_2}$$

and arc set A are constructed as follows:

1) s and t are the source and sink, respectively.
2) for each peak $p_i^j \in P^j(\sigma)$, $j = 1, 2$, $i = 1, ..., n_j$, $\sigma \in \Sigma$ the set V_j contains $f(p_j^i)$ vertices $v_j^i(1), ..., v_j^i(f(p_j^i))$. For each $v_j^i(k)$ $l(v_j^i(k)) = \sigma$, $m(v_j^i(k)) = m(p_i^j)$. For an empty peak $p_\epsilon \in P^j$, $j = 1, 2$, the set V_j contain the unique vertex v_ϵ^j with $l(v_\epsilon^j) = o$ and $m(v_\epsilon^j) = 0$.
3) For each $v \in V_1 \setminus \{v_\epsilon^1\}$ the set A contains an arc sv with $cost(sv) = 0$ and $cap(sv) = 1$. For each $v \in V_2 \setminus \{v_\epsilon^2\}$ A contains an arc vt with $cost(vt) = 0$ and $cap(vt) = 1$. There are also arcs sv_ϵ^1 and $v_\epsilon^2 t$ with $cost(sv_\epsilon^1) = cost(v_\epsilon^2 t) = 0$ and $cap(sv_\epsilon^1) = cap(v_\epsilon^2 t) = \infty$.
4) $uv \in A$ for each $u \in V_1$, $v \in V_2$ such that $|m(u) - m(v)| < \delta$ and $l(u) = l(v)$; $cost(uv) = 0$, $cap(uv) = 1$.

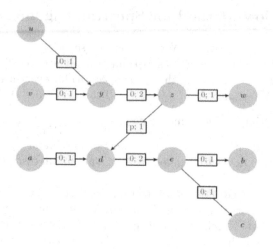

Fig. 1. Edges corresponding to relations (10),(11)

5) For every $u, v \in V_1$ and $w \in V_2$ such that
   a) $l(u) = l(v) = l(w)$,
   b) there exists $\sigma \in \Sigma$ such that $|m(u) + m(v) + m(\sigma) - m(w)| < \delta$,
   the vertex set V contains vertices $y \in V_{p_1}$ and $z \in V_{a_1}$ with $m(y) = m(z) = 0$,
   $l(y) = o$, $l(z) = l(u)\sigma$. The set A contains arcs uy, vy, yz, zw with $cost(uy) =$
   $cost(vy) = cost(yz) = cost(zw) = 0$, $cap(uy) = cap(vy) = cap(zw) = 1$,
   $cap(yz) = 2$. See Figure 1. The subgraph $N[u, v, w, y, z]$ induced by vertices
   u, v, w, y, z will be referred as left fork.
6) Analogously, for every $a \in V_1$ and $b, c \in V_2$ such that
   a) $l(a) = l(b) = l(c)$,
   b) there exists $\sigma \in \Sigma$ such that $|m(a) - m(b) - m(c) - m(\sigma)| < \delta$,
   the set V contains vertices $d \in V_{a_2}$ and $e \in V_{p_2}$ with $m(d) = m(e) = 0$,
   $l(e) = o$, $l(d) = \sigma l(b)$. The set A contains arcs ad, de, eb, ec with $cost(ad) =$
   $cost(de) = cost(eb) = cost(ec) = 0$, $cap(ad) = cap(eb) = cap(ec) = 1$,
   $cap(de) = 2$. See Figure 1. Further the subgraph $N[a, b, c, d, e]$ will be referred
   as right fork.
7) For vertices $u \in V_{a_1}$, $v \in V_{a_2}$ the set A contains an arc uv with $cost(uv) = p$
   and $cap(uv) = 1$, if $l(u) = l(v)$. See Figure 1.
8) For every $u \in V_1$ and $v \in V_2$ such that
   a) $l(u) = l(v)$,
   b) there exists $\sigma_1, \sigma_2 \in \Sigma$ such that $|m(u) - m(\sigma_1) - m(v) + m(\sigma_2)| < \delta$,
   the set V contains vertices $y \in V_{d_1}$ and $z \in V_{d_2}$ with $m(y) = m(z) = 0$,
   $l(y) = l(z) = \sigma_1\sigma_2$. The set A contains arcs uy, yz, zv with $cost(uy) =$
   $cost(yz) = cost(zv) = 0$, $cap(uy) = cap(zv) = 1$, $cap(yz) = 0$. See Figure 2.
9) for all distinct vertices $y, a \in V_{d_1}$, $z, b \in V_{d_2}$ such that $yz, ab \in A$, $cap(yz) =$
   $cap(ab) = 0$ and $l(y) = l(b)$, the set A contains arcs yb, az with $cost(yb) =$
   $cost(az) = \frac{q}{2}$, $cap(yb) = cap(az) = 1$. See Figure 2.

10) For every $v \in V_1$ there exists an arc vs with $cost(vs) = g$ and $cap(vs) = 1$.

Let $x : A \to \mathbb{N}, a \mapsto x_a$ is a flow in the network N. Problem 2 could be formulated as the following variant of the network flow problem:

$$\text{minimize} \sum_{a \in A} cost(a) x_a \tag{15}$$

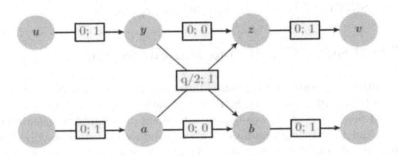

Fig. 2. Edges corresponding to relations (12),(13)

subject to

$$\sum_{uv \in A} x_{uv} - \sum_{vw \in A} x_{vw} = 0, \quad v \in V \setminus \{s, t\}; \tag{16}$$

$$\sum_{sv \in A, v \neq v_e} x_{sv} = |V_1| - 1; \tag{17}$$

$$x_{uy} - x_{vy} = 0, \quad y \in V_{p_1}; \tag{18}$$

$$x_{eb} - x_{ec} = 0, \quad e \in V_{p_2}; \tag{19}$$

$$x_{uy} - x_{zv} = 0; \quad yz \in A, cap(yz) = cost(uy) = cost(zv) = 0 \tag{20}$$

$$0 \leq x_a \leq cap(a), \quad a \in A. \tag{21}$$

This formulation differs from the classical network flow problem formulation by additional constraints which require flow to be equal on some prescribed pairs of arcs.

Arcs from 4) provide the possibility of match between peaks with close masses with 0 penalty. Vertices and arcs from 5)-7) and constraints (18)-(19) allow to match peaks with pairs of peaks according to relations (10),(11). The capacities of arcs defined in 5)-7) are chosen in such a way that if flow goes through the left fork, then it should also go through the right fork indicating the same mutation, thus forcing a fulfillment of requirement (iii) of Problem 2. Moreover, if flow goes through some pair of forks, exactly one arc of cost p between those forks is involved, thus forcing penalty for mutation. Vertices and arcs from 8)-9) and constraints (20) play the same role for relations (12),(13). Constraint (17) for total size of the flow ensures that every peak is either matched or penalized for mismatch, which is encoded by arcs from 10). Moreover, arcs from 10) ensure that the problem (15)-(21) always has a feasible solution. (16) and (21) are standard flow conservation and capacity constraints.

If P^1 and P^2 are samples of single genomes with isolated SNPs, then the number of SNPs could be estimated as $|\{a \in A : x_a > 0, cost(a) = p\}|$.

4   Test Results, Conclusions and Future Work

The algorithm was tested on simulated data. For this, 80 pairs of sequences of lengths 40-60bp with 2-4 isolated SNPs were randomly generated. For each position one of possible symbols was chosen with equal probability to generate first sequence, and then random mutations were introduced on the prescribed positions to generate the second sequence. MS profiles of generated sequences were simulated using masses $m(A) = 329.21$ DA, $m(T) = 306.17$ DA, $m(G) = 345.21$ DA, $m(C) = 305.18$ DA. The ILP formulation (15)-(21) was solved using GNU Linear Programming Kit (GLPK) (http://www.gnu.org/software/glpk/) on a computer with two 2.67GHz processors and 12 GB RAM. Since ILP solution is usually time-consuming, the time limit 30 seconds per problem was established. For 90% (72 of 80) of test instances ILP was solved within the time limit. For all that instances the numbers of SNPs were estimated correctly. Running times for ILP solution in that cases varies from 0.491 seconds in average with the standard deviation 0.968 seconds for 40bp sequences to 3.434 seconds with the standard deviation 5.824 seconds for 60bp sequences.

Thus the proposed approach enables an accurate comparison of MS profiles and provides a direct evaluation of genetic distances between DNA molecules without invoking sequences. It is potentially more accurate than the approaches based on the direct peak matching, and, moreover, in many cases allows to estimate the actual number and types of SNPs.

The proposed spectral alignment method is expected to be highly effective in evaluating genetic relatedness among viral samples and identifying transmission clusters in viral outbreaks. The reasons behind this presumption is based on the fact, that simple Hamming distance between samples could be calculated using a special case of our model with $p = q = \infty$. Hamming distance (which corresponds to the score function (3)) was shown to effectively separate transmission clusters [4][8]. Thus, the developed model allows for generating a large spectrum of distances in addition to the special case and as such offers a more general framework for measuring genetic distances using MS profiles.

The ILP-based approach to solving the problem (15)-(21) is time-consuming. Therefore more computationally effective approaches may be required to handle larger samples. It is expected that direct applications of network flow-based methods, Lagrangian relaxations or other methods should dramatically increase performance of the algorithm. The generalizations of relations (10)-(13) in order to obtain a model allowing for estimation of the actual number of mutations in highly heterogeneous samples is an important direction for the future research.

References

1. Böcker, S.: SNP and mutation discovery using base-specific cleavage and MALDI-TOF mass spectrometry. Bioinformatics 19(suppl. 1), i44–i53 (2003)
2. Böcker, S.: Sequencing from Compomers: Using Mass Spectrometry for DNA de novo Sequencing of 200+ nt. Journal of Computational Biology 11(6), 1110–1134 (2004)
3. Böcker, S., Kaltenbach, H.-M.: Mass spectra alignments and their significance. Journal of Discrete Algorithms 5(4), 714–728 (2007)
4. Dimitrova, Z., Campo, D.S., Ramachandran, S., Vaughan, G., Ganova-Raeva, L., Lin, Y., Forbi, J.C., Xia, G., Skums, P., Pearlman, B., Khudyakov, Y.: Evaluation of viral heterogeneity using next-generation sequencing, end-point limiting-dilution and mass spectrometry. Silico Biology 11, 183–192 (2011/2012)
5. Ehrich, M., Böcker, S., van den Boom, D.: Multiplexed discovery of sequence polymorphisms using base-specific cleavage and MALDI-TOF MS. Nucleic Acids Res. 33(4), e38 (2005)
6. Ehrich, M., Nelson, M.R., Stanssens, P., Zabeau, M., Liloglou, T., Xinarianos, G., Cantor, C.R., Field, J.K., van den Boom, D.: Quantitative high-throughput analysis of DNA methylation patterns by base-specific cleavage and mass spectrometry. Proceedings of the National Academy of Sciences of the United States of America 102, 15785–15790 (2005)
7. Ganova-Raeva, L., Ramachandran, S., Honisch, C., Forbi, J.C., Zhai, X., Khudyakov, Y.: Robust Hepatitis B Virus Genotyping by Mass Spectrometry. J. Clin. Microbiol. 48(11), 4161 (2010)
8. Ganova-Raeva, L., Dimitrova, Z., Campo, D.S., Yulin, L., Ramachandran, S., Xia, G.-L., Honisch, C., Cantor, C., Khudyakov, Y.: Detection of hepatitis C virus transmission using DNA mass spectrometry. J. Infect Dis. (January 31, 2013)
9. Kirpekar, F., Nordhoff, E., Larsen, L.K., Kristiansen, K., Roepstorff, P., Hillenkamp, F.D.: sequence analysis by MALDI mass spectrometry. Nucleic Acids Research 26, 2554–2559 (1998)
10. Lefmann, M., Honisch, C., Böcker, S., Storm, N., von Wintzingerode, F., Schlötelburg, C., Moter, A., van den Boom, D., Göbel, U.B.: Novel Mass Spectrometry-Based Tool for Genotypic Identification of Mycobacteria. Journal of Clinical Microbiology, 339–346 (January 2004)
11. Mäkinen, V.: Peak alignment using restricted edit distances. Biomolecular Engineering 24(3), 337–342 (2007)
12. Pevzner, P.A., Dancik, V., Tang, C.L.: Mutation-tolerant protein identification by mass-spectrometry. Journal of Computational Biology 7, 777–787 (2000)
13. Pusch, W., Kraeuter, K.O., Froehlich, T., Stalgies, Y., Kostrzewa, M.: Genotools SNP manager: a new software for automated high-throughput MALDI-TOF mass spectrometry SNP genotyping. Biotechniques 30, 210–215 (2001)
14. Rees, J.C., Voorhees, K.J.: Simultaneous detection of two bacterial pathogens using bacteriophage amplification coupled with matrix-assisted laser desorption/ionization time-of-flight mass spectrometry. Rapid Commun. Mass Spectrom 19, 2757–2761 (2005)

160 P. Skums et al.

15. Sampath, R., Hall, T.A., Massire, C., Li, F., Blyn, L.B., Eshoo, M.W., Hofstadler, S.A., Ecker, D.J.: Rapid identification of emerging infectious agents using PCR and electrospray ionization mass spectrometry. Ann. NY Acad. Sci. 1102, 109–120 (2007)
16. Sjoholm, M.I., Dillner, J., Carlson, J.: Multiplex detection of human herpesviruses from archival specimens by using matrix-assisted laser desorption ionization-time of flight mass spectrometry. J. Clin. Microbiol. 46, 540–545 (2008)
17. Stanssens, P., Zabeau, M., Meersseman, G., Remes, G., Gansemans, Y., Storm, N., Hartmer, R., Honisch, C., Rodi, C.P., Böcker, S., van den Boom, D.: High-Throughput MALDI-TOF Discovery of Genomic Sequence Polymorphisms. Genome Res. 14(1), 126–133 (2004)
18. Tost, J., Schatz, P., Schuster, M., Berlin, K., Gut, I.: Analysis and accurate quantification of CpG methylation by MALDI mass spectrometry. Nucleic Acids Res. 31, e50 (2003)
19. von Wintzingerode, F., Bocker, S., Schlotelburg, C., Chiu, N.H., Storm, N., Jurinke, C., Cantor, C.R., Gobel, U.B., van den Boom, D.: Base-specific fragmentation of amplified 16S rRNA genes analyzed by mass spectrometry: a tool for rapid bacterial identification. Proc. Natl. Acad. Sci. USA 99, 7039–7044 (2002)
20. Yang, H., Yang, K., Khafagi, A., Tang, Y., Carey, T.E., Opipari, A.W., Lieberman, R., Oeth, P.A., Lancaster, W., Klinger, H.P., Kaseb, A.O., Metwally, A., Khaled, H., Kurnit, D.: Sensitive detection of human papillomavirus in cervical, head/neck, and schistosomiasis-associated bladder malignancies. Proc. Natl. Acad. Sci. USA 102, 7683–7688 (2005)

A Context-Driven Gene Prioritization Method for Web-Based Functional Genomics

Jeremy J. Jay[1], Erich J. Baker[2], and Elissa J. Chesler[1]

[1] The Jackson Laboratory, Bar Harbor, ME 04605, USA
[2] Baylor University, Waco, TX 76798, USA

Abstract. Functional genomics experiments often result in large sets of gene centered results associated with biological concepts such as diseases. Prioritization and interpretation of these results involves evaluation of the relevance of genes to various annotations or associated terms and is often executed through the use of prior information in biological databases. These diverse databases are frequently disconnected, or loosely federated data stores. Consequently, assessing the relations among biological entities and constructs, including genes, gene products, diseases, and model organism phenotypes is a challenging task typically requiring manual intervention, and as such only limited information is considered. Extracting and quantifying relations among genes and disease related concepts can be improved through the quantification of the entire contextual similarity of gene representations among the landscape of biological data. We have devised a suitable metric for this analysis which, unlike most similar methods requires no user-defined input parameters. We have demonstrated improved gene prioritization relative to existing metrics and commonly used software systems for gene prioritization. Our approach is implemented as an enhancement to the flexible integrative genomics platform, GeneWeaver.org.

1 Introduction

High-throughput functional genomics experimental techniques have made it possible to rapidly generate vast amounts of genomic data in the context of disease related inquiry. Thousands of potential gene-disease associations must be prioritized to identify viable candidates for experimental validation and translation. Evaluation of the disease implications of gene lists and gene networks that result from genomic experimentation can be an inefficient, complex task due to the current separation of biological data stores, where typical queries must overcome barriers imposed by heterogeneous data frameworks.

There are numerous approaches to gene prioritization for functional genomics. (27; 10). Several approaches estimate the similarity of a set of candidate genes to those with known disease-gene associations (15; 2; 20; 8; 7; 26). They make use of a variety of data sources including literature, sequence, gene expression, protein domains, or annotations to curated ontology associations such as the Gene Ontology (GO) or Human Phenotype (HP) ontology(3; 25). Many resources are

Z. Cai et al. (Eds.): ISBRA 2013, LNBI 7875, pp. 161–172, 2013.
© Springer-Verlag Berlin Heidelberg 2013

designed with a focus on a single data source or pivot point, and provide meaningful results when the density of biological associations within the data source is high (19; 22; 21).

The density and quality of available data for gene prioritization continues to improve. However, curated biological associations, such as ontology annotations from individual hypothesis driven experiments, remain sparse, while the dense data afforded by functional genomics analysis is noisy and gathered in limited contexts. Manually curated gene annotations do not typically involve data surveys of all genes or processes; rather, depth of knowledge is created around specific areas of interest or well-supported hypotheses, creating an uneven landscape highlighting particular genes or gene products and specific aspects of disease function. There are limited empirical associations among the vast majority of genes and diseases. Efforts like GeneWeaver build empirical context for relevant genes through a framework where gene-disease relationships are self-organized into a hierarchical format; however, even these emergent structures cannot accurately provide prioritization among the many possible relationships(4).

Contextual information about a disease or gene has been shown to improve gene-disease associations, but typically relies on very limited data such as co-occurrence or co-expression information (15; 14). Contextual information can also include topics that are often concurrently studied, such as comorbid diseases, symptoms or other conditions that have irregular associations with the disease, but are highly relevant to its study. We describe an efficient method for incorporating diverse data context into gene-disease similarity measurement designed for use in web-based genomics analysis tools, such as GeneWeaver. Our parameter free design ensures users can get good results without mastering the method or underlying data. We evaluate a dataset consisting of associations among Entrez Gene, PubMed and Medical Subject Headings against existing similarity quantification metrics and related bioinformatics resources.

2   Background

2.1   Counting-Based Similarity Metrics

There are several metrics for quantifying the strength of a relationship between two concepts, such as a gene and a disease, based on the similarity of the set of entities associated with each. These include the widely used Jaccard Similarity Coefficient and the Rand Index(17; 23). For the Rand Index, the value is the sum of positive and negative match pairs between the two sets of associations, A and B, divided by the total number of possible matches (Eq. 1). The Jaccard Coefficient performs a similar calculation, but only includes true positive matches, withholding the true negative matches from numerator and denominator (Eq. 2). The Rand Index is ideal for measuring correspondence of two sets when knowledge is complete (in other words, that all negative associations are known definitively). The Jaccard coefficient is better suited to a genome-wide analysis because there are typically a high number of negatives (many of them presumably false) within genomic studies and few true negative assertions in the literature (16). Thus, any metric that rewards negative matches, including the widely used hypergeometric test, is upwardly biased and can be misleading in the context of functional genomics.

$$RandIndex(A, B) = \frac{TP + TN}{TP + TN + FP + FN} \tag{1}$$
$$= \frac{|U \setminus (A \cap B)|}{|U|}$$

$$Jaccard(A, B) = \frac{TP}{TP + FP + FN} \tag{2}$$
$$= \frac{|A \cap B|}{|A \cup B|}$$

Many biological annotations are organized into structured vocabularies, providing subsuming terms and annotation sets which are not accounted for by simple set-based counting methods. A method that extends Jaccard to account for these subsuming terms is referred to as SimUI (13). Given a set of terms A, this method collects the closure set (all subsuming terms, denoted A^+) to include more general descriptors in the comparison (Eq. 3). For example, a "DNA binding" gene and a "RNA binding" gene would not match using the Jaccard Coefficient or Rand Index, but with SimUI they would be able to match on the subsuming "nucleotide binding" term. This allows a more accurate depiction of the similarity of terms, but requires more knowledge and human intervention to collect all possible concepts and relationships into the structured vocabularies.

$$SimUI(A, B) = \frac{|A^+ \cap B^+|}{|A^+ \cup B^+|} \tag{3}$$

Simple intersection counting-based metrics like these work well for small and dense data sets, but large and sparse data sets have fewer possible overlaps and many more empty pairings, producing a limited resolution for comparison. Additionally, a human can easily determine that "nucleotide binding" is a more generic term than "RNA binding", but these counting-based metrics will weight them equally when a comparison is done. Techniques that quantify the amount of information conveyed by a term can distinguish terms such as these.

2.2   Information Content

The Information Content (IC) of a concept is based on the probability of a concept's occurrence within a document corpus (24). Initially this corpus was defined by prose such as biomedical abstracts, but it can be generalized further to include any knowledge base (KB) of associations. Intuitively, concepts that occur more frequently, such as "Child", in the KB provide less informative annotation than those that are infrequent, such as "Quintuplets", measured using Equation 4).

To compare two genes, we can now look at the difference between the IC values of their relations (Eq. 5) using a method called SimGIC(21).

$$IC(t) = -log\left(\frac{|KB \cap t|}{|KB|}\right) \tag{4}$$

$$SimGIC(A, B) = \frac{\sum_{x \in (A^+ \cap B^+)} IC(x)}{\sum_{y \in (A^+ \cup B^+)} IC(y)} \tag{5}$$

3   Materials and Methods

In the present study, we develop a method for unsupervised context-based analysis that handles multiple diverse data sources, is easily maintainable to add new data sources and keep them up-to-date, and is performant enough to execute queries in real time. These attributes are critical for integration of the method with the type of real time, web-based tools used by bench biologists, including systems like GeneWeaver.org

To provide flexibility in the types of features that can be used to characterize context, the available data formats and structures are decomposed to a graph format for analysis. Nodes consisting of entities such as genes, terms, concepts, publications, etc. are connected by edges representing associations found in the various data repositories. In addition, closure inferences are automatically extended into the graph structure to simplify later processing steps.

3.1   Data Sources

The National Center for Biotechnology Information houses multiple databases containing gene identifiers (Entrez Gene), publication abstracts (PubMed), and controlled vocabulary terms (Medical Subject Headings, MeSH). Both genes and MeSH terms are associated to hundreds of thousands of publications through automated and manual processes. All of these data are freely accessible through the NCBI FTP site and e-utilities for use in offline analyses.

Human Entrez Gene associations to PubMed identifiers were made using the gene2pubmed flat file downloaded from the NCBI FTP site on 21 Jan 2012. To retrieve MeSH associations to PubMed, the PubMed IDs found in the gene2pubmed file were then fetched through the NCBI e-utilities API. In total, there were 818,887 Gene to PubMed associations, and 6,918,405 MeSH to PubMed assocations fetched, linking 31,308 Entrez Gene IDs, 18,588 MeSH terms, and 368,357 PubMed identifiers into the dataset.

To evaluate the prioritized gene rankings from each method, a gold standard enables a test for precision and recall of true positive associations. For this we used two publicly available databases of curated human disease gene associations. One, the Online Mendelian Inheritence in Man (OMIM) project, is an extensively curated catalog of Human Genes and Genetic Disorders (12). This resource has been used to validate previous prioritization methods (7; 8; 2), and a comprehensive mapping from OMIM terms to MeSH terms is publicly available and maintained (9). However, the OMIM data is somewhat sparse, covering only 546 MeSH disease terms and 1,244 associations, because the resource is intended as more of an encyclopedia than a gold standard. The second resource is the Genetic Association Database (GAD), which contains a larger collection of disease gene associations covering 1,489 MeSH terms and 13,357 total associations, but has less descriptive text than the OMIM collection(5).

3.2   Methods

To develop a quantitative estimate for contextual relevance to an entity, we begin with some observations about the IC (See Eq. 4). First, a concept with low IC has a high occurrence rate in the knowledge base, and highly informative concepts (high IC value) have a lower occurrence rate. In terms of contextual pertinence, concepts found at each extreme of this spectrum provide little additional information to any particular idea. This appears counter-intuitive to the phrase "high information content" until one observes that high IC terms impart a highly restricted subset of the knowledge base (the "information" contained is the subset, not any individual feature). Thus the best terms for imparting contextual information are those with moderate IC, as they can restrict the subset enough to drop spurious results, but do not restrict it enough to discard all data with meaningful signal. A second observation is that concepts that are relevant to each other often co-occur, and will share a high similarity value through non-semantic similarity methods. Combining these two observations, we present the Context Content (CC) metric, Eq. 6.

$$CC(t, z) = log\left(1.0 + e^{-IC(t)} \cdot SimGIC(nbrs(t), nbrs(z))\right) \qquad (6)$$

The prior usage of the term "moderate IC" can also be interpreted as "moderate prevalence," and is easier to work with due to the unbounded nature of IC values. Note that in practice, prevalence and similarity, as measured by SimGIC, measure inversely differing values (a high prevalence is observed when significance values are low) and converge only in the special case where the context term covers all items in the data set. As a result, if prevalence is low or similarity is low, then the term in question should not weigh significantly in the context. By taking the product of these terms, we are in effect weighting the prevalence by the similarity to the context term. Because each of these values range from 0.0 - 1.0, their product occupies the same range. We then add one and take the logarithm in order to map the values into a positive range with similar characteristics to the information content.

We must take one further step to apply this method to all possible entities in our data. The SimGIC method only works when there are shared neighbors between two concepts. This works well for Genes and MeSH terms because they can use shared PubMed publications, however PubMed publications do not share neighbors with either Gene or MeSH terms (they are directly associated). An arithmetic mean of associated CC values would negate much of the benefit of context by equally weighting low-CC terms and high-CC terms; instead, we selected a sum of squares average to bias higher CC values without adding additional algorithmic complexity (Eq. 7).

$$CC(p, z) = \sqrt{\frac{\sum_{n \in nbrs(p)} CC(n, z)^2}{|nbrs(p)|}} \qquad (7)$$

Armed with this information, we can determine the strength of a gene's relationship to a disease by calculating the weight of its neighbors' associations to the disease. We build a metric similar to SimGIC by substituting the CC for the IC and again using a sum of squares approach to allow high-CC terms to influence the result more (Eq. 8). Where a higher-IC term may have provided a better score previously, if that same term does not have high relevance to the disease of interest, it will not contribute prominently to the ranking of a gene.

$$SimGCC(A, B, z) = \frac{\sqrt{\sum_{x \in (A^+ \cap B^+)} CC(x, z)^2}}{\sqrt{\sum_{y \in (A^+ \cup B^+)} CC(y, z)^2}} \qquad (8)$$

The end result of this process is that gene rankings will be less influenced by associations that have little relevance to the disease of interest, and likewise less influenced by very specific associations with little corroborating evidence.

3.3   Implementation

In order to implement this method efficiently, a number of data processing steps were performed. First, as previously described, the various data sets were fetched from their respective repositories and converted into a graph format of nodes and edges. To aid data updates and queries, this graph is loaded into a high-performance datastore called Redis (1). When ready to perform an analysis, the contents of the datastore can be interactively accessed or dumped to a file. The simplicity of this arrangement allows new data to be added easily.

To perform a global analysis of the ranked similarity of all genes to a term, our implementation iterates over each disease, calculates the CC for all nodes in the graph, and then calculates and outputs the SimGCC (and other methods) scores for every gene to the selected disease. These scores are then sorted to produce a gene prioritization for the disease.

Significant speedups over a naïve implementation have been achieved with a few modifications. First, individual nodes in the dataset graph are mapped onto distinct integers, which are constructed such that a simple bitmask can be used to determine the node's data partition (1=Gene, 2=MeSH, 3=PubMed). By mapping entities to integers, memory usage is significantly decreased and cache consistency is improved, in addition to making node identifier comparisons significantly faster. Second, many of the underlying equations used require the intersection and/or union of two sets of neighbors. After the initial loading stage, these neighbor relationships do not change, so they are stored in sorted order in main memory. This allows for a straightforward $O(n)$ mergesort-like algorithm to iterate all common and distinct neighbors efficiently. Finally, due to the embarrassingly parallel nature of this algorithm and the low output synchronization necessary, significant real-time speedups were obtained through the use of shared memory and multithreading – a technique well suited to modern multicore architectures. The graph structure, neighbor lists and IC values can be easily shared across processor cores since they do not change during the lifetime of the analysis. The only thread-local storage allocations required are $O(n)$ on the total number of nodes in the dataset. This allows an optimized implementation to use all available processor cores with a parallel for loop, achieving a nearly linear speedup.

4   Results and Discussion

The Receiver Operating Characteristic (ROC) plots reveal improved recall for genes associated to a selection of MeSH terms (Inset, Table 1), in addition to an overall improvement when averaged across all MeSH Mental Disorders terms in the GAD gold standard. In both instances SimGCC consistently outperforms the other ranking methods tested. Detailed gene-level rankings for OMIM disease genes also show that the improvement extends past the top 20 genes, which often consist of the highly published and highly associated disease genes (Table 1).

There are a number of online gene prioritization tools, enabling us to compare the performance of our approach to existing methods. Similar to SimGCC, many methods aggregate multiple data sources to build a quantitative measure of gene relevance. Unlike SimGCC, many of them begin the process with a set of user-defined training genes. SimGCC saves its users this added step of collecting training genes, but at the expense of hindering a power user with a specific set of genes or blocking the study of an unannotated disease.

Three tools were selected because they have a similar objective of ranking gene-disease relations, are readily available online, and have input and output formats amenable to comparison. One of the earliest attempts at gene prioritization was started over a decade ago with the Genes2Diseases project(20). It combines measures derived from MeSH Chemicals and Diseases, the Gene Ontology, RefSeq, and PubMed using a product of Jaccard scores and manually defined weights. Another method named ENDEAVOUR prioritizes genes using a large concert of data sources and metrics individually tailored to each data set, and then uses rank-order statistics and manually defined weights to combine the many rankings into a single final aggregate ranking (2). Because the data sources are separated, they can be enabled or disabled at will by the user, which allows more fine-grained control of the types of data the user wants to use. The last method in this comparison is ToppGene, which uses functional associations and protein-protein interactions to rank specific features and build a statistical model for prioritization (8). Like ENDEAVOUR, individual data sources can be enabled/disabled at will. Of the three methods, it is the most similar to SimGCC because of its use of feature-level relevance measures. However, like the other methods it has the drawback that these individual features cannot be correlated to each other in any way.

Table 1. Selected OMIM disease gene rankings and ROC Plots. Rankings show that SimGCC repeatedly ranks known genes higher. ROC plots of the first 1000 genes show better recall values for SimGCC using the larger GAD gold standard.

MeSH Disease	Entrez ID	simgcc	simgic	jaccard	rand
Alcoholism	ADH1B	1	1	1	27
Alcoholism	ADH1C	2	3	3	23
Alcoholism	GABRA2	8	8	8	110
Alcoholism	HTR2A	26	28	29	537
Alcoholism	TAS2R16	152	174	180	396
Alcoholism	**avg**	**37.8**	**42.8**	**44.2**	**218.6**
Alzheimer Disease	A2M	1	3	3	132
Alzheimer Disease	APP	3	1	1	279
Alzheimer Disease	PSEN1	10	6	6	1454
Alzheimer Disease	ACE	29	72	83	2510
Alzheimer Disease	APBB2	32	73	87	218
Alzheimer Disease	SORL1	43	80	97	1661
Alzheimer Disease	HFE	63	177	231	1324
Alzheimer Disease	NOS3	76	235	265	3609
Alzheimer Disease	BLMH	98	257	285	328
Alzheimer Disease	PLAU	126	307	314	1527
Alzheimer Disease	MPO	132	333	338	1446
Alzheimer Disease	AD5	351	630	648	1816
Alzheimer Disease	AD6	421	659	702	2944
Alzheimer Disease	AD9	1286	1987	3540	3524
Alzheimer Disease	PAXIP1	1692	2154	2649	2483
Alzheimer Disease	**avg**	**290.9**	**464.9**	**616.6**	**1683.7**
Autistic Disorder	EN2	2	40	48	154
Autistic Disorder	CNTNAP2	25	117	121	716
Autistic Disorder	MET	50	156	168	1000
Autistic Disorder	SHANK2	442	569	716	648
Autistic Disorder	**avg**	**129.8**	**220.5**	**263.2**	**629.5**
Schizophrenia	COMT	1	1	1	418
Schizophrenia	DRD3	5	4	4	277
Schizophrenia	NRG1	6	5	5	432
Schizophrenia	HTR2A	13	14	13	603
Schizophrenia	DAO	17	15	14	211
Schizophrenia	AKT1	18	50	51	529
Schizophrenia	DTNBP1	21	25	24	1595
Schizophrenia	DISC1	25	26	25	1365
Schizophrenia	DAOA	44	42	44	1696
Schizophrenia	MTHFR	58	77	91	1789
Schizophrenia	CHI3L1	63	74	76	171
Schizophrenia	PRODH	71	73	75	690
Schizophrenia	SCZD2	234	332	350	748
Schizophrenia	SCZD1	251	333	349	743
Schizophrenia	RTN4R	293	331	335	1602
Schizophrenia	SCZD6	325	376	382	924
Schizophrenia	SYN2	373	515	532	845
Schizophrenia	SCZD7	423	481	490	668
Schizophrenia	SCZD3	481	631	656	744
Schizophrenia	DISC2	622	566	583	1313
Schizophrenia	SCZD8	643	712	737	971
Schizophrenia	APOL4	708	702	734	949
Schizophrenia	APOL2	730	753	795	1307
Schizophrenia	SCZD11	964	1496	1748	1713
Schizophrenia	SCZD12	1038	1530	1747	1720
Schizophrenia	**avg**	**297.1**	**366.6**	**394.4**	**960.9**

Alcoholism

Alzheimer Disease

Autistic Disorder

Schizophrenia

Tobacco Use Disorder

Overall

method
— jaccard
— rand
— simgcc
— simgic

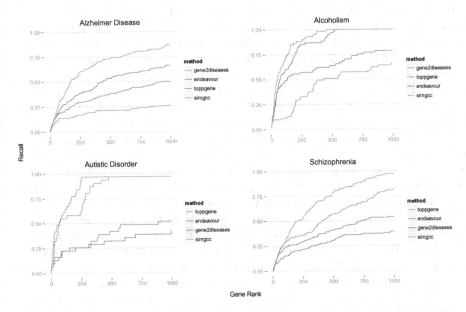

Fig. 1. External Method Comparison. Selected OMIM Disease genes were used to train recent gene prioritization methods, and then compared to SimGCC using ROC curves from the GAD gold standard. The training genes were then prepended to the results from ToppGene, ENDEAVOUR, and Genes2Diseases since these methods do not include them in rankings.

Fig. 2. SimGCC Example Application. Results from this analysis were used to highlight Phenome Map nodes in red based on their contextual intersection with Alcoholism. Darker red indicates a higher contextual association, pink indicates a small association, and green nodes indicate no contextual association could be made.

To compare SimGCC to these online gene prioritization tools, a few well-studied diseases were selected. Training genes for each method were taken from the OMIM gold standard because of its higher curation stringency. When technically feasible, all Human Entrez Gene IDs were ranked by each method. However, ToppGene was unable to rank the entire set without crashing, so a subset consisting of all the genes scored above 0.0 by the Jaccard, SimGIC, and SimGCC methods was used as the input test set. The size of this subset varied by disease tested, but was typically 20-50% of the 31,308 Entrez Gene IDs available.

Results from each method were collected and converted into a standard format. Unlike the results from SimGCC for which no training set is required, the results of these methods do not contain the genes from the provided training set, so would automatically be at a disadvantage in the prior ROC analysis. To alleviate this issue in a conservatively biased way, we prepended all of the OMIM training genes to the beginning of the rankings for each of the three online methods before determining ROC curves, in effect giving them a left biased ROC curve. The larger GAD gold standard powered the ROC curves, since it encompasses the OMIM data. Even with the training gene handicap, SimGCC is able to outperform these methods on 3 of the 4 selected diseases (Fig. 1).

A hierarchical intersection of 31 sets of genes empirically related to alcoholism was generated using the Phenome Map tool in GeneWeaver (4). This graph features individual gene sets in the lowest level, and higher order intersections of the gene sets at higher levels (Fig. 2). Red nodes contain genes with contextual similarity to Alcoholism based on SimGCC results. A previously published analysis of these 31 gene sets reveals that highly connected genes are not currently annotated to alcoholism (6), but the present study demonstrates that with inclusion of contextual information, some of the highly connected genes are found in an alcoholism relevant context.

5   Conclusion

We have described a novel method for extracting and quantifying contextual information content and demonstrated its application on a literature-centered human data set. Because this method requires no training set, it can be readily applied in new areas of investigation for which true positive data is sparse. By building upon the information content and extending similarity methods to assess the value of related associations, SimGCC was shown to provide a meaningful improvement over existing classification methods.

The extent of improvement made by SimGCC is dependent on the nature of the terms being compared and the state of data for each. The greatly improved performance for both 'Alzheimer's Disease' and 'Schizophrenia' show the value of contextual information in neurological disorders because high comorbidity and/or genetic relationships to similar disorders provide rich context in which to find similar genes. The smaller performance increase for 'Alcoholism' illustrates the heterogeneity inherent in this disorder (and many behavioral disorders) due to the effects of diverse sources of genetic predisposition and environmental variation (18). Finally, the research landscape for 'Autistic Disorder', with very large studies of many genes and very small studies of few genes, provides little contextual information for gene prioritization. As in the discussion of the CC metric (Eq. 6), data restrictions that are too loose or too narrow cannot provide enough information about individual genes or terms to extract contextual relevance.

Context similarity can be a powerful extension to the limitations of direct relational queries or bipartite gene-function prioritization familiar to bioinformatics inquiry. SimGCC facilitates analysis beyond filtered searches and into quantitative assessment of the contextual relationships of biological entities across data partitions. The SimGCC metric can be readily adapted to weighted relations, such as gene expression and other functional genomics experimental results. Integrating this method into the model organism databases, the Neuroinformatics Framework(11), and a wealth of other federated biological data resources can enable automatic discovery of related entities across data resources.

Acknowledgement. The authors gratefully acknowledge funding from NIH AA 18776.

References

[1] Redis, http://redis.io
[2] Aerts, S., Lambrechts, D., Maity, S., Van Loo, P., Coessens, B., De Smet, F., Tranchevent, L.C., De Moor, B., Marynen, P., Hassan, B., et al.: Gene prioritization through genomic data fusion. Nature biotechnology 24(5), 537–544 (2006)
[3] Ashburner, M., Ball, C.A., Blake, J.A., Botstein, D., Butler, H., Cherry, J.M., Davis, A.P., Dolinski, K., Dwight, S.S., Eppig, J.T., Harris, M.A., Hill, D.P., Issel-Tarver, L., Kasarskis, A., Lewis, S., Matese, J.C., Richardson, J.E., Ringwald, M., Rubin, G.M., Sherlock, G.: Gene ontology: tool for the unification of biology. Nature Genetics 25(1), 25–29 (2000)
[4] Baker, E.J., Jay, J.J., Bubier, J.A., Langston, M.A., Chesler, E.J.: GeneWeaver: a web-based system for integrative functional genomics. Nucleic Acids Research (November 2011)
[5] Becker, K.G., Barnes, K.C., Bright, T.J., Wang, S.A.: The genetic association database. Nature Genetics 36(5), 431–432 (2004)
[6] Bubier, J., Chesler, E.: Accelerating discovery for complex neurological and behavioral disorders through systems genetics and integrative genomics in the laboratory mouse. Neurotherapeutics, 1–11 (2012)
[7] Chen, C., Mungall, C.J., Gkoutos, G.V., Doelken, S.C., Köhler, S., Ruef, B.J., Smith, C., Westerfield, M., Robinson, P.N., Lewis, S.E., Schofield, P.N., Smedley, D.: MouseFinder: candidate disease genes from mouse phenotype data. Human Mutation 33(5), 858–866 (2012)
[8] Chen, J., Bardes, E.E., Aronow, B.J., Jegga, A.G.: ToppGene suite for gene list enrichment analysis and candidate gene prioritization. Nucleic Acids Research 37, W305–W311 (2009)
[9] Davis, A.P., Wiegers, T.C., Rosenstein, M.C., Mattingly, C.J.: MEDIC: a practical disease vocabulary used at the comparative toxicogenomics database. Database: The Journal of Biological Databases and Curation 2012 (February 2012)
[10] Fernald, G.H., Capriotti, E., Daneshjou, R., Karczewski, K.J., Altman, R.B.: Bioinformatics challenges for personalized medicine. Bioinformatics 27(13), 1741–1748 (2011)

[11] Gardner, D., Akil, H., Ascoli, G.A., Bowden, D.M., Bug, W., Donohue, D.E., Goldberg, D.H., Grafstein, B., Grethe, J.S., Gupta, A., Halavi, M., Kennedy, D.N., Marenco, L., Martone, M.E., Miller, P.L., Müller, H., Robert, A., Shepherd, G.M., Sternberg, P.W., Van Essen, D.C., Williams, R.W.: The neuroscience information framework: a data and knowledge environment for neuroscience. Neuroinformatics 6(3), 149–160 (2008)
[12] McKusick-Nathans Institute of Genetic Medicine, J.H.U.B.: Online mendelian inheritance in man, OMIM®, http://omim.org
[13] Gentleman, R.: Visualizing and distances using GO (2005), http://bioconductor.fhcrc.org/packages/2.11/bioc/vignettes/GOstats/inst/doc/GOvis.pdf
[14] Hibbs, M.A., Hess, D.C., Myers, C.L., Huttenhower, C., Li, K., Troyanskaya, O.G.: Exploring the functional landscape of gene expression: directed search of large microarray compendia. Bioinformatics 23(20), 2692–2699 (2007)
[15] Homayouni, R., Heinrich, K., Wei, L., Berry, M.W.: Gene clustering by latent semantic indexing of MEDLINE abstracts. Bioinformatics 21(1), 104–115 (2005)
[16] Hubert, L., Arabie, P.: Comparing partitions. Journal of Classification 2(1), 193–218 (1985)
[17] Jaccard, P.: Étude comparative de la distribution florale dans une portion des alpes et des jura. Bulletin de la Société Vaudoise des Sciences Naturelles 37, 547–579 (1901)
[18] Kreek, M., Nielsen, D., LaForge, K.: Genes associated with addiction: alcoholism, opiate, and cocaine. NeuroMolecular Medicine 5(1), 85–108 (2004)
[19] Lord, P.W., Stevens, R.D., Brass, A., Goble, C.A.: Investigating semantic similarity measures across the gene ontology: the relationship between sequence and annotation. Bioinformatics 19(10), 1275–1283 (2003)
[20] Perez-Iratxeta, C., Bork, P., Andrade, M.A.: Association of genes to genetically inherited diseases using data mining. Nature Genetics 31(3), 316–319 (2002)
[21] Pesquita, C., Faria, D., Bastos, H., Falcão, A., Couto, F.: Evaluating go-based semantic similarity measures. In: Proc. 10th Annual Bio-Ontologies Meeting, pp. 37–40 (2007)
[22] Pesquita, C., Faria, D., Bastos, H., Ferreira, A.E., Falcão, A.O., Couto, F.M.: Metrics for GO based protein semantic similarity: a systematic evaluation. BMC Bioinformatics 9(S5), S4 (2008)
[23] Rand, W.M.: Objective criteria for the evaluation of clustering methods. Journal of the American Statistical association, 846–850 (1971)
[24] Resnik, P.: Using information content to evaluate semantic similarity in a taxonomy. In: Proceedings of the 14th International Joint Conference on Artificial Intelligence, pp. 448–453 (1995)
[25] Robinson, P.N., Köhler, S., Bauer, S., Seelow, D., Horn, D., Mundlos, S.: The human phenotype ontology: A tool for annotating and analyzing human hereditary disease. The American Journal of Human Genetics 83(5), 610–615 (2008)
[26] Tiffin, N., Adie, E., Turner, F., Brunner, H.G., van Driel, M.A., Oti, M., Lopez-Bigas, N., Ouzounis, C., Perez-Iratxeta, C., Andrade-Navarro, M.A., Adeyemo, A., Patti, M.E., Semple, C.A.M., Hide, W.: Computational disease gene identification: a concert of methods prioritizes type 2 diabetes and obesity candidate genes. Nucleic Acids Research 34(10), 3067–3081 (2006)
[27] Tranchevent, L., Capdevila, F.B., Nitsch, D., Moor, B.D., Causmaecker, P.D., Moreau, Y.: A guide to web tools to prioritize candidate genes. Briefings in Bioinformatics 12(1), 22–32 (2011)

Exploiting Dependencies of Patterns in Gene Expression Analysis Using Pairwise Comparisons*

Nam S. Vo and Vinhthuy Phan**

Department of Computer Science, The University of Memphis,
Memphis, TN 38152, USA
{nsvo1,vphan}@memphis.edu

Abstract. In using pairwise comparisons to analyze gene expression data, researchers have often treated comparison outcomes independently. We now exploit additional dependencies of comparison outcomes to show that those with a certain property cannot be true patterns of genes' response to treatments. With this result, we leverage p-values obtained from comparison outcomes to predict true patterns of gene response to treatments. Functional validation of gene lists obtained from our method yielded more and better functional enrichment than those obtained from the conventional approach. Consequently, our method promises to be useful in designing cost-effective experiments with small sample sizes.

Keywords: gene expression, pairwise comparison, sample size, partially ordered set.

1 Introduction

Various methods using pairwise comparisons (e.g. [4, 7, 11, 13]) have been introduced to analyze gene expression data. The goal is to compare the expression of genes at various time points, or when treated with various chemical compounds. In case of studies involving multiple treatments of chemical compounds, researchers are interested in understanding not only the effects of certain drugs (compared to untreated) but also the differences and similarities among the drugs themselves. In the simplest case, one would like to know the effects of treatment t versus untreated (control). Pairwise comparisons would identify genes that are up-regulated or down-regulated by treatment t. But as the number of treatments increases, it becomes harder to interpret patterns observed from pairwise comparisons. Additionally, instead of making $\binom{n}{2}$ measurements (n is the number of treatments), to save cost, researchers such as Sutter et al. [13] made only n measurements and then used *post hoc* computations to determine the outcomes of $\binom{n}{2}$ comparisons. In exploring dependencies among $\binom{n}{2}$ comparisons under a

---

* The rights of this work are transferred to the extent transferable according to title 17 U.S.C. 105.
** Corresponding author.

Z. Cai et al. (Eds.): ISBRA 2013, LNBI 7875, pp. 173–184, 2013.
© Springer-Verlag Berlin Heidelberg 2013

simple model of comparison, Longacre et al. [10] demonstrated that only $2n - 1$ measurements would be needed to compute results of all $\binom{n}{2}$ comparisons.

Researchers have recognized that each treatment requires many samples measured independently to account for biological variation. Having a sufficient *sample size* for accurate measure of gene expression is critical for both microarray and RNA-seq technologies [8, 5]. Statistical methods such as [9, 15], calculate a sample size with respect to a prescribed portion of type I/type II error rates, or to ensure that a given proportion of genes are significantly differentially expressed. Consequently, a characteristic of these high-throughput technologies is that one sample size is calculated for all genes; measurements taken from 5 microarrays would mean than all studied genes have exactly 5 samples. Having one sample size for all genes means that observed patterns of highly-variantly-expressed genes might not be accurate.

We introduce a method that employs pairwise comparisons and *post hoc* calculations (similarly to [13, 11]) to address the problem of having one sample size for all genes in gene expression analysis. The first novel aspect of our method is in the characterization of pattern of response to treatments of each gene as a *strict partially ordered set* (poset). Using this characterization, the main result indicates that, under some reasonable assumptions, true response patterns must be *linearly orderable* posets and that many observed patterns might not be linearly orderable because of small sample sizes. Consequently, we devise a strategy to predict most likely (least erroneous) linearly orderable extensions of those patterns that are not linearly orderable. The rationale is that these linearly orderable extensions are most likely true patterns. We validated this strategy by scoring functional enrichment of resulting gene lists. As a result, this approach can help design more effective and efficient experiments.

2   Method

2.1   Gene Response Patterns Obtained from Pairwise Comparisons

Suppose that we are given a set of significantly expressed genes and must determine the patterns of response of these genes; gene expression may be collected using procedures such as Kruskal-Wallis with permutation resampling [3] and false discovery rate control [1, 2]. The approach of using pairwise comparisons to characterize the pattern of gene response to treatments has been long established [10, 4, 7, 11, 13, 14]. In the simplest case, we have 2 treatment groups, for example, a *control group* c and a treatment group t. To determine which genes are up-regulated ($t > c$), down-regulated ($t < c$) or unaffected by t, we compare two groups of samples $\{c_1, \cdots, c_r\}$ and $\{t_1, \cdots, t_s\}$. When there are many treatments, we seek to understand the effects being treated versus untreated as well as differences and similarities among different treatments. This makes it harder to describe and analyze patterns of comparison outcomes.

Additionally, researchers showed that in order of obtain accurate patterns, sample sizes (r and s) must be sufficiently large to account for technical and biological variations; this is necessary for both microarray [8] and RNA-seq [5] technologies. Statistical tests such as MannWhitney U test or Wilcoxon rank-sum test can be used to compare two groups of samples. The basic idea is to order samples from both groups in a non-decreasing order and assign each sample the corresponding rank in this order. The observed sum of ranks, w_t, of samples in the t group will tell us if samples in t are statistically larger or smaller than those in c. More specifically, if the p-value[1] (i.e. the probability of observing a rank-sum for t that is at most w_t) is less than a false positive rate α, then the alternative hypothesis $H_1 : t < c$ is accepted. Conventionally, α is set to be 0.05 or even 0.01; the smaller α is, the less likely we make a mistake.

In general, given expression data of a given gene for two treatment groups A and B, comparing A versus B yields one of three possible outcomes (patterns):

$A \prec B$: the gene responds more significantly to A than to B if the pairwise test $H_0 : A = B$ is rejected in favor of $H_1 : A < B$.

$B \prec A$: the gene responds more significantly to B than to A if the pairwise test $H_0 : A = B$ is rejected in favor of $H_1 : A > B$.

$A \sim B$: If H_0 is accepted because the sample sizes are too small to sufficiently determine the order at the given α.

Note that the outcome of a comparison is a function of two things: (1) the sample sizes of the two groups, and (2) α. Larger α's yield more outcomes of types $A \prec B$ and $B \prec A$, but having larger α's also means higher error rates. We will show how to vary α intelligently to predict true patterns.

2.2   Comparison Outcomes Are Strict Partially Ordered Sets (Posets)

Given a gene g treated with n measurements for n treatments, define the pattern P_g of g to be the collective outcomes obtained from $\binom{n}{2}$ *post hoc* comparison tests on all different treatment pairs. Each comparison yields one of three outcomes as described above. Note that P_g is a (strict) partially ordered set, with respect to the relation $\prec$ defined by the comparison test procedure. This is because a valid test procedure, e.g. the Wilcoxon rank-sum test, yields outcomes that are:

1. Antisymmetic: either $A \prec B$ or $B \prec A$, but not both, may be obtained.
2. Transitive: if $A \prec B$ and $B \prec C$ are obtained, then $A \prec C$ must be obtained.

Our goal is to exploit the structure of P_g to determine if it is accurate and if not, what the true pattern might be. To do this, we make a few assumptions. First, given any two treatments A and B, a gene must either (1) it must respond more (higher expression) to B than to A, or (2) to more to A than to B, or (3) identically to both. Second, for the first two cases, we can determine the true response pattern ($A \prec B$ or $B \prec A$) with a sufficiently large (but finite) sample size; and for the third case, no matter how large the sample size, only the outcome $A \sim B$ can be obtained.

---

[1] These p-values can be calculated exactly using a recursive procedure [12].

These assumptions mean that when the outcome $A \sim B$ is observed, either there are too few samples or the gene responds identically to both treatments. For only 2 treatments, it is impossible to tell which is the scenario. But for 3 or treatments, the following lemma can help identify the two cases.

Lemma 1. *Given treatments A, B, C, suppose Δ is the pattern defined by the following outcomes: $A \prec C, A \sim B, B \sim C$. Then Δ is not the true pattern of response to the 3 treatments due to an insufficiently small sample size.*

Proof. Suppose, to the contrary, that the number of samples are sufficiently large and the pattern Δ is a true pattern. As the number of samples is sufficient, we are able to resolve all differences among treatments if such differences exist. And if no difference exists between two treatment groups, then they must be identical. Therefore, the outcomes $A \sim B$ and $B \sim C$ mean that $A \equiv B$ and $B \equiv C$ (i.e. the gene responds identically to A and B, and to B and C, respectively). But, this would imply that $A \equiv C$, which is a contradiction to the observed outcome $A \prec C$. Therefore, Δ is not true, and further we need more samples to determine accurately either the outcome A versus B, or B versus C, or both.

2.3   Linearly Orderable Patterns

The same reasoning as in Lemma 1 shows that if a pattern P_g – obtained from pairwise comparing n treatment groups – contains Δ as a sub-pattern, then P_g is not a true pattern and more samples would be needed to obtain the true response pattern of g. This can be seen more easily in a visual representation. Fig 1a shows Δ. Fig 1b shows a pattern that contains Δ as a sub-pattern, which involves treatments a, b, and e. Thus, we can conclude, based on Lemma 1, that the pattern in Fig 1b is not a true response pattern (for any gene).

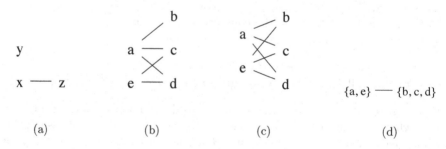

(a)	(b)	(c)	(d)

Fig. 1. Response patterns as posets. (a) $\Delta = \{x \prec z, x \sim y, y \sim z\}$ cannot be a true response pattern. (b) A not-true pattern as it contains Δ. (c) A linearly orderable pattern. (d) the linear order of the pattern in (c).

We cannot, however, disqualify the pattern shown in Fig 1c as untrue. The following definition helps to characterize patterns that are possibly true.

Definition 1. *A pattern P based on n elements $t_1, \cdots, t_n$ is* **linearly orderable** *if $\forall i, j$ such that $t_i \sim t_j$, $G(t_i) = G(t_j)$ and $L(t_i) = L(t_j)$.*

where $G(t_i) = \{t_k | t_i \prec t_k\}$ is the set of elements "larger" than t_i, and similarly $L(t_i) = \{t_k | t_k \prec t_i\}$ is the set of elements "smaller" than t_i.

Lemma 2. *A pattern is linearly orderable if and only if it does not contain $\triangle$.*

Proof. If a pattern P contains $\triangle$, e.g is the pattern in Fig 1a. We see $x \sim y$, but $\{z\} = G(x) \neq G(y) = \emptyset$, implying P is not linearly orderable. Conversely, suppose that P does not contain $\triangle$, then given any pair of x and y such that $x \sim y$, there cannot exist a z such that $x \prec z$ and $y \sim z$. This implies $G(x) = G(y)$. By symmetry, we can show that $L(x) = L(y)$. Thus, P is linearly orderable.

Lemmas 1 and 2 implies that

Theorem 1. *True patterns must be linearly orderable.*

If an observed pattern P is not linearly orderable, then it does not represent a true response. Although it must be linearly orderable, without additional samples, we cannot know what the true pattern is. If, however, we assume that already observed outcomes of types $A \prec B$ are correct (they are with high probabilities), then additional samples do not change these and consequently the true pattern must be an *extension* of P. To be precise, we introduce a definition.

Definition 2. *Q is a* **linearly orderable extension** *of P if (1) Q is linearly orderable, and (2) $\forall i, j$ if P contains $t_i \prec t_j$, then Q also contains $t_i \prec t_j$.*

For instance, if the pattern in Fig 1c is observed, then the true pattern are likely among its 39 linearly orderable extensions. (Additional samples can result in any of 3 linearly orderable sets out of $\{a, e\}$; and any of 13 linearly orderable sets out of $\{b, c, d\}$. This gives a total of 3*13=39 combinations.)

2.4   Determining True Response Patterns Using *ad hoc* Thresholds

As true patterns must be linearly orderable, a linearly orderable pattern obtained with least assumptions might most likely be true patterns. To see the important role of α, recall that in determining the outcome for A versus B, if $H_1 : A < B$ is accepted with p-value $\leq \alpha$, then the outcome is $A \prec B$. Otherwise, if $H_1 : B < A$ is accepted with p-value $\leq \alpha$, then the outcome is $B \prec A$. Otherwise, the outcome is $A \sim B$. For example, if the p-value for hypothesis $H_1 : A < B$ is found to be 0.07, and $\alpha = 0.1$, then the outcome $A \prec B$ is obtained. On the other hand, if $\alpha = 0.05$, then the outcome $A \sim B$ is likely obtained.

Conventionally, pairwise comparisons are tested independently and α is set to a fixed small number (e.g. 0.05) to reduce false positives. But, if we consider altogether the collective outcomes of $\binom{n}{2}$ pairwise comparisons, we can exploit the fact that true response patterns must be linearly orderable. Thus, instead of using a global, fixed α for all genes, we could gain more by using an *ad hoc* α that works best for each individual gene.

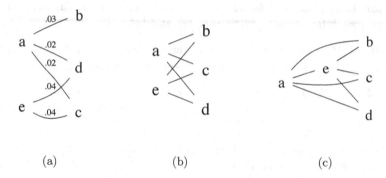

(a)   (b)   (c)

Fig. 2. (a) Pattern shown in Fig 1b with outcomes labeled by p-values, (b)-(c) Two of 39 possible linearly orderable extensions of the pattern shown in (a)

An example shown in Fig 2 helps explain our strategy. Fig 2a shows the pattern previously shown in Fig 1b with some additional information. Each outcome of type $A \prec B$ is labeled with the p-value at which the outcome is obtained. For example, the outcome $a \prec b$ is obtained with p-value 0.03, and $a \prec d$ is obtained with p-value 0.02. With $\alpha = 0.05$, the pattern in Fig 2a is observed. It is not linearly orderable and hence is not a true pattern; the true pattern is likely among the 39 linear orderable extensions of this pattern. *But which one among these is most likely the true pattern?*

Consider a hypothetical scenario in which the p-value for $H_1 : e < b$ is 0.09. If this p-value is the smallest p-value that is greater than 0.05, we can deduce that the pattern in Fig 2b is most likely to be true. Thus, setting $\alpha = 0.09$, we would observe correctly this linearly orderable pattern. In addition, suppose that the p-value for $H_1 : a < e$ is 0.07. Then, the pattern in Fig 2c is more likely to be true. Essentially, our strategy, as described in Algorithm 1, is to **find the minimal threshold α that yields a non-trivial linearly orderable pattern.**

For a given gene g, this procedure increments the *ad hoc* threshold α_g (initially set at α_{min}) until a non-trivial linearly orderable pattern is obtained. The procedure returns both α_g and the linearly orderable pattern. If there are too few samples, it is not possible to determine the order of any treatment pair. For large numbers of treatments, it is necessary to start with a very small α_{min} to control false positives. Techniques to control false positives and false discovery rates such as Bonferrroni or Benjamini-Hochberg [1, 2] require hypothesis tests be independent. These methods, unfortunately, do not apply because as we have shown these tests' outcomes are highly dependent of each other.

3   Experimental Results

3.1   Data and Method of Validation

To validate our method, we used a gene expression data set (GEO accession number GSE8880 [11]), which came from a controlled study of samples of livers of Sprague-Dawley rats treated with either control diet or one of three

---

Algorithm 1. GetPattern(gene g)

1: Given each pair of treatment groups A and B for the gene g, compute two p-values
   p_1 and p_2 for the alternative hypotheses $H_1 : A < B$ and $H_2 : B < A$.
2: Sort L, the list of these $2 \cdot \binom{n}{2}$ p-values, in increasing order (duplicates removed).
3: $\alpha_{\min} = 0.05$.
4: **if** $\alpha_{\min} > L[m]$, where m is the last index of L **then**
5:    **return** $(\emptyset, L[m])$
6: **else**
7:    j be the index such that $L[j] \leq \alpha_{\min} < L[j+1]$; or $j = 1$ if $\alpha_{\min} < L[1]$
8: **end if**
9: **for** $i = j$ to m **do**
10:    $\alpha_g = L[i]$
11:    Let P^{α_g} be the pattern obtained from having the threshold α_g.
12:    **if** P^{α_g} is not $\emptyset$ and is linearly orderable **then**
13:       **return** (P^{α_g}, α_g)
14:    **end if**
15: **end for**
16: **return** $(\emptyset, \alpha_g)$

---

chemopreventive compounds with well understood pharmacological activities, 5,6-benzoflavone (BNF), 3H-1,2-dithiole-3-thione (D3T) and 4-methyl-5-pyrazinyl-3H-1,2-dithiole-3-thione (OLT). There were 5 samples in each of 4 treatment groups. Pairwise comparisons were performed to determine patterns of response for 1737 significantly differentially expressed genes.

To compare our approach (Algorithm 1) of using an *ad hoc* threshold for each gene against the conventional approach of having a fixed global threshold α, we placed genes into *lists* that are labelled by those genes' patterns; genes having the same patterns are placed into the same lists. We did not use any conventional clustering methods (e.g. hierarchical clustering or k-mean) because we did not want to introduce additional biases from any clustering method.

Thus, we have two sets of poset-labelled gene lists. The first set was produced by observing patterns using a fixed threshold α (0.05). We observed that 1252 (72%) genes acquired 45 linearly orderable patterns and were placed into 45 gene lists; 485 (28%) genes acquired 69 non-linearly orderable patterns and were placed into 69 gene lists. The second set of clusters were produced by observing linearly orderable patterns using *ad hoc* α for each individual gene (Algorithm 1). Using this method, all 1737 significantly differentially expressed genes acquired a total of 55 linearly orderable patterns and were placed into 55 gene lists. These 55 linearly orderable patterns include all 45 linearly orderable patterns observed with fixed α at 0.05, plus 10 new patterns.

We evaluated and compared the functional enrichment of the two sets of gene lists using DAVID [6], which is a resource aimed at systematically extracting biological meaning from large gene or protein lists. DAVID integrates biological information from all major public bioinformatics resources. We used the Gene Functional Classification tool of DAVID to extract highly-enriched clusters from each gene list. To quantify the degree of enrichment, we counted the number of functionally enriched clusters and calculated the total as well as average enrichment score that DAVID returned. We expect biologically meaningful assignments of patterns to genes will yield many and better functionally enriched clusters.

3.2   Analysis of Linearly Orderable Extensions

As described, fixing α at 0.05 resulted in 485 genes acquiring 69 non-linearly orderable patterns. By relaxing α, our method reassigned the pattern of each of these 485 genes to its most likely linearly orderable extension. We observe that many linearly orderable extensions can be obtained by modestly raising α beyond 0.05. Shown in Fig 3a, with $\alpha = 0.05$, patterns of 72% of genes linearly orderable; with $\alpha \leq 0.075$, patterns of 84% of genes were linearly orderable; with $\alpha \leq 0.15$, 97% of genes had linearly orderable patterns.

Fig 3a shows an important aspect of our method. Conventionally, it would be unthinkable to set α at something higher than 0.15 as there would be too many false positives. Our method allow flexible α, but in a strategically *progressive* manner. For this data set, patterns for 72% genes were determined with α set at 0.05. Patterns for the next 12% (for a total of 84%) of genes were determined with α at 0.075. And the next 13% (total of 97%) were determined with α at 0.15. Thus, although our method considers large values of α's, this is done only for very few genes whose patterns remain not-linearly orderable at smaller values.

To compare the structural difference between patterns not linearly orderable when α is 0.05 and their linearly orderable extensions (as obtained by our method), we define the difference between two pattern P and Q as $d(P, Q) = \sum_{i=1}^{\binom{n}{2}} \delta(p_i, q_i)$, where $\delta(p_i, q_i)$ is 0 if the ith outcome of P and Q is the same, and 1 if it is different. For example, if P and Q are the patterns shown in Fig 1b and 1c, then $d(P, Q) = 1$ as they differ only in the comparison of b and e.

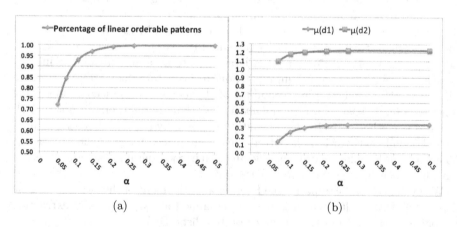

Fig. 3. (a) fraction of linearly orderable patterns at increasing values of α. (b) structural difference between patterns acquired at $\alpha = 0.05$ and at higher values.

Fig 3b shows the structural difference (on average) between a pattern observed at $\alpha = 0.05$ and its linearly orderable extension at increasing values of α. This difference is denoted in the figure as $\mu(d1)$. As this number is well below 0.4 for all values of α, we see that on average a pattern is only very slightly different from its linearly orderable extension. This number, however, does not tell the whole story, because 72% of genes acquired linearly orderable patterns at $\alpha = 0.05$ (meaning they are trivially their own linearly orderable extensions).

Thus, we proceed to analyze patterns that were not linearly orderable at $\alpha = 0.05$. The structural difference between these patterns and their linearly orderable extensions is denoted in the figure as $\mu(d2)$. We see that this number is roughly 1.2 at all values of α. This means that on average adding roughly 1 outcome (of type $A \prec B$) to these patterns would make them linearly orderable.

Given that patterns determined with α fixed at 0.05 and patterns determined by our method do not differ structurally very much, *is there any resulting biological difference?* We examine this question next.

3.3   Functional Analysis Using DAVID

Validating against Random Assignment of Patterns. As described in Algorithm 1, patterns not linearly observable with α fixed at 0.05 are extended to linearly orderable patterns. This is equivalent to taking these patterns and strategically assign them with certain linearly orderable patterns. We demonstrate that this strategy is biologically meaningful by comparing it to a random strategy. Suppose that our method assigned all 485 not-linearly-orderable patterns to k linearly orderable patterns $P_1, \cdots, P_k$; this would yield k corresponding gene lists $L_1, \cdots, L_k$ (L_i includes all genes having pattern P_i). To compare, we created k lists $R_1, \cdots, R_k$ by randomly placing genes into these lists in such a way that P_i has the same number of genes as R_i (i.e. $|P_i| = |R_i|$) for all i.

Then, the gene lists $L_1, \cdots, L_k$ and $R_1, \cdots, R_k$ would be given to DAVID for functional analyses. (Technically, these lists only consist of genes whose patterns are not linearly orderable at $\alpha = 0.05$. Together with these, we also gave to DAVID the lists of genes whose patterns are linearly orderable at $\alpha = 0.05$.) In response, for each gene list, DAVID would return a non-negative number enriched clusters; each of these enriched cluster is also given an enrichment score.

Table 1 shows the results of comparing our method against the random assignment strategy. We see that compared to the random strategy, the gene lists based on linearly orderable extensions yield more and larger enriched clusters

Table 1. Our method versus random assignment. μ_R and σ_R are mean and standard deviation of the random assignment strategy.

	Our method	μ_R	σ_R
No. of enriched clusters	23	16.60	1.90
No. of genes in enriched clusters	119	88.90	7.87
Total enrichment score	46.71	28.70	4.86

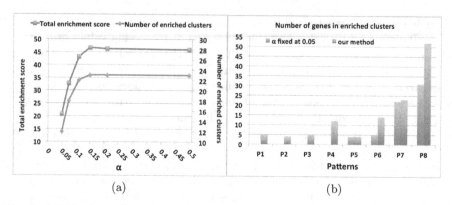

Fig. 4. (a) Total enrichment scores and number of enriched clusters for gene lists obtained at various values of α. (b) Number of genes in enriched clusters obtained by our method vs. fixing α at 0.05.

as well as higher total enrichment score. (We averaged the score of the random strategy over 20 different runs.) The fact that our method performed about 3 standard deviations better than random suggests that the strategy of assigning linearly orderable extensions is biologically meaningful.

Comparing with Assignment of Patterns with Fixed α. Ultimately, we expect that allowing a flexible α to obtain linearly orderable patterns yields better prediction of true response patterns. As mentioned in Section 3.1, fixing α at 0.05 yielded 45 gene lists, each of which corresponded to a linearly orderable pattern, and 69 gene lists, each of which corresponded to a not-linearly orderable pattern. On the other hand, Algorithm 1 yielded 55 gene lists corresponding to linearly orderable patterns. If a gene list is biologically meaningful, it should be functionally enriched, and we would expect such result from DAVID.

We found that for gene lists obtained with α fixed at 0.05, DAVID retrieved 12 enriched clusters, with a total enrichment score approximately 20.8. Meanwhile, as shown in Fig 4a, for gene lists obtained from our method, DAVID retrieved up to 23 enriched clusters with a total enrichment score of 46.70. Thus, gene lists produced by our method are twice as enriched. Additionally, Fig 4b shows functional enrichment of 8 gene lists, corresponding to 8 linearly orderable patterns. Genes having these 8 patterns were observed both with $\alpha = 0.05$ and with our method. The figure shows that the reassignment of genes to these 8 patterns (Algorithm 1) increased functional enrichment noticeably.

Fig 4a also shows the number of enriched clusters and total enrichment scores for gene lists obtained at various α ranging from 0.05 to 0.5. We can see that gene lists obtained with $\alpha \leq 0.15$ were most enriched. This suggests that genes whose patterns are not linearly orderable at $\alpha \leq 0.15$ did not add to the functional enrichment of the existing gene lists, and might be removed from consideration. With other data sets, it might be necessary to impose an upper bound on α, e.g. 0.15, to reduce false positives. This result suggests that doing so does not negatively affect functional enrichment very much if at all.

4   Conclusion and Discussion

The proposed approach to gene expression data analysis exploits dependencies in outcomes of pairwise comparing gene response to treatments. We showed that observed patterns are true if and only they are *linearly orderable*. For genes that have too few samples, linearly orderable extensions of their observed patterns might be inferred as true patterns. This process is accomplished by allowing the error rate α to be flexible instead of being fixed at a constant. We demonstrated that a small α (≤ 0.15) are sufficient to obtain linearly orderable patterns for most (97%) genes. Importantly, we demonstrated that this method yielded more and larger enriched gene lists in comparison to conventional analysis that holds α at a constant value of 0.05, thus validating the effectiveness of the method.

The method should be useful in helping designing cost-effective experiments. Statisticians continually introduce sample-size calculations that guarantee low type I & type II errors while retrieving as many significantly differentially expressed genes as possible. And yet, there is an inherent characteristic in high-throughput technologies that measure expressions of tens of thousands of genes in one batch: *one sample size is used for all genes*. Thus, large sample sizes yield accurate patterns for many genes, but are expensive. Small sample sizes, however, do not yield true patterns for highly variantly expressed genes.

This method suggests middle ground, whereby we only need to have a reasonable sample size that yields true patterns for most genes. For highly-variantly expressed genes, their patterns are supposedly not linearly orderable. Nevertheless, their linearly orderable extensions can be predicted using the proposed technique. These linearly orderable extensions are more likely to be true patterns of response. Using functional enrichment as the basis of biological validation, we have shown that this technique has a better chance of making correct predictions.

Finally, partially ordered sets (posets) as patterns can serve as a *pre-filtering* step for subsequent clustering analysis. They can also serve as meaningful labels of gene lists, as we did in this paper. These posets describe precisely interactions of how genes having such patterns respond to treatments. They provide a level of expressiveness beyond the simplistic upregulation/downregulation description.

Acknowledgement. We would like to thank the anonymous reviewers for their insightful comments.

References

1. Benjamini, Y., Hochberg, Y.: Controlling the false discovery rate: a practical and powerful approach to multiple testing. J. R. Statist. 57(1), 289–300 (1995)
2. Benjamini, Y.: Discovering the false discovery rate. Journal of the Royal Statistical Society: Series B (Statistical Methodology) 72(4), 405–416 (2010)
3. Davidson, A.C., Hinkley, D.V.: Bootstrap methods and their application. Cambridge University Press, Cambridge (1997)

4. Geman, D., d'Avignon, C., Naiman, D., Winslow, R.: Classifying gene expression profiles from pairwise mrna comparisons. Statistical Applications in Genetics and Molecular Biology 3(article19) (2004)
5. Glaus, P., Honkela, A., Rattray, M.: Identifying differentially expressed transcripts from rna-seq data with biological variation. Bioinformatics 28(13), 1721–1728 (2012)
6. Huang, D., Sherman, B., Lempicki, R.: Systematic and integrative analysis of large gene lists using david bioinformatics resources. Nature Protocols 4(1), 44–57 (2008)
7. Hulshizer, R., Blalock, E.M.: Post hoc pattern matching: assigning significance to statistically defined expression patterns in single channel microarray data. BMC Bioinformatic 8, 240 (2007)
8. Lee, M.L., Kuo, F.C., Whitmore, G.A., Sklar, J.: Importance of replication in microarray gene expression studies: statistical methods and evidence from repetitive cdna hybridization. Prot. Natl. Acad. Sci. 97(18), 9834–9839 (2000)
9. Lin, W.J., Hsueh, H.M., Chen, J.J.: Power and sample size estimation in microarray studies. BMC Bioinformatics 11, 48–48 (2010)
10. Longacre, A., Scott, L., Levine, J.: Linear independence of pairwise comparisons of dna microarray data. J. Bioinform. Comput. Biol. 3(6), 1243–1262 (2005)
11. Phan, V., George, E.O., Tran, Q.T., Goodwin, S., Bodreddigari, S., Sutter, T.R.: Analyzing microarray data with transitive directed acyclic graphs. Journal of Bioinformatics and Computational Biology 7(1), 135–156 (2009)
12. Ross, M.S.: Simulation, 3rd edn. Academic Press, San Diego (2002)
13. Sutter, T.R., He, X.R., Dimitrov, P., Xu, L., Narasimhan, G., George, E.O., Sutter, C.H., Grubbs, C., Savory, R., Stephan-Gueldner, M., Kreder, D., Taylor, M.J., Lubet, R., Patterson, T.A., Kensler, T.W.: Multiple comparisons model-based clustering and ternary pattern tree numerical display of gene response to treatment: procedure and application to the preclinical evaluation of chemopreventive agents. Mol. Cancer Ther. 1(14), 1283–1292 (2002)
14. Tran, Q.T., Xu, L., Phan, V., Goodwin, S., Rahman, M., Jin, V., Sutter, C.H., Roebuck, B., Kensler, T., George, E.O., Sutter, T.R.: Chemical genomics of cancer chemopreventive dithiolethiones. Carcinogenesis 30(3), 480–486 (2009)
15. van Iterson, M., 't Hoen, P.A., Pedotti, P., Hooiveld, G.J., den Dunnen, J.T., van Ommen, G.J., Boer, J.M., Menezes, R.X.: Relative power and sample size analysis on gene expression profiling data. BMC Genomics 10(1), 439 (2009)

Cloud Computing
for *De Novo* Metagenomic Sequence Assembly

Xuan Guo, Xiaojun Ding, Yu Meng, and Yi Pan

Department of Computer Science
Georgia State University
Atlanta, GA, 30303
pan@cs.gsu.edu

Abstract. In metagenomics, the population sequencing is an approach to recover the genomic sequences in the genetically diverse environment. Combined with the recently developed next generation sequencing platform, mategenomics data analysis has greatly enlarged the size of sequencing datasets and decreased the cost. The complete and accurate assembly of sequenced reads from an environmental sample improves the efficiency of genome functional and taxonomical classification. A common bottleneck of the available tools is the high computing requirement for efficiently assembling vast amounts of data generated from large-scale sequencing projects. To address these limitations, we developed a parallel strategy to accelerate computation and boost accuracy. We also presented an instance of this strategy for a state-of-the-art assembly tool, Genovo, on Apache hadoop platform. As a demonstration of the capability of our approach, we compared the performance of our method to two other short read assembly programs on a series of synthetic and real datasets created using the 454 platform, the largest of which has 683k reads. Under the parallel strategy, the ability of reconstruction of bases outperformed other tools both on speed and several assembly evaluation metrics.

Keywords: Metagenome, NGS, Sequence Assembly, Cloud Computing.

1 Introduction

Metagenomics is the study of all micro-organisms coexistent in an environment area, including environmental genomics, ecogenomics or community genomics. In the past, microbial genomic studies usually focused on one single individual bacterial strain which is suitable to be separately cultivated [1]. Due to experimental limitations, most of the microbes could not be isolated in laboratory conditions and, in fact, all micro-organisms in a habitat have various functional effects on one another and their hosts, such as human gut [2] and larger ecosystems [3]. Furthermore, recent research [4] [5] has shown the strong association between common diseases and the diversity of microbes in humans, like inflammatory bowel disease and gastrointestinal disturbance. Population sequencing is an essential tool for recovering the genomic sequences in a genetically diverse

Z. Cai et al. (Eds.): ISBRA 2013, LNBI 7875, pp. 185–198, 2013.
© Springer-Verlag Berlin Heidelberg 2013

environmental sample, which is known as metagenomics and also known as environmental genomics or community genomics. These studies are fundamental for identifying and discovering novel genes, studying ecosystems by utilizing other bioinformatics tools, like multiple sequence alignment services [6] and hence advancing our systemic understanding of biological processes and communities.

With high-throughput next generation sequencing (NGS) technology, like the 454 technology, it becomes possible to give solutions for metagenomics analysis[7]. The high coverage of fragmented reads increased the size of sequencing datasets for large projects. Current methods for metagenomics sequencing recovery are deficient in both the scale of data they can handle and the quality of assembly contigs, particularly for terascale metagenomics projects. For example, 454 Roche GS Titanium system could collect 200 Mbp to 300 Mbp data within five hours. As far as we know, there are a few *de novo* assemblers aimed at metagenomics [8], like MetaVelet [9], Genovo [10] and MetaIDBA [11], and these assemble tools have been plagued by the issues mentioned above. Our approach was to utilize recent high-performance computing (HPC) technology–cloud computing. One of the advantages of cloud based technologies is that execution environment and experiment conditions can be easily and completely customized; the large distributed infrastructures could be accessed by non-distributed computing experts. Moreover, since the infrastructure is rented on a pay-per-use basis, when the experiments have finished, immediate access and release to required resources become possible without planning beforehand. For the parallel assembly procedure, we designed a clustering and merging strategy to execute adjusted Genovo software on the cloud environment.

In this paper, we designed a cloud based *de novo* metagenomic assembly algorithm to address the aforementioned problems. The rest of the paper is organized as follows: a briefly introduction of MapReduce model and sequencing assembly problem was given in Section 2; our framework of the cloud based assembly algorithm was described in Section 3 with the design of Map-Reduce model implantation; experiment results of the reads clustering was listed in Section 4; Section 5 gave our conclusion and future works.

2   Preliminaries

In this section, we briefly reviewed the MapReduce model, and some basic concepts, notations of the short reads *de novo* assembly problem for metagenomes.

2.1   The Map-Reduce Model

Map-Reduce is a programming model for processing and generating large data sets in a distributed computing environment. It was first proposed and implemented by Google. Some instances of MapReduce implementations, like Apache Hadoop, removed the burden of managing I/O explicitly and hidden its low-level details from programmers. Thus, map-reduce applications are amenable to a broad variety of real-world tasks and make them more accessible to users who are not HPC experts. Dean and Ghemawat [12] illustrated the MapReduce programming model as follows:

– A map function first takes a collection of user defined key/value pairs as input, executes a designed processes and outputs a collection of key/value pairs. A hidden sorting procedure will arrange pairs according to the order of their keys and allocate them to reducers.
– A reduce function receives a collection of key/value pairs, which share the same key, to conduct some calculations and finally yields results collected by the MapReduce framework.

Figure 1 shows the data flow and different phases of the MapReduce framework with only one round MapReduce iteration. In our design of parallel *De Novo* assembly strategy, multiple rounds of MapReduce iteration were employed, which meant the consequent task directly used the result from the previous job.

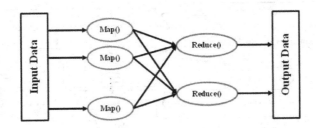

Fig. 1. The MapReduce programming model

2.2 *De Novo* Assembly of Metagenomes

The *De Novo* sequencing assembly problem is defined to find the shortest common superstring of a given set of sequences as follows: Given strings $\{s_1, s_2, \cdots\}$, find the shortest string T such that every s_i is a substring of T. For the metagenomes assembly, the target is to improve the accuracy of genetic functional and taxonomical classification by assembled contigs of genomes of multiple species Generally, there are four aspects which are required to pay attention to in metagenomes assembly: (1) the components of an environmental sample are complicated with related species different from high to low taxonomies; (2) genome alternations are uncertain, like horizontal gene transfer which is widespread among co-existent species; (3) the coverages of genomes are various owning to the unequal abundance of distinct species; (4) the number of reads in one dataset is in millions. The intuitive tactic to import those algorithms based on overlap-layout to parallel mode is to split and balance input short reads according to the ranking of overlap, which indicates the possibility of whether they come from the same part of the consensus sequence, then to execute transitional assembly software on each partition. In the present study, we addressed the problem of *de novo* short read metagenomes assembly by a parallel strategy to adjust and improve overlap-layout based method, which can be framed in MapReduce model to speed up the computation and enhance the assembly quality.

3   Algorithm for *De Novo* Assembly

In this section, we presented a parallel *De Novo* assembly method, which is a cloud based algorithm with four phases: weight-edge construction (Phase I) to separated sequenced reads into small groups using 4 Map-Reduce jobs, clustering (Phase II) to hierarchically cluster reads as potential contigs, parallel-Genovo (Phase III) to yield a set of candidate contigs by executing adjusted Genovo and post-merge (Phase IV) designed to combine the candidate contigs together if their overlaps meet a predefined requirement. Figure 2 outlines the four phases of our method.

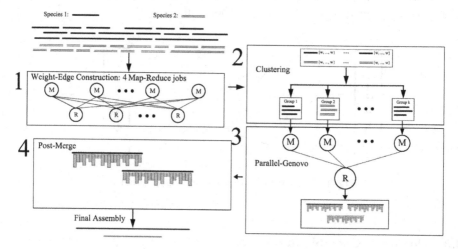

Fig. 2. The pipeline of Cloud *De Novo* assembly is divided into four major phases: Weight-Edge Construction, Clustering, Parallel-Genovo and Post-Merge.

3.1   Phase I: Weight-Edge Construction

Our algorithm for weight-edge construction was adapted from two alignment methods: ZEBRA algorithm [13], a fast scan rule to skip non-match reads, the local alignment algorithm, like Smith-Waterman (SW) dynamic programming algorithm [14]. Two steps were designed in Phase I: (1) scanning, that every l-mer (short DNA sub-strings of length l) of every read additionally with its location was store in a hash-position table HP, then an approximate offset for further pair sequences alignment operation was calculated if two reads shared at least K common l-mers and (2) aligning, that the approximate entry point from scanning were used to do the SW alignment of a pair of reads. In practice, we used the weak requirement for the parameter K that at least six common 10-mers shared by two reads were filter out as candidate alignment reads. Our solution adapted from ZEBRA algorithm and SW algorithm is given in Algorithm 1, which is written in a sequential way for easy understanding. The Map-Reduce implementation with 4 round Map-Reduce jobs will be discussed in next part. A threshold τ_{align} was used to remove bad alignments:

$$\tau_{align} = max\left(s_{mis}, s_{ins}, s_{del}\right) \times l_{align} \times 2p_{err} + s_{hit} \times l_{align} \times (1 - 2p_{err})$$
$$= l_{align}(s_{hit} + 2p_{err}(max(s_{mis}, s_{ins}, s_{del}) - s_{hit}))$$

This weak threshold could effectively reject those pairs of reads whose probability that both reads were sequenced from the same segment of genome were extremely low. It promised that the rest of pairs should be further considered to be connected.

Algorithm 1. Weight-Edge Construction

Input: Reads set $R = \{r_i\}, i \in [0, |R|]$
Output: Weight array $W = \{W_i\}$ (with $i \in [0, |R|]$)
1   Maintain a hash-position table $HP = \left\{ \left\langle h_1, \left\{ p_{r_j}^{h1}, r_j, \dots \right\} \right\rangle, \left\langle h_2, \left\{ p_{r_j}^{h2}, r_j, \dots \right\} \right\rangle \dots \right\}$;
2   **for** $r_i \in R$ **do**
3   hash every constituent l-mer to a integer h and store the position of $p_{r_i}^h$ and the id of r_i in HP ;
4   **end**
5   **for** $h_j \in HP$ **do**
6   For all pairs of $\left\langle p_{r_i}^{h_j}, r_i \right\rangle, \left\langle p_{r_j}^{h_j}, r_j \right\rangle$; generate $\left\langle (r_i, r_j), p_{r_i}^{h_j} - p_{r_j}^{h_j} \right\rangle$;
7   **end**
8   Merge and count all pairs $\left\langle (r_i, r_j), count, p_{r_i}^{h_j} - p_{r_j}^{h_j} \right\rangle$;
9   **for** *every pair* $\left\langle (r_i, r_j), count, p_{r_i}^{h_j} - p_{r_j}^{h_j} \right\rangle$ **do**
10   **if** *count* $> \beta$ **then**
11   Calculate $W_{ij} = SW(r_i, r_j)$ on position $p_{r_i} - p_{r_j}$; Store W_{ij} in W_i and W_{ij} in W_j ;
12   **end**
13 **end**
14 **return** Weight array $W = \{W_i\}$;

Map-Reduce Implementation. In the Map-Reduce framework, we used the notation $< key, value >$ to denote a key/value pair. In task 1, the input was index of read, r_i, and sequence of read, rs_i; In task 2, the mapper used approximated entry point $p_{i,j}$ for each pair of read as value. In task 3, entry point information (one pair just needs one entry point) and the original sequences were treated as input. In task 4, it regenerated pairs of reads as key to guild the sequence of these two reads to the same reducer and implemented SW algorithm to obtain the similarity scores for the phase II. The similarity score was calculated as Equation 1. $l_{hit}, l_{mis}, l_{ins}, l_{del}$ were the count of the four base alignments.

$$score = s_{hit}l_{hit} + s_{mis}l_{mis} + s_{ins}l_{ins} + s_{del}l_{del} \qquad (1)$$

Task 1:
    Input: $< r_i, rs_i >$
    Map: Generate hash values of all l-mers in r_i; For each h emit $< h, p_{r_i}^h, r_i >$.
    Reduce: $< h, [IR_i, \cdots, IR_j] >$. (with $IR_i = (p_{r_i}^h, r_i)$)

Task 2:

Input:$< h, [IR_i, , IR_j] >$

Map: For pair $< IR_i, IR_j >, i \neq j,$, emit $< (\min(r_i, r_j), \max(r_i, r_j)), p_{i,j} >$

Reduce: $< (\min(r_i, r_j), \max(r_i, r_j)), \left\{ p_{i,j}^0, p_{i,j}^1, \ldots, p_{i,j}^m \right\} >$; if $m \geq min_{count}$,

emit $< (\min(r_i, r_j), \max(r_i, r_j)), p_{i,j} >$ (with $p_{i,j}$ is the approximate start point).

Task 3:

Input: $< r_i, rs_i >$, $< \min(r_i, r_j), \{ \max(r_i, r_j), p_{i,j} \} >$ and $< \max(r_i, r_j), \min(r_i, r_j) >$

Map: For pair $< r_i, rs_i >$, emit each input entry as it is;

For pair $< (\min(r_i, r_j), \max(r_i, r_j)), p_{i,j} >$,

emit $< \min(r_i, r_j), \{ \max(r_i, r_j), p_{i,j} \} >$ and $< \max(r_i, r_j), \min(r_i, r_j) >$

Reduce: Emit $< \min(r_i, r_j), \left\{ \max(r_i, r_j), p_{i,j}, rs_{\min(r_i, r_j)} \right\} >$ and

$< \max(r_i, r_j), \left\{ \min(r_i, r_j), rs_{\max(r_i, r_j)} \right\} >$

Task 4:

Input: $< \min(r_i, r_j), \left\{ \max(r_i, r_j), p_{i,j}, rs_{\min(r_i, r_j)} \right\} >$ and $< \max(r_i, r_j), \left\{ \min(r_i, r_j), rs_{\max(r_i, r_j)} \right\} >$

Map: Emit $< (\min(r_i, r_j), \max(r_i, r_j)), \left\{ p_{i,j}, rs_{\min(r_i, r_j)} \right\} >$ and

$< (\max(r_i, r_j), \min(r_i, r_j)) \left\{ rs_{\max(r_i, r_j)} \right\} >$

Reduce: Banded align r_i, r_j from $p_{i,j}$: If $score_{i,j} > \tau_{align}$, emit $< (\min(r_i, r_j), \max(r_i, r_j)), score_{i,j} >$

3.2 Phase II: Clustering

The basic idea in the clustering was to partition the set of reads into disjoint subsets and maintain the property that reads in the same subset possess larger weights among them than the reads from different subsets. The weight was coming from local alignment in the previous phase. Our method were presented in Algorithm 2. When two reads were assigned into one subsets, it can be considered as one read was extended on either its $3'$ end or $5'$ end. Thus two things were required to address: (1) when to stop extension and (2) what kind of clusters can be further combined. For the first question, we exerted a threshold weight τ_c on the extension of reads, based on optimal stopping theory (OST). There was a $1/e$-law of best choice in OST to determine the worst case; here e was the *Euler's constant* and τ_c was the worst weight among the first $1/e$ part of extended reads. The first $1/e$ of extensions of $1/e + 1$ reads would be accepted. The process of extension would be terminated when the first subsequent combination whose weight was less than τ_c and we store the current cluster c as a candidate for next step. For the second question, a user defined upper bound of clusters N was given at the begin of clustering, which was related to the hardware information of clusters, like the number of computing nodes and the type of CPU they equipped. Each read in $R = \{r_1, r_2, \ldots, r_{|R|}\}$ was appended a weight set W_i, $i \in [1, 2, \ldots, |R|]$. The weight set $W_i = \{w_{ij}\}$ consisted of the score of alignment of read r_i to r_j. For each r_i, w_{i*} stood for the high score in W_i. First, a list of weight L composed of largest weight from each read r_i was produced and sorted descendent (line 1). τ_c was determined by *count*, which was less than $(1/e \times n_{ave})$ (line 2-10). Each cluster may have different τ_c. We tended to allocate the rest of read, i.e. not belonging to any candidate cluster, to the current cluster, if they met the specific value of τ_c (line 11-19). The forcing combination further merged the candidate clusters to reduce the number of cluster for satisfying the pre-defined N (line 20-25). In this phase, each read was merged only once, so the complexity of phase II in computing clusters is $O(|R|)$. In our experiment,

---

Algorithm 2. Clustering

Input: A family of Weight $W = \{W_i\}$ (with $i \in [0, |R|]$); Designed number of clusters N;
Output: Cluster $C = \{c_1, c_2, \ldots, c_N\}$ (with $c = \{r_i\}$)

1 Generate a list of weight $L = \left\{w_{ij}^*\right\}$, where w_{ij}^* is the largest weight for read r_i and it incidents to read r_i and read r_j; $k \leftarrow 0$;

2 **while** $\exists r \notin c_{k'}, k' \in [0, |C|]$ **do**

3 $\quad$ Pop the largest w_{ij}^* in L; $c_k = \cup \{r_i, r_j\}$, $W_{c_k} = \cup W_i$, update W_j; $C = \cup c_k$, $count \leftarrow 1$; $\tau_c \leftarrow -\infty$;

4 $\quad$ **while** $count < (1/e \times n_{ave})$ **do**

5 $\quad\quad$ Get the largest $w_{i'j'}^*$ from W_{c_k};

6 $\quad\quad$ **if** $\tau_c > w_{i'j'}^*$ **then**

7 $\quad\quad\quad$ $\tau_c \leftarrow w_{i'j'}^*$

8 $\quad\quad$ **end**

9 $\quad\quad$ Extend c_k by read $r_{i'}$, $r_{j'}$ and weight $w_{i'j'}^*$; Update list L, $W_{j'}$ and W_{c_k} ;

10 $\quad$ **end**

11 $\quad$ **while** $|W_{c_k}| > 0$ **do**

12 $\quad\quad$ Get the largest $w_{i'j'}^*$ from W_{c_k};

13 $\quad\quad$ **if** $w_{i'j'}^* < \tau_c$ and j is a index of read **then**

14 $\quad\quad\quad$ Stop extend current cluster c_k break ;

15 $\quad\quad$ **else**

16 $\quad\quad\quad$ Extend c_k by read $r_{i'}$, $r_{j'}$ and weight $w_{i'j'}^*$; Update list L, $W_{j'}$ and W_{c_k} ;

17 $\quad\quad$ **end**

18 $\quad$ **end**

19 **end**

20 **while** $|C| > N$ and $|L| > 0$ **do**

21 $\quad$ Pop the largest $w_{c_x c_y}^*$ in L ;

22 $\quad$ **if** $|c_x| + |c_y| < n_{up}$ **then**

23 $\quad\quad$ $c_x \leftarrow c_x \cup c_y$; update W_{c_x}, W_{c_y} and L

24 $\quad$ **end**

25 **end**

26 **return** $C = \{c_1, c_2, \ldots, c_N\}$;

---

it showed that the clustering can be done very fast (less than 120 seconds) on one computing node and the clustering is implemented sequentially.

3.3 Phase III: Parallel-Genovo

Our algorithm for parallel-Genovo was adapted from the metagenomes *De Novo* assembly approach Genovo [10]. The reason for us to choose Genovo was that it can reconstruct more bases and produce a assembly with better quality especially for the low-abundance dataset comparing to other methods. Genovo was an instance of iterated conditional modes (ICM) algorithm, which sequentially performed a random walk on states corresponding to different assemblies for maximizing local conditional probabilities. The drawback of Genovo was the computation time. Take middle size of metagenomic dataset for example, which had $220k$ reads of 400 bp average length, Genovo consumed more than 37 hours. The Parallel-Genovo was modified to run on hadoop platform and to produce a set of contigs with the count information with the structure as shown in Figure 3. The count information were coming from those reads covering the current base. The frequency was also employed to compute the post-merge in Phase IV.

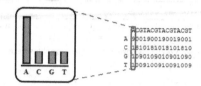

Fig. 3. Count of each deoxyribonucleic acid of base on contig

The parallel-Genovo took the clustered reads as input. Each cluster was fetched by a mapper in our Map-Reduce version parallel-Genovo and all mappers were simultaneously performing the previous described three steps by certain times of iterations. The reduce phase of Map-Reduce simply gathered the reconstructed contigs with count information on each base.

3.4   Phase IV: Post-Merge

The post-merge phase was designed to further merge contigs for achieving higher quality of contigs. In this phase, two general questions were required to address beforehand: (1) identify and correct indel bases on overlapping ends of two contigs; (2) remove chimeric reads, i.e. a prefix or a suffix matching distant locations in the genome [16]. For identifying indel bases, we treated the process of sequencing as a binomial distribution with error rate p_{err} for miss matching, insertion and deletion for each base. The random variable x was the number of count of insertion or deletion base, while the total trials n_t was the largest total count on that base or neighbours of that base. The probability to give an error with exactly x insertion or deletion was $Pr(X = x) = \binom{n_t}{x} p_{err}^x (1 - p_{err})^{n_t - x}$. The null hypothesis that the original sequence did have the insertion or deletion could be tested according to the significance level of the possibility $Pr(X = x)$. In our implementation, we set the level of significance to be 5%. For removing chimeric reads, we aligned two ends of contigs without those bases whose total counts of four bases were equal to one. Thus correct ends of contigs can be aligned for merging when the scores of alignment excessed a certain threshold. Since the end of one contig can be considered into two state, either keeping or removing bases with counts equal to one, four different aligning patterns need to be calculated. A list L of values of alignment scores and merging locations were stored and continuously updated along with the procedure. In L, a contig only kept the best alignment and merging information which gave the highest alignment score among four aligning patterns. The process of post-merge was given in Algorithm 3. It took contigs of the count information as input and generated the final assembly as output. A list L was constructed at the initiation, where t_i, t_j were index of contigs (line 1). The score $Score_{ijk}$ was a ratio of the original alignment score to the length of that alignment. In our alignment score matrix setting, when τ_{merge} was equal to 0.4, the algorithm can obtain the best result. Along with the process of merging of contigs, the list L would be updated until no more score was larger than τ_{merge} (line 2-7).

Algorithm 3. Post-Merge

Input: Contigs Set $T = \{t_i\}$, $i = [1, |T|]$; Merge Threshold τ_{merge};
Output: Merged Contigs Set $T = \left\{t'_i\right\}$, $i = \left[1, |T'|\right]$

1 Generate a list L of potential merge location, $L = \left\{(l^k_{t_i}, l^k_{t_j}, Score_{ijk})\right\}$;
2 **while** $min(L) > \tau_{merge}$ **do**
3   Pop the $(l^k_{t_i}, l^k_{t_j}, Score_{ijk})$ in L with the largest $Score$;
4   Merge Contig t_i and t_j as new Contig t'_i ;
5      $T = T - \{t_i, t_j\} + \left\{t'_i\right\}$;
6   Update L ;
7 **end**
8 return Merged Contigs Set $T = \left\{t'_i\right\}$, $i = \left[1, |T'|\right]$

---

4   Results

4.1   Compared Methods and Experimental Environment

To assess the efficiency of our cloud-based assembly approach, we performed a set of experiments on simulated and real metagenomic datasets which were sequenced using the 454 machine to the other two tools, the MetaVelvet [9] and the Genovo [10]. Both tools were designed for short metagemomic reads assembly. Note that MetaVelvet was designed for the Illumina Genome-Analyzer platform, but supported the 454 reads as well and both of them were freely available on line. There two methods were used under their default parameters setting in our experiments. After running on a set of reads, the assembly output, a list of contigs (sequences), were evaluated for completeness and correctness in several metrics: the number of contigs, the total length of contigs (TLC), the longest length of contigs (LLC), N50, N90 and the coverage of genomes (CG), that it was calculated as the union of all matching segments with the length more than 400bp on the reference. In addition to these, the running time of genovo and our method were also listed for illustrating the speed-up.

We deployed our cloud based *de novo* assembler on a ten nodes hadoop cluster where each node was equipped a 16 GB main memory. The cluster also had a 625 GB secondary storage under the control of the HDFS. Additionally, each computing node had a quad-core Opteron(tm) processor 2376 with 2.3Ghz and was interconnected through Network File System. The typical hadoop configurations were left without any changes, including the default 64 MB block size, 3 HDFS replication factor, one master node for controlling and the rest of nodes serving as workers for executing computations and storing data blocks.

4.2   Experiment on a Single Sequence Dataset

First we benchmarked the evaluation of our approach on a single sequence dataset, *E. coli* whose reference strand can be accessed from NCBI short read archive. This dataset was sequenced by the 454 Titanium and the total length of the genome of *E. coli* was 4.6 Mb, which contained 110k reads with the average length 351 bp. As shown in Table 1, our method outperformed both the Genovo and the MetaVelvet on every evaluation metric.

Table 1. Comparing the Methods on a Single Sequencing Task

	Our method	Genovo	MetaVelvet
No. contigs	88	93	412
TLC (kb)	4587	4687	4676
LLC(kb)	281.3	205.7	186.5
N50 (kb)	94.2	87.6	10.4
N90 (kb)	32.9	25.9	3.7
CG(%)	89.0	88.4	87.9

4.3   Experiments on Multi-species Simulated Datasets

We used the software, MetaSim [17], to artificially construct two metagenome sequence reads datasets with the different complexity. According to the simulation conducted in [8], we also selected 112 different species from the NCBI bacteria genome library. The first dataset, named LC, had only two dominant organisms which were strongly related taxonomically; the second dataset, named HC, did not have a dominant organism, that is all the species had equal weight for obtaining a similar coverage rate. The summary of these two datasets are shown in Table 2. We used the coverage of contigs (CC) as the evaluation metric instead of the coverage of genome, by calculated the union of all matching intervals on that contig with the length longer than 500bp aligning against the reference. The summary of the statistics of final assemblies of two simulated datasets are listed in Table 3. We can obviously see that our method generated better assembly results than those from Genovo and MetaVelvet. On both datasets, MetaVelvet did not perform very well, because the total contigs' lengths were too small compared to our method and Genovo. This was the main reason that we did not choose the MetaVelvet to parallel on cloud. The difference between our method and the original sequential algorithm Genovo was not too much. But we could still see some improvement on N50 and N90. However, our method gave higher coverage on contigs and the total contigs length and the number of contigs were subtle. In terms of higher contigs coverage, N50 and N90, a lower number of contigs may mean a better assembly result.

Table 2. Summary of the simulated and real datasets used in this study

Dataset	Number of species	Number of base pairs	Number of reads
LC	112	89.5Mb	220288
HC	112	92.8Mb	220288
Mb	NA	32.2Mb	135205
NTS	NA	254.5Mb	683082

Table 3. Summary of the assembly statistics of the simulated dataset

Assemblers		Our method	Genovo	MetaVelvet
LC	No. contigs	9004	10315	1430
	TLC (kb)	13808.0	13612.6	958.6
	LLC (kb)	172.3	153.1	1.4
	N50 (kb)	1.6	1.2	0.6
	N90 (kb)	0.7	0.6	0.5
	CC (%)	94.7	84.7	97.2
HC	No. contigs	29352	31743	3827
	TLC (kb)	21193.2	22932.2	2746.4
	LLC (kb)	4.2	3.8	1.7
	N50 (kb)	0.8	0.7	0.6
	N90 (kb)	0.6	0.5	0.5
	CC (%)	91.1	82.3	90.6

4.4   Experiments on Real Metagenomic Datasets

We also conducted comparison on two dataset from real metagenomic projects. The first dataset, named Microbes (Mb), was sampled from the Rios Mesquites stromatolites in Cuatro Cienagas, with accession numbers SRR:001043 in NCBI Short Read Archive. The number of total reads in this samples was about 135k. Another real dataset, hot springs microbial with name NTS, was from Los Alamos National Laboratory, and it was available at . http://metagenomics.anl . gov/linkin.cgi?project=223. The summary of these two real metagenomic datasets were listed in Table 2. The statistic analysis appears in Table 4. Once again, our method achieved the best results on both real datasets among three assemblers. To obtain a comprehensive comparison on the real datasets, we adopted the same evaluation metric utilized in [10], i.e. BLAST-score-per-base (Bspb). The calculation process of Bspb was to first BLAST the contigs to the GenBank and use the outcome BLAST hits to compile a pool of genomes which best represented the consensus sequences among the results. Each hit represented one GenBank sequence. Then it filtered out those non-significant BLAST hits whose E-value were less than 10^{-9}. The computation of Bspb was simply to divide the BLAST score of one hit by the half of the length of that hit. The highest Bspb reached near 2.0 in our study. Thus a BLAST profile can be plotted by moving the threshold of Bspb from low to high and from which it can be easy to find out how many bases could be covered by one method. Based on this definition, Figure 4(a) and Figure 4(b) show the BLAST profile for two real datasets. In the Figure 4(a), our method covered more bases than the others along the decreasing of alignment threshold on the dataset of Microbes. The MetaVelvet nearly covered nothing when the threshold raised to 0.8. Similarly our method gained more covered bases on the whole range of align threshold, which meant that it produced contigs with higher quality.

Table 4. Summary of the assembly statistics of the real datasets

Assemblers		Our method	Genovo	MetaVelvet
Mb	No. contigs	50	47	7
	TLC (kb)	58.9	54.7	4.2
	LLC (kb)	4.1	3.8	0.7
	N50 (kb)	1.5	1.2	0.6
	N90 (kb)	0.7	0.6	0.5
NTS	No. contigs	98044	101783	4040
	TLC (kb)	67982.2	69789.2	2434.2
	LLC (kb)	10.5	8.7	1.3
	N50 (kb)	0.7	0.6	0.6
	N90 (kb)	0.6	0.5	0.5

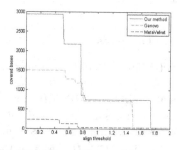 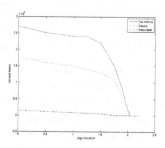

(a) BLAST profile on Microbes for three methods

(b) BLAST profile on NTS for three methods

Fig. 4. BLAST-score-per-base profiles for two real datasets

4.5   Time Comparison

We were also interested to know how much speed up has been achieved by our framework compared to Genovo, which was shown in Table 5. We did not consider the MetaVelvet here since MetaVelvet could not give us satisfactory results. In Table 5, our method was quite faster than Genovo. For the *E. coli*, our method did not explore the power of all computing nodes (only ten cores were used), due to the size of background contigs were relatively small (from 100 to 200) and in Phase II, only coarse partition was execrated. Similar results came from the Microbes. For the other two simulated datasets, the speed-ups were significant, owning to the number of contigs were actually larger than the

Table 5. Comparing the Methods on a Single Sequencing Task

Dataset	E. coli	LC	HC	Mb	NTS
Our method	1027s	253m	301m	1723s	667m
Genovo	9750s	2271m	2485m	4248s	232h

number of CPU cores used in our framework. For the dataset of NTS, the size of reads is more than 600k and Genovo shown its weakness to deal with the huge metagenomic data since it consumed lots of time on solving the assembly. It was because every read will be checked on the consensus sequence in the Genovo, and the length of consensus sequences were linear to the number of reads. Therefore, the time complexity would be more than $O(n^2)$ in reality. On the contrary, with our parallel strategy, the input data can always be kept in a small size and as a result, it gained a great speed up on large datasets.

5 Conclusion

Sequence assembly of large-scale environmental samples is considered a difficult and challenging problem in metagenomics, especially with the advance in sequencing technology, which leads to high coverage of genomes and huge numbers of reads. Our work was motivated by the computational and accurate issues of existing methods when they were applied to large scale metagenomics projects. Our approach was implemented based on the MapReduce model by the widely available Hadoop platform. By utilizing the parallel strategy, our framework has both accelerated and improved sequence assembly results on simulated and real metagenomics datasets.

Acknowledgments. This study is supported by the Molecular Basis of Disease (MBD) at Georgia State University.

References

1. Wu, X., Cai, Z., Wan, X.F., Hoang, T., Goebel, R., Lin, G.: Nucleotide composition string selection in hiv-1 subtyping using whole genomes. Bioinformatics 23(14), 1744–1752 (2007)
2. Gill, S.R., Pop, M., DeBoy, R.T., Eckburg, P.B., Turnbaugh, P.J., Samuel, B.S., Gordon, J.I., Relman, D.A., Fraser-Liggett, C.M., Nelson, K.E.: Metagenomic analysis of the human distal gut microbiome. Science 312(5778), 1355–1359 (2006)
3. Venter, J.C., Remington, K., Heidelberg, J.F., Halpern, A.L., Rusch, D., Eisen, J.A., Wu, D., Paulsen, I., Nelson, K.E., Nelson, W., Fouts, D.E., Levy, S., Knap, A.H., Lomas, M.W., Nealson, K., White, O., Peterson, J., Hoffman, J., Parsons, R., Baden-Tillson, H., Pfannkoch, C., Rogers, Y.H., Smith, H.O.: Environmental genome shotgun sequencing of the sargasso sea. Science 304(5667), 66–74 (2004)
4. Qin, J.: A human gut microbial gene catalogue established by metagenomic sequencing. Nature (2009)
5. Khachatryan, Z.A., Ktsoyan, Z.A., Manukyan, G.P., Kelly, D., Ghazaryan, K.A., Aminov, R.I.: Predominant Role of Host Genetics in Controlling the Composition of Gut Microbiota. PLoS ONE 3(8), e3064 (2008)
6. Nguyen, K.D.: On the edge of web-based multiple sequence alignment services. Tsinghua Science and Technology 17(6), 629–637 (2012)
7. Turnbaugh, P.J.: A core gut microbiome in obese and lean twins. Nature (2009)

8. Pignatelli, M., Moya, A.: Evaluating the Fidelity of De Novo Short Read Metagenomic Assembly Using Simulated Data. PLoS ONE 6(5), e19984 (2011)

9. Namiki, T., Hachiya, T., Tanaka, H., Sakakibara, Y.: Metavelvet: an extension of velvet assembler to de novo metagenome assembly from short sequence reads. In: Proceedings of the 2nd ACM Conference on Bioinformatics, Computational Biology and Biomedicine, BCB 2011, pp. 116–124. ACM, New York (2011)

10. Laserson, J., Jojic, V., Koller, D.: Genovo: de novo assembly for metagenomes. In: Berger, B. (ed.) RECOMB 2010. LNCS, vol. 6044, pp. 341–356. Springer, Heidelberg (2010)

11. Peng, Y., Leung, H.C.M., Yiu, S.M., Chin, F.Y.L.: Meta-idba: a de novo assembler for metagenomic data. Bioinformatics 27(13), i94–i101 (2011)

12. Dean, J., Ghemawat, S.: Mapreduce: simplified data processing on large clusters. Commun. ACM 51(1), 107–113 (2008)

13. Grillo, G., Attimonelli, M., Liuni, S., Pesole, G.: Cleanup: a fast computer program for removing redundancies from nucleotide sequence databases. Computer Applications in the Biosciences: CABIOS 12(1), 1–8 (1996)

14. Smith, T., Waterman, M., Fitch, W.: Comparative biosequence metrics. Journal of Molecular Evolution 18, 38–46 (1981)

15. Lander, E.S., Waterman, M.S.: Genomic mapping by fingerprinting random clones: A mathematical analysis. Genomics 2(3), 231–239 (1988)

16. Lasken, R., Stockwell, T.: Mechanism of chimera formation during the multiple displacement amplification reaction. BMC Biotechnology 7(1), 19 (2007)

17. Richter, D.C., Ott, F., Auch, A.F., Schmid, R., Huson, D.H.: MetaSim–A Sequencing Simulator for Genomics and Metagenomics. PLoS ONE 3(10), e3373 (2008)

Protein Closed Loop Prediction
from Contact Probabilities

Liang Ding[1,*], Joseph Robertson[2,*],
Russell L. Malmberg[2,3], and Liming Cai[1,**]

[1] Department of Computer Science
[2] Institute of Bioinformatics
[3] Department of Plant Biology,
University of Georgia, GA 30602, USA
{lding,josephr}@uga.edu, russell@plantbio.uga.edu, cai@cs.uga.edu

Abstract. According to the theory of Trifonov [1] a subset of closed loops found in globular proteins, descriptively named loop-n-lock (LNL) structures, are returns of a polypeptide chain trajectory to close contact with itself. These closed loops are distinguished by a characteristic length, 20 to 35 residues, and a "lock" that stabilizes the loop. Evidence supports their contention that globular proteins are composed of linear combinations of such closed loops. Occupying an intermediate position in protein structure hierarchy i.e., between secondary structure and domain, makes these loops good candidates for units of a hierarchical folding model.

In this study we investigated the potential utility of such closed loops in protein structure prediction. We first proposed a method to predict closed loops using sequence information, specifically interaction potential and secondary structure potential. As far as we know, this is the first program to predict closed loops solely based on sequence information. Then to support their use in a hierarchical folding model, we explored the placement of secondary structure elements with respect to closed loops. Our investigations showed that 80% of secondary structures do not cross a closed loop boundary, lending support to the hypothesis that the closed loop is a higher level intermediate structure of proteins. Moreover, we explored the relationship between closed loops and aligned segments from protein chain pairs from the Dali data set . The results showed protein pairs which exhibited significant structural similarity were more likely to have conserved closed loops.

1 Introduction

Protein structure is typically described in terms of the four levels first proposed in the early fifties by Linderstrom-Lang [2]: primary, secondary, tertiary and quaternary. However, this hierarchy is now considered overly simplified, hence

---

* These authors, listed alphabetically, contributed equally.
** To whom correspondence should be addressed.

Z. Cai et al. (Eds.): ISBRA 2013, LNBI 7875, pp. 199–210, 2013.
© Springer-Verlag Berlin Heidelberg 2013

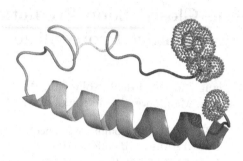

Fig. 1. A loop-n-lock in protein chain 1a0cA with two secondary structures α helix and β strand, and the "lock", or contacting, residues shown as filled spheres. The figure was generated using PyMOL (http://www.pymol.org/).

the addition of intermediate levels of structure e.g., super-secondary and domain. Such intermediate structure levels may alleviate the issue known as Levinthal's paradox [3] i.e., the time to fully sample conformation space yields folding times many orders of magnitude greater than those observed. That is, substructures composed of relatively short contiguous sequences i.e., tens of residues, have only a relatively small conformational space to explore, and once established significantly restrict the conformation space for the entire chain.

One such structure type, namely loop-n-lock (LNL), was proposed by Trifonov's group [1]; it is a closed loop, typically consisting of 20 to 35 amino acid residues, stabilized by van der Waals, as well as other, interactions among residues at its C and N termini [12]. These substructures, an example of which is shown in figure 1. are hypothesized to have their origin in the physical characteristics of the polypeptide chain, in particular the length at which it becomes highly probable that the chain will make contact with itself. Additionally, under the hypothesis, these loops are proposed to represent the original functional protein units, and that globular protein domains [13], and thus entire globular proteins, as found now are derived from the concatenation of such loops e.g. ,by duplications, and latterly also deletions.

More detailed models of protein structure hierarchy clearly can facilitate more efficient and accurate protein structure prediction, hence such intermediate structures could have significant utility. However, the added complexities introduced are a potential liability for computational efficiency, thus the need for easily computed features, based on local information. These benefits are readily apparent, and here derive from the characteristics of LNL structure composition, in that they are composed of contiguous short sub-sequences, and there are detectable signals of the "locks" in the sequences. Thus straightforward approaches to prediction, based solely on sequence analysis for interaction potential between pairs of residues, may be used to select candidate loops. To this end we propose a method based on contact prediction to generate candidate loop elements, which may then be processed algorithmically to find an optimal non-conflicting subset. As we know, this is the first program to predict closed loops from sequence.

Although such predictions in themselves are not generally sufficiently accurate to call a majority of loops in any given instance, if a subset with relatively high confidence can be identified, they may be used as anchors. Even a small number of such candidates could help refine the selection of the remaining candidates for inclusion in a putatively optimal subset.

With reference to the proposed origins and evolutionary significance of such loops, we also argue that deletions and insertions should be enriched for unit sizes and positions corresponding to LNL sizes and positions in alignments between orthologous proteins. Further, the positions of LNLs should be conserved for structurally similar, orthologous pairs e.g., from the Dali data set, and in particular the locks, due to their higher conservation [4]. These assertions are also examined here.

2   Methods

2.1   Contact Probability Based Closed Loop Prediction

In this section, we describe a new method to decompose a protein sequence into closed loops, which as far as we know, is the first approach to predicting closed loop structures solely from sequence information. The basic idea is that a weighted candidate set containing all plausible closed loops is first constructed. Next, a greedy strategy is used to select the best subset by selecting the locally optimal loop with largest weight. Due to the known major limitation of such strategies for selecting global optima i.e., becoming trapped in a local minimum, we also tested a dynamic programming based approach. Results from both were then compared with decompositions generated by DHcL [7] to determine which performed better. Note that for the purposes of this investigation it is assumed DHcL generates the correct set of closed loops for each of the known protein structures in its database.

We first give a brief definition of contact map: a square matrix where rows and columns represent sequence positions, where protein residue-residue contact, often using C_α-C_α distance as a proxy, is shown by placing a one in each cell where the distance is below a threshold, and a zero otherwise. For two residues i and j in a protein sequence with length n, and $1 \leq i, j \leq n$, if the two residues are closer than a predetermined threshold (e.g., 8Å), the element of the contact map matrix at position i, j is 1, and 0 otherwise. This idea is trivially extended to represent predicted contact by instead filling the cells with probabilities. Such matrices have formed the basis for several distanced-based algorithms for reconstructing protein 3D structures.

For our investigation we chose NNcon [5] to generate the contact potentials maps, based on its strong performance in the Eighth Critical Assessment of Techniques for Protein Structure Prediction (CASP8). By default NNcon generates maps based on thresholds of 8Å and 12Å. However, because close contacts are the determinant of a closed loop, we used only 8Å maps.

Once the map for a given polypeptide was generated, we obtained the weighted candidate set by including all the chain segments having $15 - 45$ residues, whose contact potential between the two ends was greater than a chosen threshold. Three different weight schemes were used to generate candidate sets: single weight, sum weight, and median weight. Let M be the contact map generated by NNcon, then $M[i, j]$ represents the contact potential between residues i and j. The single weight method simply set $M[i, j]$ as the weight of closed loop candidate $[i, j]$, while the sum weight was set to $\sum_{i-2 \leq i \leq i+2, j-2 \leq j \leq j+2} M[i, j]$, and if we let sw be the sum weight, then the median weight was defined as $sw/25$. For the generated closed loop candidates, secondary structures of lock parts are also considered as an additional factor of weight based on the fact that lock parts of majority of closed loops are within some secondary structure (either α-helix or β-sheet)[4]. We used a linear equation $W_1 \cdot M[i, j] + W_2 \cdot (S[i] + S[j])$, where $S[t]$ is the potential that position t is within some secondary structure, to incorporate secondary structure factor into the weights of the candidates.

Our greedy strategy iteratively selects the closed loop candidate with the greatest weight, while simultaneously excluding all overlapping candidates. The algorithm halts when either no more candidates can be inserted into the solution set, or all the candidates have been inserted. Suppose a polypeptide sequence has length n, then according to the definition of the closed loop, it can have at most $n/15$ closed loops. Thus the running time is bounded by $O(n^2)$.

We now briefly introduce the interval graph, which is a graph theoretical construct derived from a multiset of intervals on the real line. It has one vertex for each interval in the multiset, and edges between all pairs of vertices corresponding to intervals that intersect. Given the closed loop candidate set, it is easy to see that each candidate can be represented by an interval on the real line, and since, according to Trifonov *et al*'s theory [1], globular proteins are composed of linear combinations of closed loops, the closed loops prediction problem can be reduced to the maximum weight independent set problem on an interval graph. We implemented a version the dynamic programming algorithm by Hsiao [10], with time complexity $O(n \log n)$, to compute a globally optimal solution.

Candidate sets, once generated, were then processed to select a nonconflicting subset using either the greedy algorithm defined above, or a dynamic programming approach utilizing Hsiao's algorithm. To estimate prediction accuracy we selected two sets of protein chains. The first set contains 722 random X-ray crystallographic polypeptide chain models derived from more than $74,000$ proteins of Protein Data Bank. To reduce the effects of sequence similarities, we chose an ASTRAL [8] sequence subset with 802 protein chains as the second test set. Each sequence in the second test set is a representative of a superfamily in SCOP [6]. Both test sets exclude proteins with duplicated or missing atoms, and structural gaps. We compared the prediction results with the closed loops decompositions from DHcL. The average prediction accuracy (APA) was used to provide a measure of correspondence with DHcL predictions. Let n be the number of sequences to predict. For i-th sequence, let TP_i and FN_i be the number of true positive and false negative results respectively. Then the sensitivity for the prediction of the i-th sequence is $TP_i/(TP_i + FN_i)$. And the APA among n sequences is defined as $(1/n) \cdot \sum_{1 \leq i \leq n} TP_i/(TP_i + FN_i)$.

Hereafter, when referring to a closed loop structure the notation $[s_i^g, e_i^g]$ will be used, where "s" and "e" are the start and end positions of the loop having index "i" in the set identified by "g". So, assuming we had a predicted closed loop $[s_i^p, e_i^p]$, it is a true prediction if there was a closed loop $[s_i^d, e_i^d]$ in the result set from DHcL whose start and end positions satisfied $s_i^p - 5 \leq s_i^d \leq s_i^p + 5$ and $s_i^p - 5 \leq s_i^d \leq s_i^p + 5$ respectively. In the above inequalities, the error gap was defined to be 5, because the locks stabilizing closed loops are typically formed from 3-5 contacting residues at each end.

2.2   Closed Loop Conservation for Structurally Similar Proteins

In this section we describe tests to support the hypothesis that closed loops could act as a higher level structure, relative to secondary structure, in a hierarchical folding model for proteins. In the first, we investigated the relationship of closed loops with aligned segments from polypeptide chain pairs from the Dali data set [9] in the same structural neighborhood. In the second, described in section 2.2, we explored the connection between closed loop gaps and pairwise alignment gaps.

Closed Loop Matching. Conservation of closed loop structures for orthologous, structurally similar proteins would lend support to the hypothesis that they are a higher level structure in a hierarchical model of protein folding. We performed two experiments based on 5000 randomly selected pairs from the Dali data set. In the first experiment, for each Dali pair (P, Q), we treated the first chain P as the basis. Suppose f is a function reflecting the Dali pairwise alignment which aligns chain P to chain Q. Then for each closed loop (s_i^P, e_i^P) of the first chain, we calculate the corresponding aligned position $(f(s_i^Q), f(e_i^Q))$ of the second chain. Two segments (s_i, e_i) and $(f(s_i), f(e_i))$ are considered matching closed loops if either $(f(s_i), f(e_i))$ is a closed loop or $(f(s_i), f(e_i))$ contains a closed loop. In Table 6 the matching described above is called strict matching.

In addition to the strict matching, we defined less stringent matching criteria, based on the definition of closed loops, which as defined, have at most 5 close contacts among residues at their two ends. On that basis we used 5 as the cutoff for what we termed "loose matching". So formally, given (s_i, e_i) and $(f(s_i), f(e_i))$, they are loosely matched, if there is a loop with start between $f(s_i) - 5$ and $f(s_i) + 5$ and end between $f(e_i) - 5$ and $f(e_i) + 5$. As a control, we randomly generated 5000 pairs with which to compare the results obtained using the 5000 Dali pairs, and analyzed them using the same criteria.

Closed Loop Unit Test. To assess the utility of closed loops as a unit for structural alignment of proteins, we investigated the relation between closed loop gaps and Dali pairwise alignment gaps. A closed loop gap is defined as the polypeptide segment between two continuous closed loops. Let n be the number of gaps and $E = \{(e_1^1, e_2^1), \cdots, (e_1^n, e_2^n)\}$ contain the end points of the gaps derived from Dali pairwise alignment. We calculate the probability that an end point of E is included in a closed loop gap. This experiment was based on

Table 1. Average accuracies of predictions for the random test set and the ASTRAL derived test set based on both the greedy method and the dynamic programming method. Three weight calculations used: single, sum and median.

	Avg Pred. Acc.	Single	Sum	Median
Random Set	Greedy	52.76%	52.14%	54.19%
	Dynamic Progr.	58.31%	56.62%	56.46%
ASTRAL Set	Greedy	46.06%	46.84%	47.38%
	Dynamic Progr.	50.14%	50.72%	50.65%

Table 2. Average accuracies of the predictions for other error gaps of the ASTRAL derived test set

Avg Pred. Acc.	Error Gap 4	Error Gap 3	Error Gap 2
Greedy	42.42%	33.64%	23.78%
Dynamic Progr.	44.75%	35.55%	24.78%

5413 polypeptide chains. The probability that such gaps will align by chance is calculated by dividing the total length of closed loop gaps by the total length of all the sequences, which is less than 31%.

2.3　Closed Loops and Secondary Structures

This test was intended to help assess LNL structures as units in a structure hierarchy, specifically their relation to secondary structure elements. In brief, this consisted of simply checking both ends of secondary structure elements against LNLs' positions to determine the proportions of secondary structure elements that do, and do not, straddle LNL boundaries.

3　Results

3.1　Contact Probability Based Closed Loop Prediction

Tests on randomly selected data utilizing both the greedy and dynamic programming approaches yielded similar results for the three weighting schemes, as can be seen in Table 1, which shows the average prediction accuracies(APA). Using the second test set, derived from ASTRAL with low sequence identity, the results achieved by same methods are almost comparable to the random set, shown in Table 1. The APAs derived with other error gaps settings are shown in table 2. Two examples of well predicted polypeptide chains are shown in Table 3, along with the corresponding results from DHcL.

Average prediction accuracies for three weight values and the percentages of the total they represent are given in Table 5. In addition to overall accuracies, we determined the proportions of chains with one or more LNL with weight

Table 3. The closed loops of protein chain 1aw2A and 1cm7B: a comparison with DHcL's results. Protein chain 1aw2A has 8 closed loops, 7 of them are predicted correctly. Protein chain 1cm7B has 11 closed loops, 8 of them are predicted correctly.

Protein	DHcL loops	Len.	Pred. loops	Len.	DHcL loops	Len.	Pred. loops	Len.
1aw2A	7-39	33	4-37	34	39-62	24	37-61	25
	64-92	29	61-92	32	92-124	33	92-124	33
	—	–	124-164	41	133-172	40	—	–
	169-212	44	164-208	45	211-232	22	208-231	24
	231-251	21	231-246	16				
1cm7B	6-39	34	8-39	32	—	–	44-75	32
	46-90	45	—	–	—	–	71-107	37
	112-133	22	107-135	29	—	–	136-151	16
	144-165	22	148-163	16	—	–	162-191	30
	176-192	17	—	–	191-222	32	189-223	35
	—	–	223-243	21	226-249	24	—	–
	—	–	243-269	27	268-282	15	266-281	16
	282-300	19	281-297	17	—	–	293-309	17
	306-322	17	305-321	17	325-358	34	321-354	34

Table 4. This table shows the percentage of protein chains containing one or more LNL with weight at least that given. These results come from the random set and the ASTRAL set.

Weight	Wt(>0.4)	Wt(>0.2)	Wt(>0.15)	Wt(>0.1)
Random Set	64.69%	98.35%	99.31%	99.58%
ASTRAL Set	44.65%	71.71%	75.12%	82.21%

equal to or higher than the threshold for the random test set and the SCOP test set with the results given in Table 4. For example, the second column for the ASTRAL set in Table 4 shows that 44.65% of the protein chains in the SCOP test set have at least one LNL with weight greater than 0.4.

Both the random test set and the ASTRAL based set showed similar relationships between the weights and accuracies of the predictions, which appears to be approximately linear. Figure 2 show the accuracies for selected weight values from each of the three weighting schemes for the random set. Both methods perform relative better for protein chains which have LNL with higher weight. However, for protein chains which have LNL with higher weight, the dynamic programming method performs much better than the greedy method. For example, for protein chains which don't have LNL with weight greater than 0.4, the dynamic programming method has APA 32.25%, whereas the greedy method only has APA 14.39%.

Besides interaction potential, the secondary structures factor of lock parts are considered. We trained this factor using the ASTRAL set. The prediction results show at best 1.5% improvement in terms of average prediction accuracy by considering secondary structure factor, which agrees the definition of the

Table 5. This table shows the percent of closed loops (corresponding to total number of closed loops) with weight greater than the thresholds in the corresponding columns

Random Set	Weight Setting	Wt(>0.4)	Wt(>0.2)	Wt(>0.15)	Wt(>0.1)
	Single Wt	11.13%	35.20%	39.69%	44.23%
	Sum Wt	7.93%	16.24%	21.29%	30.23%
	Median Wt	7.96%	16.56%	21.76%	31.34%
ASTRAL Set	Single Wt	11.24%	24.58%	28.28%	31.39%
	Sum Wt	8.18%	14.05%	16.51%	22.03%
	Median Wt	8.17%	14.26%	16.91%	22.54%

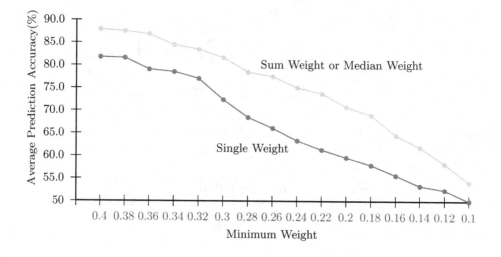

Fig. 2. Relationship between average prediction accuracy and the minimum weight of the closed loop candidates for the three weighting methods: cyan for the prediction results using single weight, and yellow for sum weight and median weight. The results are from the random test set.

closed loops. This explains interaction potential of the lock part is the key factor of the closed loops and secondary structure potential is less important than interaction potential.

3.2   Closed Loop Conservation for Structurally Similar Proteins

Figure 3 shows the probability both end points of a Dali pairwise alignment gap are included in some closed loop gap, plotted against minimum gap lengths for Dali pairwise alignments.

Table 6. Comparison of closed loops matching between Dali pairs and random pairs in terms of total number of closed loops

Dali Pairs	Total CLs	Match CLs	Percent
Strict Match	11418	5527	48.40%
Loose Match	11418	8420	73.74%
Rand. Pairs	Total CLs	Match CLs	Percent
Strict Match	4014	1782	44.39%
Loose Match	4014	2960	73.74%

Table 7. Comparison of closed loops matching between Dali pairs and random pairs in terms of Dali alignment. In the 3-rd column, aligned pairs (0.7) shows the number of pairs with 70% of closed loops matching. Similarly, aligned pairs (0.5) is the number of pairs with 50% of closed loops matching.

	Dali Pairs	Aligned Pairs (0.7)	Percent (0.7)	Aligned Pairs (0.5)	Percent (0.5)
Strict Match	2738	633	23.11%	1503	54.89%
Loose Match	2738	1509	55.11%	2302	84.07%
	Rand. Pairs	Aligned Pairs (0.7)	Percent (0.7)	Aligned Pairs (0.5)	Percent (0.5)
Strict Match	5000	117	2.34%	371	7.42%
Loose Match	5000	260	5.20%	607	12.14%

3.3   Closed Loops and Secondary Structures

Results for the secondary structure were derived from analysis of closed loop positions with respect to placement of secondary structure elements for 25,869 X-ray crystal protein models. The results show that almost 80% of closed loops entirely contain one or more secondary structure elements. And the percent is irrelevant to the length of the closed loops.

4   Discussion

Three things were investigated in this work: prediction potential for closed loop structures, whether loop conservation correlated with fold conservation, and evidence to support the use of the closed loop in a hierarchical model. All of these were aimed at determining the utility of these structures for improving protein folding models. Any structure to be used must have some foundation in biology which is provided by the work of Trifonov et al [1]. for their hypotheses concerning the closed loops, and supported by analysis of experimental results. If they are to be part of a hierarchical model, then they must fit sensibly into a hierarchy, and the work of Berezovsky et al. has supported this, as well as one part of this work. More importantly, there must be some way to predict them based on computational analysis of sequence information.

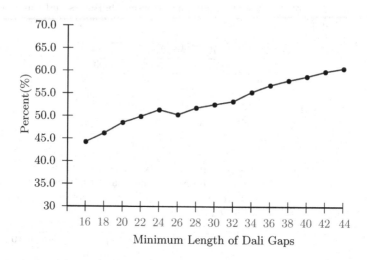

Fig. 3. The minimum length of Dali gap shows a linear relationship with the probability that both end points of a Dali pairwise alignment gap are included in some closed loop gap. Total number of testing chains is 5,413. Horizontal coordinates represent the minimum length of Dali gaps, which are the cutoffs for determining gaps. Vertical coordinates represent the percentages calculated by dividing the number of end points of Dali pairwise alignment gaps which are included in the gaps of closed loops, by the total number of end points.

No matter how much potential a new structural definition has for improving protein folding models, it is no use unless there is a straightforward prediction method. Our results show prediction accuracy of about half of true closed loops (averages 54% and 47% for random and SCOP tests, compared with DHcL). As the biases that DHcL have, our results cannot clearly identify the boundries of the true closed loops. Hence this to some extent affected the overall prediction accuracy. Despite of it, our approach yields a subset of high accuracy closed loops with convincing agreement between the predicted closed loops and those generated by DHcL. Correlating the results in Figure 2, those LNLs with weight greater than 0.18 have average prediction accuracy greater than 72%. We believe that sufficient to improve structure prediction in itself, as even a relatively small number of good predictions could provide sufficient information to usefully restrict the solution space to be explored. From tables 1 median weight setting has the best average prediction accuracy, the dynamic programming method performs almost 10% better than the greedy strategy. Our observation was hat to achieve the globally optimal solution, the dynamic programming approach tended to decompose long closed loops, i.e., those with length greater than 35, into two or three shorter loops, in contrast to DHcL. Therefore, other objective functions which could better address closed loops with large potentials should be incorporated in the future.

In addition to the primary goal of prediction, tests for enrichment for conserved loops in structurally similar orthologous pairs (Dali), and for support of closed loop as a higher order structure, i.e., super-secondary, were made. As would be expected, pairs with conserved structure, shared more conserved loops than those without. Compared to the results obtained for the Dali pairs, the random pairs showed no evidence of conservation, in particular note the last column in Table 7, where it can be seen that 84.07% of Dali pairs have about half of their closed loops matching, whereas for random pairs, the proportion is only 12.14%. Further testing to better quantify this difference is needed, but the results here are sufficiently promising to warrant the additional work. Finally, in support of the closed loop's candidacy for super-secondary structure, we found that over 80% of closed loops had no secondary structure element only partially contained. This too should be further examined, in particular with reference to the assertion that the locks are more highly conserved, and tend to be in secondary structure elements themselves [4].

5 Conclusion

Overall the results support the inclusion of the closed loop structure as a putative intermediate structure in a hierarchical model of protein structure. The supporting evidence from the conservation and secondary structure element containment warrants further investigation. LNLs good prediction potential makes them also good candidates for inclusion in computational models, where this is critical. Additionally, analysis by one group [4] showed that consideration of LNL structures yields better agreement with experimental evidence relating to folding order and nucleation, lending further support for their use in folding models.

In considering the LNL as a unit of structure, the enrichment for insertions and deletions within Dali gaps supports closed loops' potential as units for protein structure prediction. That is, the closed loops could replace secondary structures to do insertion or deletion in many current prediction implementations.

Acknowledgments. We thank Xingran Xue and Yingfeng Wang of the RNA Informatics Lab for sharing literature and data. This work is supported in part by NSF IIS grant (No: 0916250).

References

1. Trifonov, E.N., Berezovsky, I.N.: Proteomic Code. Molecular Biology 36(2), 315–319 (2002)
2. Linderstrom-Lang, K.U.: Proteins and Enzymes. In: Lane MedicalLectures. University Series, Medical Sciences, vol. 6. Stanford University Publications, Stanford University Press (1952)
3. Levinthal, C.: How to Fold Graciously. Mossbauer Spectroscopy in Biological Systems. In: Proceedings of a Meeting Held at Allerton House, Monticello, Illinois, pp. 22–24 (1969)

4. Chintapalli, S.V., Yew, B.K., Illingworth, C.J.R., Upton, G.J.G., Reeves, P.J., Parkes, K.E.B., Snell, C.R., Reynolds, C.A.: Closed Loop Folding Units from Structural Alignments: Experimental Foldons Revisited. J. Comput. Chem. 31, 2689–2701 (2010)
5. Tegge, A.N., Wang, Z., Eickholt, J., Cheng, J.: NNcon: Improved Protein Contact Map Prediction Using 2D-Recursive Neural Networks. Nucleic Acids Research 37, w515–w518 (2009)
6. Murzin, A.G., Brenner, S.E., Hubbard, T., Chothia, C.: SCOP: a structural classification of proteins database for the investigation of sequences and structures. Jounral Molecular Biology 247, 536–540 (1995)
7. Koczyk, G., Berezovsky, I.N.: Domain Hierarchy and closed Loops (DHcL): a server for exploring hierarchy of protein domain structure. Nucl. Acids. Res. 36, W239–W245 (2008)
8. Chandonia, J.M., Hon, G., Walker, N.S., Lo Conte, L., Koehl, P., Levitt, M., Brenner, S.E.: The ASTRAL compendium in 2004. Nucleic Acids Research 32, D189–D192 (2004)
9. Holm, L., Rosenstrom, P.: Dali server: conservation mapping in 3D. Nucl. Acids Res. 38, W545–W549 (2010)
10. Hsiao, J.Y., Tang, C.Y., Chang, R.S.: An efficient algorithm for finding a maximum weight 2-independent set on interval graphs. Information Processing Letters 43(5), 229–235 (1992)
11. Kogan, S.B., Kupervasser, O.: Domain Hierarchy of Protein Loop-Lock Structure (DHoPLLS): a server for decomposition of a protein structure on set of closed loops. CoRR. abs/1106.1356 (2011)
12. Aharonovsky, E., Trifonov, E.N.: Sequence Structure of van der Waals Locks in Proteins. Journal of Biomolecular Structure Dynamics 22(5), 545–553 (2005)
13. Berezovsky, I.N.: Discrete structure of van der Waals domains in globular proteins. Protein Engineering 16(3), 161–167 (2003)

A Graph Approach to Bridge the Gaps in Volumetric Electron Cryo-microscopy Skeletons

Kamal Al Nasr, Chunmei Liu, Mugizi Robert Rwebangira, and Legand L. Iii Burge

Department of Systems and Computer Science
Howard University
Washington, DC 20059
{kalnasr,chunmei,rweba,blegand}@scs.howard.edu

Abstract. Electron Cryo-microscopy is an advanced imaging technique that is able to produce volumetric images of proteins that are large or hard to crystallize. *De novo* modeling is a process that aims at deriving the structure of the protein using the images produced by Electron Cryo-microscopy. At the medium resolutions (5 to 10Å), the location and orientation of the secondary structure elements can be computationally identified on the images. However, there is no registration between the detected secondary structure elements and the protein sequence, and therefore it is challenging to derive the atomic structure from such volume data. The skeleton of the volume image is used to interpret the connections between the secondary structure elements in order to reduce the search space of the registration problem. Unfortunately, not all features of the image can be captured using a single segmentation. Moreover, the skeleton is sensitive to the threshold used which leads to gaps in the skeleton. In this paper, we present a threshold-independent approach to overcome the problem of gaps in the skeletons. The approach uses a novel representation of the image where the image is modeled as a graph and a set of volume trees. A test containing thirteen synthesized images and two authentic images showed that our approach could improve the existent skeletons. The percent of improvement achieved were 117% and 40% for Gorgon and MapEM, respectively.

1 Introduction

Electron Cryo-microscopy (CryoEM) technique is an advanced imaging technique that aims at visualizing and interpreting unstained nanostructures from biological complexes such as viruses [1-4]. Unlike X-ray Crystallography or Nuclear Magnetic Resonance (NMR), CryoEM is able to produce volumetric images of proteins that are poorly soluble, large, and hard to crystallize. Due to various experimental difficulties, many proteins have been resolved to the medium resolution range (5-10Å). Many volumetric images (henceforth affectionately referred to as density maps) of large protein complexes have been generated [3, 5-9].

In *de novo* modeling, the atomic structure of the protein is derived using the information obtained from the 3-D density map and the 1-D structure of the protein. At medium resolution range, the atomic structure of the protein can't be derived directly

Z. Cai et al. (Eds.): ISBRA 2013, LNBI 7875, pp. 211–223, 2013.
© Springer-Verlag Berlin Heidelberg 2013

from the density map. Fortunately, other information can be used in *de novo* modeling. The location and the orientation of major secondary structure elements on the density map (SSEs-V) such as helices and β-sheets are detectable [10-14]. On the other hand, the location of secondary structure elements on sequence (SSEs-S) can be predicted [15, 16] (see Fig. 1). The early step in *de novo* modeling is to find the order and direction of assigning SSEs-S to SSEs-V. Such a problem called topology determination problem. Topology determination is challenging and is proven to be NP-Hard [17]. The total number of possible topologies is $\binom{M}{N}N!\,2^N$, where M is the number of SSEs-S and N is the number of SSEs-V. To derive the backbone of the protein, the correct topology of the SSEs has to be determined first and then the backbone of the protein can be derived for further optimization [18-20].

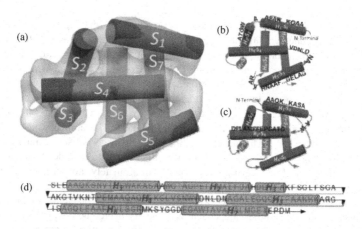

Fig. 1. Helical SSEs-V and the topologies. (a) The density map (gray) was simulated to 10Å resolution using protein 1FLP (PDB ID). The seven SSEs-V were detected using SSETracer. Two alternative topologies are shown. A portion of the sequence is threaded to the SSEs-V and the virtual loops (in blue) in the correct topology in (b) and a wrong topology in (c). (d) The SSEs-S of the protein are marked as H_1 to H_7.

Many *de novo* modeling tools use CryoEM volumetric skeleton [20, 21]. The CryoEM volumetric skeleton (henceforth affectionately referred to as skeleton) is a compact density volume that represents the shape of the original density map in a simplified form. The skeleton is used to reduce the search space of the topology problem or to derive the final atomic structure of the protein. When the detected SSEs-V are overlaid with the skeleton the connection relationship among them is reviewed.

Many methods have been developed to extract the skeleton of a 3-D object [22-27]. Many of 3-D skeletonization algorithms still have limitations when the noise is present [28]. Three tools were developed to extract the skeleton of the density maps: binary skeletonizer [29], grayscale skeletonizer [30] and interactive skeletonizer [31]. Binary skeletonizer is composed of two algorithms: iterative thinning and skeleton pruning [29]. On the other hand, the grayscale skeletonizer [30] is generated by applying the binary skeletonization on a range of segmentations at different thresholds. The threshold used to extract the skeleton still plays a major role in the final quality of the skeleton in both methods. Remarkably, no single threshold can be used to capture all features of the density map. When a less selective density threshold is used, more misleading connections appear in the skeleton (Fig. 2b). In contrast, the use of more selective threshold will result in discontinuities (Fig. 2a).

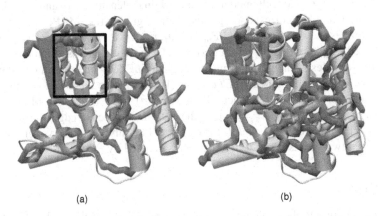

(a)   (b)

Fig. 2. An example of a skeleton (in green) produced by Gorgon for the authentic density map at 6.8Å resolution (EMDB ID 5100) and the corresponding Protein "Scorpion Hemocyanin resting state" (PDB ID 3IXV). The red cylinders represent helical SSEs-V. (a) The skeleton is extracted at 1.2 threshold and one gap is shown in the black box. (b) The skeleton is extracted at 1.1 threshold. The SSEs-V are over-connected. More outgoing connections can be visually seen clearly.

The quality of the density map obtained from the CryoEM experiments determines the quality of the skeleton. When the resolution of the density map is high, the skeleton is well resolved. At the medium resolution, the skeleton can be ambiguous. The skeleton sometimes suffers from the problem of incompleteness. Skeleton incompleteness, or skeleton gaps, can be defined as the absence of the skeleton connectivity when a real connection exists either on the density map or the structure of the protein. In this paper, we present a graph-based approach to help overcome the incompleteness problem of the skeleton using the original density map. The approach relies on a novel representation of the density map, where the map is modeled as a graph and a number of volume trees.

2   Materials and Methods

2.1   Basic Notions

The density map and the skeleton are examples of volumetric images. The volume image defined on an orthogonal grid, $\mathbb{Z}^3$. In the grid cell model, the cells of a cube in a 3-D volume are 3-D voxel locations with integer coordinates. The voxel p can be referred to by its orthogonal location (x, y, z). The 6-neighborhood of voxel p is a set defined with $N_6(p) = \{(x', y', z'): |x - x'| + |y - y'| + |z - z'| \leq 1\}$. The value saved in the cell corresponding to voxel p represents the associated magnitude of the electron density of the protein at that location and is denoted by $d(p)$. The voxel p is called *end voxel* if only one voxel with density greater than zero can be found in $N_6(p)/\{p\}$.

Let MAP be the grid cell model of the original density map and let $MAP_G = (V_m, E_m)$ denote the corresponding undirected graph for MAP, where $V_m = \{v: d(v) > 0\}$ is the set of nodes and $E_m = \{(v_1, v_2): v_1 \in N_6(v_2), v_1 \neq v_2\}$ is the set of undirected edges. In this paper, the terms node and voxel are used interchangeably to refer to the nodes of the graph. Similarly, let $SKELETON$ be the grid cell model of the original skeleton and let $SKELETON_G = (V_s, E_s)$ denote the corresponding undirected graph for $SKELETON$ where $V_s = \{v: d(v) > 0\}$ is the set of nodes and $E_s = \{(v_1, v_2): v_1 \in N_6(v_2), v_1 \neq v_2\}$ is the set of undirected edges. In the graph model, voxel p is called *end voxel* if it is connected to only one other voxel. Finally, a path between two voxels p and q in the graph is denoted by $P_{p,q}$.

2.2   Method

A possible solution to the problem of incompleteness is to develop a skeletonizer that uses different local thresholds. Instead, we present a threshold-independent approach to overcome the incompleteness problem of the skeletons extracted from the density maps. The first step in our approach is to preprocess MAP in order to keep only voxels with high density in a small neighborhood. We observe that such voxels are good representatives for local regions. We apply the concept of a screening filter called local-peak-counter (LPC) proposed in sheettracer [32]. The LPC rewards voxels with high local density values and thereby tolerates the variations in the magnitude of densities throughout the density map. In LPC, for each voxel p, the average density of a cube centered at p and with edge length of seven is calculated. Those voxels in the cube with density value greater than the calculated average have their counter incremented. At the end of counting, each voxel with density magnitude greater than 99% of the average densities calculated for the 343 cubes formed around it is saved in a new grid model called $PEAKS$. The voxels saved in $PEAKS$ are the strongest voxels that can be found locally in the density map. Each voxel is a peak of its local neighborhood. The intensity value of the peak voxels is set to the LPC counter while it is set to zero for other voxels.

The process of gap bridging starts by locating the *end voxels* in $SKELETON_G$. *End voxels* are good candidates because they are more likely to be around the gaps. Voxel p is out of our interest if $|N_6(p)/\{p\}| > 1$, which means that p is located on a continuous path of the skeleton. The essential idea used in this approach is to locate the local volume peaks around the *end voxels* in $SKELETON_G$. To do so, the map around that *end voxels* will need to be split into volumes that satisfy certain properties. Volume-based split was used by helix tracer [12]. The insight of the split process is to recognize the clusters of voxels that are of high local density. The split of the $PEAKS$ can be accomplished by building the corresponding directed graph $PEAKS_G = (V_p, E_p)$, where $V_p = \{v: d(v) > 0\}$ is the set of nodes and $E_p = \{(v_1, v_2): v_2 \in$

$N_6(v_1), d(v_2) = \max_{v \in N_6(v_1)} d(v), v_1 \neq v_2\}$ is a set of directed edges from the voxel to the highest-density voxel in its 6-adjacent neighborhood. $PEAKS_G$ is a directed-acyclic-graph and it is, if a linear asymptotic running time function is used to invert the direction of edges, a forest of trees is produced. The root of each tree is the voxel with the highest density in the volume tree. For the voxel p, let the volume tree contains p be denoted by $VOLTREE(p)$. Given $PEAKS_G$ and any voxel $p \in V_p$, the construction of $VOLTREE(p)$ is simple and asymptotically linear. Fig. 3 depicts an example of $PEAKS_G$ (in pink) for the authentic density map (EMDB ID 5030).

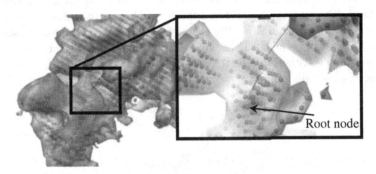

Fig. 3. Local peaks and volume trees. The graph of local peaks is shown in pink for the authentic density map 5030 (EMDB ID). The graph of skeleton voxels is also shown in blue. The detected *end voxels* are colored in red. In the augmented portion of the map (boxed), some dashed lines are drawn to show some examples of close voxels found for some pendant voxels in the volume tree. The root of the volume tree is colored in black (also is marked).

Let $ENDv$ denote the *end voxels* determined for $SKELETON_G$. Some of these *end voxels* may not be part of $PEAKS_G$. In such case, a nearby voxel that belongs to $SKELETON_G$ and $PEAKS_G$ is picked up and marked as a new *end voxel*. After $ENDv$ is found, a list of candidate voxels, denoted by $CLOSE^i$ for the *end voxel* $i \in ENDv$ is determined. The voxels that belong to $CLOSE^i$ are good representatives for the high local density volumes surrounding the gap. To find voxels in $CLOSE^i$, the process starts by constructing the volume tree $VOLTREE(i)$ and then marking the pendant nodes of the tree. For each pendant node v, the closest voxel $q \in V_p$ that is not part of $VOLTREE(i)$ is added to $CLOSE^i$ and $VOLTREE(q)$ is constructed. If two voxels in $CLOSE^i$ belong to the same volume tree, the farthest should be removed from $CLOSE^i$. Furthermore, for each voxel $q \in CLOSE^i$, the voxel q should be removed from $CLOSE^i$ if there is a path of length less than ε between the voxel i and any node in $VOLTREE(q)$ in $SKELETON_G$. The value of ε used in this paper is 15Å.

The final phase of our approach is to connect each *end voxel* i with the candidate voxels in $CLOSE^i$. The voxels in $CLOSE^i$ are called candidates because not necessarily all of them will be connected with i. The phase involves two steps. The first step is to find a candidate path to bridge the gap. For each $q \in CLOSE^i$, we find the path $P_{i,r}$ where r is the root voxel of $VOLTREE(q)$ in $PEAKS_G$. The path $P_{i,r}$ actually is a concatenation of the three subpaths $P_{i,v}, P_{v,q}$, and $P_{q,r}$, where v is the pendant node of

$VOLTREE(i)$ that is closest to q. The paths $P_{i,v}$ and $P_{q,r}$ can be calculated directly from $PEAKS_G$ if the direction is ignored. The subpath $P_{v,q}$ is the part of the path that is located at the gap. To calculate $P_{v,q}$, the approach finds a path between v and q in MAP_G in a greedy manner. A best first search is applied to calculate the path with the highest local density. The counters used in LPC previously are used in such a search. When $P_{i,r}$ is calculated, the process stops if for any voxel $p \in P_{i,r}, p \in V_s$. That means that the process successfully finds a candidate path to bridge the gap. If no voxel in $P_{i,r}$ is located on the skeleton, the method continues the search for a path of high local density starting from r and ends at any voxel of $SKELETON_G$. Again, a best first search is used. The process stops when either it reaches a voxel that is part of the skeleton or cannot find a path at all. If a path is found, it will be added to the skeleton and $SKELETON_G$ will be updated. The bridged skeleton of the authentic map shown in Fig. 3 is overlaid with the original skeleton in Fig. 4. Fig. 5 illustrates the main steps of the proposed algorithm.

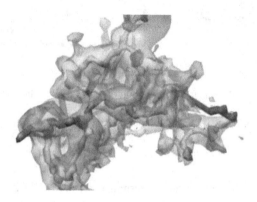

Fig. 4. The bridged skeleton. The original skeleton is shown in green. The generated bridged skeleton after applying our approach is shown in purple. The original density map is also shown (transparent gray).

3   Results

A set of fifteen density maps and their associated skeletons were used to evaluate the performance of our approach. Thirteen of the density maps were synthesized to 10Å resolution using the structure of the protein and the *molmap* command in Chimera package [33]. Two other density maps (EMDB ID 5100 with 6.8Å resolution and EMDB ID 5030 with 6.4Å resolution) are the authentic data downloaded from the EMDB. The backbone structures (3IXV_A and 3FIN_R) of the authentic density maps are available in the PDB and are aligned with the density maps in EMDB. The first 222 residues in the n-terminal of 3IXV are used in this test. The thirteen proteins selected for the synthesized maps are helical due to the fact that helices are often detected more accurately than the β-sheets in the medium resolution density maps. It is still a challenging problem to detect the SSEs-V from the medium resolution data when β-sheets are involved. Therefore, sheet-type SSEs-V are not considered in this test. For each density map, we use SSETracer [14] to detect the helical SSEs-V sticks. The true location of the helical SSEs-S was generated from the PDB file of the protein structures.

GapBridging $(MAP, SKELETON)$
Input: the grid model of the CryoEM map and skeleton
Output: the grid model of the bridged CryoEM skeleton

```
build SKELETONG
find ENDv
build PEAKSG
for each i ∈ ENDv
        if i ∉ PEAKSG
                find a nearby voxel i' ∈ PEAKSG and i' ∈ SKELETONG
                let i = i'
        endif
endfor
for each i ∈ ENDv
        find CLOSEi
        for each q ∈ CLOSEi
                let v be the pendant node of VOLTREE(i)
                let r be the root of VOLTREE(q)
                find the paths Pi,v , Pv,q and Pq,r
                let Pi,r = Pi,v ∘ Pv,q ∘ Pq,r
                if any v' ∈ Pi,r and v' ∈ SKELETONG
                        accept Pi,r and add it to SKELETONG and SKELETON
                else
                        find the path Pr,v'' from r to any node v'' ∈ SKELETONG
                        if Pr,v'' is found
                                add it to SKELETONG and SKELETON
                        endif
                endifelse
        endfor
endfor

return SKELETON
```

Fig. 5. The proposed algorithm. The pseudo code illustration of the core steps in the proposed algorithm that bridges the gaps of the skeletons extracted from CryoEM density maps.

The skeleton was obtained using Gorgon 2.1 [34]. Two kinds of the skeletons were extracted and used in testing: the binary skeleton and the grayscale skeleton. The binary skeleton was used as the base to extract the grayscale skeleton. In general, the skeletons were extracted from density maps at a threshold that visually minimizes the sheet planes in the skeleton and shows the connections between helical sticks. Gorgon 2.1 and MapEM [17, 21, 35] were used to evaluate the impact of our approach on the accuracy of final ranking of the true topology for each protein. The topologies were ranked before and after the bridging process of the gaps. The correctness evaluation of the two methods was carried out by comparing the produced topologies with the correct topology of each protein obtained from the PDB. The rank of the true topology is then reported in Table 1. A failure is reported (N/A in Table 1) if the tool cannot find the true topology within the top 35 topologies. Even though, the native topology can be found after the 35th topology, a failure is reported. Note that the gap problem may cause the problem of memory failure because of the size of the search space. When the skeleton quality is good enough, the search space of the problem is significantly reduced and large proteins can be tested. The present of gap problem negatively impacts the reduction of the search space and the entire *de novo* modeling as well.

We used the same skeleton and same helical SSEs-V sticks detected by SSETracer for both tools. A Max Euclidian Loop Distance parameter (ε) was set to 15Å. All other parameters are default parameters in Gorgon. The gap tolerance threshold used in MapEM was set to 10Å. MapEM can overcome the problem of the gap in skeletons for a particular length. The gap tolerance is the length of the gaps that MapEM can deal with. The top 35 ranked topologies were generated for each protein using Gorgon and MapEM before and after applying our approach on the skeleton. The binary skeleton used for protein 1HG5 is shown in Fig. 6. The original skeleton is extracted for 1HG5 (row 7, Table 1) at a threshold of 0.36. The skeleton consists of misleading points and gaps (circled in Fig. 6) as commonly seen in typical skeletons.

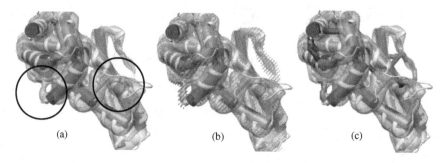

(a)   (b)   (c)

Fig. 6. The bridged skeleton for 1HG5. (a) The detected helical SSEs-V sticks (red) is superimposed on the original skeleton (green) and the synthesized original density map (gray) at 10Å resolution. Certain gaps (circled) in the skeleton are shown. (b) The local peaks generated from the original density map. (c) The bridged skeleton (purple) is superimposed on the original skeleton. The gaps found on the original skeleton are bridged and shown in purple.

We first tested the two tools using fifteen cases of two different kinds of skeletons obtained by Gorgon. There is a minor significant difference in accuracy between the two kinds of skeletons. The grayscale skeleton performs slightly better in certain cases as shown in Columns 6 and 10 of Table 1. The difference is clear when Gorgon is used as shown in Column 10. Gorgon was not able to find the true topology within the top 35 topologies for all binary skeletons as shown in column 9. On the contrary, it can predict the true topology when the grayscale skeleton is used for 40% of the set. Similarly, MapEM was able to find the native topology within the top 35 ranked topologies for ten of the fifteen test cases when a binary skeleton was used. On the other hand, it was able to find the true topology for the thirteen of the fifteen maps when the grayscale skeleton was used. MapEM performs better than Gorgon on original skeletons because it takes the gaps of the skeletons into consideration. MapEM successfully deals with gaps of length 10Å or shorter. If the gap is longer than 10Å or there is multiple small gaps on the skeleton between the SSEs-V, MapEM fails to find the true topology. For instance, MapEM fails to find the true topology for 3XIN because of two small consecutive gaps on the skeleton between the first two helices. Therefore, the performance of the two tools proves that the quality of the grayscale skeletons is better than binary skeletons.

Table 1. The accuracy of ranking true topology using density maps

No.	ID[a]	#SSE-S[b]	#SSE-V[c]	MapEM Original Skeletons		MapEM Bridged Skeletons		Gorgon Original Skeletons		Gorgon Bridged Skeletons	
				Binary[d]	Gray[e]	Binary	Gray	Binary	Gray	Binary	Gray
1	1ENK	3	3	7 (0.32)	23 (0.30)	15	15	N/A	6	5	5
2	3FIN*	4	4	1 (3.70)	9 (2.50)	1	1	N/A	3	1	1
3	3THG	4	4	7 (0.32)	2 (0.28)	3	1	N/A	1	4	1
4	1GV2	6	6	19 (0.37)	21 (0.33)	5	4	N/A	N/A	1	1
5	1FLP	7	7	1 (0.33)	1 (0.28)	2	1	N/A	1	1	1
6	3IEE	9	8	4 (0.38)	4 (0.35)	4	4	N/A	N/A	3	4
7	1HG5	11	9	1 (0.36)	2 (0.32)	1	5	N/A	1	N/A	1
8	2OVJ	12	9	N/A (0.39)	6 (0.35)	3	3	N/A	N/A	5	4
9	2XB5	13	9	11 (0.35)	9 (0.29)	13	24	N/A	1	N/A	1
10	1P5X	13	9	4 (0.40)	4 (0.37)	4	5	N/A	N/A	12	10
11	3XIN*	14	10	N/A (1.20)	13 (1.10)	3	9	N/A	N/A	N/A	N/A
12	1OAZ	18	13	N/A (0.37)	N/A (0.32)	7	3	N/A	N/A	18	10
13	1HV6	18	13	N/A (0.36)	N/A (0.29)	N/A	N/A	N/A	N/A	N/A	N/A
14	1WER	20	15	N/A (0.36)	4 (0.31)	4	1	N/A	N/A	5	1
15	3HJL	20	20	1 (0.31)	7 (0.25)	1	1	N/A	N/A	2	2

a: Protein PDB ID, proteins with * are the proteins for authentic density maps.
b: The number of actual helical SSEs-S in the protein.
c: The number of detected helical SSEs-V from the density map.
d: The rank of the true topology using binary skeleton. The value in ( ) is the threshold used to extract the skeleton.
e: The rank of the true topology using grayscale skeleton. The value in ( ) is the threshold used to extract the skeleton.

Table 1 shows the performance of the tools after applying our approach on skeletons. Our approach was applied on both skeletons and the new skeletons were tested using the same tools. Similar to the performance with the original skeletons, the tools perform better for grayscale skeletons as shown in Columns 8 and 12. Gorgon could find the true topologies of eleven of the fifteen proteins that represent 73.3% of the set when the binary skeleton was used. However, the four cases that Gorgon fails to find the true topology for were successfully bridged using our approach. 86.7% of the topologies of the set were recognized correctly when the grayscale skeleton was used as shown in Column 12. The percentage of improvement of the number of true topologies recognized correctly for the grayscale skeletons is 117% (Columns 10 and 12).

For example, the true topology of the protein 1GV2 was correctly recognized when the bridged binary and grayscale skeletons were used (ranked 1 and 1 respectively, Row 4, Table 1). However, Gorgon fails to recognize the correct topology within the top 35 topologies when the original skeletons are used (Row 4, Table 1). Likewise, the performance of MapEM is improved by 40% when the binary skeletons are used (Columns 5 and 7). No significant improvement is found in the performance of MapEM for the grayscale skeletons other than the improved rank of some proteins. For example, the rank of the true topology for 3THG was moved from the second (Column 6, Row 3) to the first (Column 8, Row 3) position for the grayscale skeleton. The enhancement is expected to be less for MapEM since it tolerates the gaps of certain length.

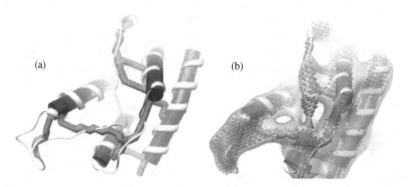

(a)   (b)

Fig. 7. Failure in topology determination. (a) The wrong skeleton connection of the protein 1HV6. The trace deviates from the structure to connect with the wrong helix. (b) The graph of the local peaks shows the relatively strong density on the correct trace.

MapEM and Gorgon fail to rank the true topology of 1HV6 (Row 13) within the top 35 topologies. In addition to the gaps, the skeletons contain wrong skeleton connections (Fig. 7). The skeletons of the protein wrongly deviate from the correct trace of the structure to reach a wrong helix (Fig. 7a). The skeletons have no *end voxel* at this region that indicates an existence of a gap. Hence, our approach will fail to bridge the trace. Interestingly, the graph of local peaks (Fig. 7b) shows some volume trees on the empty region of the skeleton. The local peaks indicate a relatively strong density on the map at the region after the skeleton deviates. Therefore, we believe that local peaks can be used to extract the skeleton of the density maps. The current skeletonizers are implemented using general thinning and pruning techniques. They may be acceptable for other domains where the quality of the descriptive skeleton is not crucial. In contrast, the connections between the SSEs-V are very important to the topology determination problem in *de novo* modeling. Thus, the quality of the skeleton becomes essential. Missing one connection may mislead the entire process. On the other hand, a vast tolerance of such errors may lead to miss the true topology in the top ranked list.

4 Conclusion

CryoEM is becoming an important structure determination technique. More density maps are being produced by the CryoEM experiments and many of them arrive at the medium resolution range. The topology of the secondary structure elements detected from the density map is a critical piece of information for deriving the atomic structures from such maps. Several tools for *de novo* prediction use the skeleton of the density map in order to reduce the search space of the topology problem or to derive the final atomic structure of the protein. The skeleton sometimes suffers from the problem of incompleteness that misleads the prediction process. In this paper, we presented a threshold-independent approach to overcome the incompleteness problem. The approach relies on a novel representation of the density map, where the map is modeled as a graph of local peaks and a set of volume trees.

We tested the approach using fifteen protein density maps in which thirteen are synthesized and two are authentic. Two kinds of the skeletons were extracted and used in testing: the binary skeleton and the grayscale skeleton. Gorgon 2.1 and MapEM were used to evaluate the impact of our approach on the accuracy of final ranking of the true topology for each protein. The topologies were ranked before and after the bridging process of the gaps. In general, the performance of the tools is better for grayscale skeletons. The test shows that our approach can improve the performance of the tools used in *de novo* modeling. The percentages of improvement are 117% and 40% for Gorgon and MapEM, respectively. Interestingly, local peaks are found to be an enhanced meter for local densities than the thinning and pruning techniques.

Acknowledgment. This work was supported by NSF Science & Technology Center grant (CCF-0939370) and NSF CAREER (CCF-0845888).

References

[1] Chiu, W., Schmid, M.F.: Pushing back the limits of electron cryomicroscopy. Nature Structural Biology 4, 331–333 (1997)
[2] Zhou, Z.H., Dougherty, M., Jakana, J., He, J., Rixon, F.J., Chiu, W.: Seeing the herpesvirus capsid at 8.5 A. Science 288(5467), 877–880 (2000)
[3] Ludtke, S.J., Song, J.L., Chuang, D.T., Chiu, W.: Seeing GroEL at 6 A resolution by single particle electron cryomicroscopy. Structure 12(7), 1129–1136 (2004)
[4] Chiu, W., Baker, M.L., Jiang, W., Zhou, Z.H.: Deriving folds of macromolecular complexes through electron cryomicroscopy and bioinformatics approaches. Current Opinion in Structural Biology 12(2), 263–269 (2002)
[5] Conway, J.F., Cheng, N., Zlotnick, A., Wingfield, P.T., Stahl, S.J., Steven, A.C.: Visualization of a 4-helix bundle in the hepatitis B virus capsid by cryo-electron microscopy. Nature 386(6620), 91–94 (1997)
[6] Zhang, X., Jin, L., Fang, Q., Hui, W.H., Zhou, Z.H.: 3.3 Å Cryo-EM Structure of a Non-enveloped Virus Reveals a Priming Mechanism for Cell Entry. Cell 141(3), 472–482 (2010)

[7] Baker, M.L., Jiang, W., Wedemeyer, W.J., Rixon, F.J., Baker, D., Chiu, W.: Ab initio modeling of the herpesvirus VP26 core domain assessed by CryoEM density. PLoS Computational Biology 2(10), e146 (2006)

[8] Martin, A.G., Depoix, F., Stohr, M., Meissner, U., Hagner-Holler, S., Hammouti, K., Burmester, T., Heyd, J., Wriggers, W., Markl, J.: Limulus polyphemus hemocyanin: 10 A cryo-EM structure, sequence analysis, molecular modelling and rigid-body fitting reveal the interfaces between the eight hexamers. Journal of Molecular Biology 366(4), 1332–1350 (2007)

[9] Villa, E., Sengupta, J., Trabuco, L.G., LeBarron, J., Baxter, W.T., Shaikh, T.R., Grassucci, R.A., Nissen, P., Ehrenberg, M., Schulten, K., Frank, J.: Ribosome-induced changes in elongation factor Tu conformation control GTP hydrolysis. Proceedings of the National Academy of Sciences of the United States of America (PNAS) 106(4), 1063–1068 (2009)

[10] Lasker, K., Dror, O., Shatsky, M., Nussinov, R., Wolfson, H.J.: EMatch: discovery of high resolution structural homologues of protein domains in intermediate resolution cryo-EM maps. IEEE/ACM Transactions on Computational Biology and Bioinformatics 4(1), 28–39 (2007)

[11] Jiang, W., Baker, M.L., Ludtke, S.J., Chiu, W.: Bridging the information gap: computational tools for intermediate resolution structure interpretation. Journal of Molecular Biology 308(5), 1033–1044 (2001)

[12] Del Palu, A., He, J., Pontelli, E., Lu, Y.: Identification of Alpha-Helices from Low Resolution Protein Density Maps. In: Proceeding of Computational Systems Bioinformatics Conference (CSB), pp. 89–98 (2006)

[13] Baker, M.L., Ju, T., Chiu, W.: Identification of secondary structure elements in intermediate-resolution density maps. Structure 15(1), 7–19 (2007)

[14] Si, D., Ji, S., Al Nasr, K., He, J.: A machine learning approach for the identification of protein secondary structure elements from cryoEM density maps. Biopolymers 97, 698–708 (2012)

[15] Jones, D.T.: Protein secondary structure prediction based on position-specific scoring matrices. Journal of Molecular Biology 292(2), 195–202 (1999)

[16] Pollastri, G., McLysaght, A.: Porter: a new, accurate server for protein secondary structure prediction. Bioinformatics 21(8), 1719–1720 (2005)

[17] Al Nasr, K., Ranjan, D., Zubair, M., He, J.: Ranking Valid Topologies of the Secondary Structure elements Using a constraint Graph. Journal of Bioinformatics and Computational Biology 9(3), 415–430 (2011)

[18] Al Nasr, K., Sun, W., He, J.: Structure prediction for the helical skeletons detected from the low resolution protein density map. BMC Bioinformatics 11(suppl. 1), S44 (2010)

[19] Lindert, S., Staritzbichler, R., Wötzel, N., Karakaş, M., Stewart, P.L., Meiler, J.: EM-Fold: De Novo Folding of α-Helical Proteins Guided by Intermediate-Resolution Electron Microscopy Density Maps. Structure 17(7), 990–1003 (2009)

[20] Lindert, S., Alexander, N., Wötzel, N., Karaka, M., Stewart, P.L., Meiler, J.: EM-Fold: De Novo Atomic-Detail Protein Structure Determination from Medium-Resolution Density Maps. Structure 20(3), 464–478 (2012)

[21] Al Nasr, K., Chen, L., Si, D., Ranjan, D., Zubair, M., He, J.: Building the initial chain of the proteins through de novo modeling of the cryo-electron microscopy volume data at the medium resolutions. In: Proceedings of the ACM Conference on Bioinformatics, Computational Biology and Biomedicine, Orlando, Florida, pp. 490–497 (2012)

[22] Khromov, D., Mestetskiy, L.: 3D Skeletonization as an Optimization Problem. In: The Canadian Conference on Computational Geometry, Charlottetown, pp. 259–264 (2012)

[23] Dey, T.K., Zhao, W.: Approximate medial axis as a voronoi subcomplex. In: Proceedings of the Seventh ACM Symposium on Solid Modeling and Applications, Saarbrücken, Germany, pp. 356–366 (2002)

[24] Foskey, M., Lin, M.C., Manocha, D.: Efficient Computation of A Simplified Medial Axis. Journal of Computing and Information Science in Engineering 3(4), 274–284 (2003)

[25] Tam, R., Heidrich, W.: Shape simplification based on the medial axis transform, pp. 481–488

[26] Tran, S., Shih, L.: Efficient 3D binary image skeletonization, pp. 364–372

[27] She, F.H., Chen, R.H., Gao, W.M., Hodgson, P.H., Kong, L.X., Hong, H.Y.: Improved 3D Thinning Algorithms for Skeleton Extraction, pp. 14–18

[28] van Dortmont, M.A.M.M., van de Wetering, H.M.M., Telea, A.C.: Skeletonization and distance transforms of 3D volumes using graphics hardware. In: Kuba, A., Nyúl, L.G., Palágyi, K. (eds.) DGCI 2006. LNCS, vol. 4245, pp. 617–629. Springer, Heidelberg (2006)

[29] Ju, T., Baker, M.L., Chiu, W.: Computing a family of skeletons of volumetric models for shape description. Computer-Aided Design 39(5), 352–360 (2007)

[30] Abeysinghe, S.S., Baker, M., Wah, C., Tao, J.: Segmentation-free skeletonization of grayscale volumes for shape understanding, pp. 63–71

[31] Abeysinghe, S.S., Ju, T.: Interactive skeletonization of intensity volumes. Vis. Comput. 25(5-7), 627–635 (2009)

[32] Kong, Y., Zhang, X., Baker, T.S., Ma, J.: A Structural-informatics approach for tracing beta-sheets: building pseudo-C(alpha) traces for beta-strands in intermediate-resolution density maps. Journal of Molecular Biology 339(1), 117–130 (2004)

[33] Pettersen, E.F., Goddard, T.D., Huang, C.C., Couch, G.S., Greenblatt, D.M., Meng, E.C., Ferrin, T.E.: UCSF Chimera—A visualization system for exploratory research and analysis. Journal of Computational Chemistry 25(13), 1605–1612 (2004)

[34] Baker, M.L., Abeysinghe, S.S., Schuh, S., Coleman, R.A., Abrams, A., Marsh, M.P., Hryc, C.F., Ruths, T., Chiu, W., Ju, T.: Modeling protein structure at near atomic resolutions with Gorgon. Journal of Structural Biology 174(2), 360–373 (2011)

[35] Al Nasr, K.: De novo protein structure modeling from cryoem data through a dynamic programming algorithm in the secondary structure topology graph. Dissertation, Department of Computer Science, Old Dominion University (2012)

Measure the Semantic Similarity of GO Terms Using Aggregate Information Content

Xuebo Song[1], Lin Li[1], Pradip K. Srimani[1], Philip S. Yu[2],
and James Z. Wang[1,*]

[1] School of Computing, Clemson University, Clemson, SC 29634-0974
{xuebos,ll,srimani,jzwang}@clemson.edu
[2] Department of Computer Science, University of Illinois, Chicago, IL 60607
psyu@uic.edu

Abstract. The rapid development of Gene Ontology (GO) and huge amount of biomedical data annotated by GO terms necessitate computation of semantic similarity of GO terms and, in turn, measurement of functional similarity of genes based on their annotations. This paper proposes a novel and efficient method to measure the semantic similarity of GO terms. This method addresses the limitations in existing GO term similarity measurement methods by using the information content of all ancestor terms of a GO term to determine the GO term's semantic content. The aggregate information content of all ancestor terms of a GO term implicitly reflects the GO term's location in the GO graph and also represents how human beings use this GO term and all its ancestor terms to annotate genes. We show that semantic similarity of GO terms obtained by our method closely matches the human perception. Extensive experimental studies show that this novel method outperforms all existing methods in terms of the correlation with gene expression data.

1 Introduction

Gene Ontology (GO) [1] describes the attributes of genes and gene products (either RNA or protein, resulting from expression of a gene) using a structured and controlled vocabulary. GO consists of three ontologies: biological process (BP), cellular component (CC) and molecular function (MF), each of which is modeled as a directed acyclic graph. In recent past, many biomedical databases, such as Model Organism Databases (MODs) [2], UniProt [3], SwissProt [4], have been annotated by GO terms to help researchers understand the semantic meanings of biomedical entities. With such a large diverse biomedical data set annotated by GO terms, computing functional or structural similarity of biomedical entities has become a very important research topic. Many researchers have tried to measure the functional similarity of genes or proteins based on their GO annotations [5–16]. Since different biomedical researchers may annotate the same or similar gene function with different but semantically similar GO terms based on

---

* Corresponding author.

Z. Cai et al. (Eds.): ISBRA 2013, LNBI 7875, pp. 224–236, 2013.
© Springer-Verlag Berlin Heidelberg 2013

their research findings, an accurate measure of semantic similarity of GO terms is critical to accurate measurement of gene functional similarities.

While those existing studies have proposed different methods to measure the semantic similarity of GO terms, they all have their limitations. In general, there are three types of methods for measuring the semantic similarity of GO terms: node-based [9, 17–19], edge-based [10, 20, 21], and hybrid [6, 11] methods. See section 2 for a brief discussion of some most representative methods and their limitations.

In this paper, we propose a novel method to measure the semantic similarity of GO terms. This method is based on two major observations: (1) In general, the dissimilarity of GO terms near the root (more general terms) of GO graph should be larger than that of the terms at a lower level (more specific terms); (2) the semantic meaning of one GO term should be the aggregation of all semantic values of its ancestor terms (including the term itself). The first observation follows the human perception of term semantic similarity at different ontology levels. The second observation agrees with how human beings use the term to annotate genes.

The rest of the paper is organized as follows. We review existing most representative methods for semantic similarity measurement of GO terms in section 2; we introduce our proposed Aggregate Information Content based approach (AIC) in section 3. Section 4 provides details of experimental evaluation of AIC, while section 5 concludes the paper with a summary of unique charateristics of AIC.

2 Related Prior Work

A large number of studies [5–14] have appeared in the literature in the last 15 years to measure the semantic similarity of GO terms. All of these methods can be broadly classified into three categories: node-based, edge-based, and hybrid methods. The three most cited representative methods [17–19] were originally designed to measure the semantic similarity of natural language terms. While each of them has its limitations they have been widely adopted by bioinformatics researchers to measure the semantic similarity of GO terms. In 2007, Wang [6] proposed a new measure of the semantic similarity of GO terms: this new hybrid method considers both the GO structure and the semantic content (biological meaning) of the GO terms in measuring the semantic similarity of GO terms, and many studies [5, 11, 15, 16] have shown the superiority of this hybrid method. Besides, it has been widely accepted by biomedical researchers [11] since it was published.

2.1 Limitations of Current Methods

Node-based measures (e.g. Resnik's [17], Lin's [18], Jiang and Conrath's [19], Schlicker's [9]) rely mainly on Information Content (IC) of the GO terms to represent their semantic values; IC of a GO term is derived from the frequency of its presence (including the presence of its children terms) in a certain corpus (e.g. SGD database, GO database). Resnik's [17] method concentrates only on the Maximum Information Contained in Ancestors (MICA) of the compared GO terms, but ignores the locations of these terms in the GO graph, e.g., a GO term's distance from the root of the ontology, and the semantic impact of other ancestor terms. A term's distance to the root of the ontology shows the specialization level of this term in human perception. If a term is far from the root in the ontology, it means biomedical researchers know more details about this term and the meaning of the term is more specific. On the other hand, if a term is closer to the root of the ontology, it means the term is a more general term, such as cellular process or metabolic process, which does not provide too much details about the related biomedical entities. Ignoring the specialization level of a term is the principal reason that the semantic similarity obtained by these methods is inconsistent with human perception; they suffer from "shallow annotation" problem [8, 13, 6] in which the semantic similarity of GO terms near the root of the ontology are sometimes measured very high.

Edge-based approaches [10, 20, 21] are based on the length of graph paths connecting the terms being compared. Some edge-based approaches [20] treat all edges equally, ignoring the levels of edges in the ontology. This simple edge-based approach also suffers from "shallow annotation" because based on this approach, the semantic similarity of two terms with a certain graph distance near the root would be equal to the semantic similarity of two terms with the same graph distance but away from the root. To address the "shallow annotation" problem, other edge-based methods [10, 21] assign different weights to the edges at the different levels of the ontology, assuming that the edges at the same level of the ontology have the same weight. However, the terms at the same level of the GO graph do not always have the same specificity because different gene properties demand different levels of detailed studies. It means the edges at the same level of the GO graph but in different GO branches do not necessarily have the same weights.

The hybrid method [6] considers both the GO structure and the semantics (biological meanings) of GO terms at different ontological levels. However, this method uses two semantic contribution factors, obtained from empirical study of gene classification of certain species, to calculate the semantic values of GO terms. Semantic contribution factors obtained by empirical studies on genes from certain species may not be suitable for genes of other species.

2.2   Review of Existing Representative Methods

We provide a brief overview of the four most representative methods for GO term semantic similarity measure: Method A by Resnik [17], Method B by Lin [18], Method C by Jiang and Conrath [19], and Method D by Wang et. al [6]. We use these four methods as benchmarks to evaluate the relative performance of our proposed AIC method in this paper in the next sections.

Method A. The frequency of a GO term is recursively defined as,

$$freq(t) = annotation(t) + \sum_{i \in child(t)} freq(i) \tag{1}$$

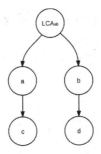

Fig. 1. GO terms at different ontology levels sharing the same LCA

where $annotation(t)$ is the number of gene products annotated with term t in the GO database. $child(t)$ is the set of children of term t. For each term $t \in T$, $p(t)$ denotes the probability that term t occurs in the GO database,

$$p(t) = freq(t)/freq(root) \tag{2}$$

Information Content(IC) of term t is defined as

$$IC(t) = -\log p(t) \tag{3}$$

Method A uses Maximum Information Contained in Ancestors (MICA) of two terms to measure the semantic similarity between them.

$$sim_{GO}(a,b) = \max_{c \in P(a,b)} IC(c) \tag{4}$$

where $P(a,b)$ denotes the set of common ancestor terms of term a and term b in the ontology graph. Based on the definition of IC in Method A (Equations 1, 2, 3), MICA often happens to be the IC value of the Least Common Ancestor LCA of terms a and b.

The principal limitation of method A derives from the fact that it considers only MICA of two terms while ignoring the distances of the two terms to their LCA and the semantic contribution of other ancestor terms. For example, terms a and b have the same LCA with terms c and b in the partial GO graph shown in Figure 1. Using method A, the semantic similarity between term a and b would be equal to the semantic similarity between term c and d, inconsistent with human perception.

Methods B & C. Method B is based on the ratio between IC values of two terms and that of their MICA; the semantic similarity between two terms a and b is defined as,

$$sim_{GO}(a,b) = \frac{2 * max_{c \in P(a,b)} IC(c)}{IC(a) + IC(b)} \tag{5}$$

Method C introduces the concept of term distance into the semantic similarity calculation. The intuition is that two terms closer in the GO graph should be more similar than two terms farther in the GO graph. The distance between two terms a and b is defined as

$$Dis_{GO}(a, b) = IC(a) + IC(b) - 2 * \max_{c \in P(a,b)} IC(c) \qquad (6)$$

The semantic similarity of two terms a and b are then defined as

$$sim_{GO}(a, b) = \frac{1}{1 + Dis_{GO}(a, b)} \qquad (7)$$

Note: Methods B and C ameliorated the principal limitation of Method A by implicitly considering the graph distance of the two terms in the semantic similarity measure. Consider the example in Figure 1; $sim_{GO}(c, d)$ should be less than $sim_{GO}(a, b)$ according to human perception because the graph distance between c and d is greater than the graph distance between a and b. Since term a is a parent of term c, we have $freq(a) > freq(c)$ and $p(a) > p(c)$ (Equations 1 and 2). According to the definition of IC in Equation 3, we have $IC(c) > IC(a)$. Similarly, we have $IC(d) > IC(b)$. Therefore, the semantic similarity values obtained by both methods B and C are consistent with human perception in this aspect.

However, it is possible that a GO term has multiple parent terms with different semantic relations; using MICA alone does not account for multiple parents. Also, two terms at a higher level (more general terms) of GO graph should be, as is perceived by humans, semantically more dissimilar than two terms with the same graph distance at a lower level (more specific terms). Because methods B and C do not consider the specialization level of two terms' LCA in the semantic similarity measure, the semantic similarity values obtained by these two methods may still be inconsistent with the human perception as demonstrated in our experiment in Section 4.

Method D. Method D attempts to address the shortcomings of other existing methods by aggregating the semantic contributions of ancestor terms in the GO graph. The S-value of GO term t related to term a (where term t is an ancestor of term a) is defined as,

$$S_a(t) = \begin{cases} 1 & \text{if } t = a \\ \max\{w_e * S_a(t') \mid t' \in \text{children of } t\} & \text{if } t \neq a \end{cases} \qquad (8)$$

where w_e is the semantic contribution factor of an edge. Then the semantic value (SV) of a GO term a is,

$$SV(a) = \sum_{t \in T_a} S_a(t) \qquad (9)$$

where T_a is the set of GO terms in DAG_a (Directed Acyclic Graph consisting all the ancestors of the term including the term itself). Finally, the semantic similarity between two GO terms a, b is defined as,

$$sim_{GO}(a, b) = \frac{\sum_{t \in T_a \cap T_b} (S_a(t) + S_b(t))}{SV(a) + SV(b)} \qquad (10)$$

where $S_a(t)$ is the S-value of GO term t related to term a and $S_b(t)$ is the S-value of GO term t related to term b. While this method combines both the semantic and the topological information of GO terms to address weaknesses of methods A, B and C, it still suffers from two disadvantages. First, it needs to use a semantic contribution factor value (weight) empirically obtained from gene classification to calculate the semantic values of GO terms. Using a semantic contribution factor obtained from the classification of genes from certain species may not be suitable for measuring the functional similarity of genes in other species. Second, some biomedical studies need to obtain the similarity matrix for a large group of GO terms or genes. Dynamically calculating the semantic values of GO terms is time consuming and may result in a long user response time, which will be shown in our experimental studies.

3   Proposed Aggregate Information Content Based Method (AIC)

We address the limitations of the existing methods using an aggregate information content approach.

3.1   GO Similarity

This *aggregate information content* based similarity measurement method (Method AIC) considers the aggregate contribution of the ancestors of a GO term (including this GO term) to the semantics of this GO term, and takes into account how human beings use the terms to annotate genes. We use a term's IC value, as defined before (Equations 1, 2, 3), to represent their semantic contribution values. Given the fact that terms at upper levels (more general terms) of ontology graph are less specific than those at lower levels, we define the weight of a term t as,

$$W(t) = 1/IC(t) \qquad (11)$$

We further propose a logarithmic model to normalize W(t) into a semantic weight $SW(t)$:

$$SW(t) = \frac{1}{1 + e^{-W(t)}} \qquad (12)$$

We then compute semantic value $SV(a)$ of the GO term a by adding the semantic weights of all its ancestors (i.e., aggregating semantic contribution of the ancestors).

$$SV(a) = \sum_{t \in T_a} SW(t) \qquad (13)$$

where T_a is the set of all of its ancestors including a itself. We define the semantic similarity between GO terms a and b, based on their aggregate information content (AIC), as follows.

$$sim_{GO}(a, b) = \frac{\sum_{t \in T_a \cap T_b} 2 * SW(t)}{SV(a) + SV(b)} \qquad (14)$$

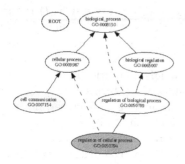

Table 1. IC values & Semantic Weights of GO terms

Go Terms	IC value	SW value
0050794	1.2931	0.6842
0007154	2.0939	0.6172
0050789	1.1339	0.7072
0065007	1.0343	0.7245
0009987	0.4346	0.9090
0008150	0	1

Fig. 2. GO Graph containing terms GO:0050794 and GO:0007154

where $SW(t)$ is the semantic weight of term t defined in Equation 12, and $SV(t)$ is the semantic value of term t defined in Equation 13. Aggregating the semantic contribution of all ancestor terms implicitly factors in the position of the term in the GO graph, and overcomes the weakness of the MICA based approaches.

We demonstrate how to use the AIC method to compute the semantic similarity between two terms, GO:0050794 and GO:0007154, shown in Figure 2.(All the similarity comparison figures showed in this paper are retrieved from the tools in [22].) First, we use the GOSim R package [23] to retrieve the IC information for all related GO terms, shown in Table 1. Second, we calculate the semantic weight for each GO term using Equation 12. Finally, we use Equation 13 and Equation 14 to get the semantic similarity of GO terms GO:0050794 and GO:0007154 as $sim_{GO}(0050794, 0007154) = 0.5828$.

3.2   Gene Similarity

There are several methods [6, 8, 12] to measure the functional similarity of gene products based on the semantic similarity of GO terms. The common methods are: MAX [6, 8] and AVE [12] methods; they define functional similarity between gene products as the maximum or average semantic similarity values over the GO terms annotating the genes respectively. In this paper, we use AVE method as follows,

$$sim_{AVE}(g_1, g_2) = \underset{\substack{t_1 \in annotation(g_1) \\ t_2 \in annotation(g_2)}}{\text{average}} sim(t_1, t_2) \qquad (15)$$

where $annotation(g)$ is the set of GO terms that annotates gene g. Although some studies [6, 8] use the MAX method to compute the functional similarity of genes, people [5] found that the AVE method is more stable and less sensitive to outliers. In addition, the AVE method is more compatible with our original objective of capturing all available information while the MAX method often ignores the contribution of other GO terms.

4 Experimental Evaluation of AIC

It is well known, as demonstrated in [7, 5, 8], that there is a high correlation between gene expression data and the gene functional similarity obtained from GO term similarities, i.e., genes with similar expression patterns should have high similarity in GO based measures because they should be annotated with semantically similar GO terms. We use the correlation of genes obtained from gene expression data to validate the gene functional similarities obtained by GO based similarity measures. As in many existing studies [13, 24–26], we use gene expression data from Spellman dataset [27], which comprises of 6178 genes, to obtain the gene correlation patterns. The gene annotation data used to calculate the gene functional similarity is obtained from the GO database (2012-07). In the next two subsections, we provide comparison of our method (AIC) with the state-of-the-art current methods: Method A [17], Method B [18], Method C [19], and Method D [6] in terms of GO term semantic similarity and gene functional similarity.

4.1 Evaluating AIC Method Using GO Term Semantic Similarity

From human perspective, we know that two GO terms at higher levels of the gene ontology should have larger dissimilarity than two GO terms with the same graph distance at lower levels. Our AIC method is compatible with this observation in that two GO terms with the same graph distance at the lower levels of the gene ontology usually share more common ancestors. Therefore, the semantic similarity of GO terms obtained by our AIC method is consistent with human perception as shown in an illustrative example from our experimental results in Figure 3 and Table 2.

Consider the two GO terms GO:0005739 and GO:0005777 as shown in Figure 3. The semantic similarity values obtained by Methods A, B, C, D and AIC are shown in Table 2. These two very specific GO terms have only one different ancestor term GO:0042579; the semantic similarity between them should be very high. However, the semantic similarity values obtained by Method A [17], Method B [18], and Method C [19] fail to exhibit this expected behavior while Method D [6] and the proposed AIC method correctly exhibit this expected behavior. This observation reinforces our previous contention that use of MICA alone in computing similarity is not sufficient because of loss of important information.

Now, we check whether all these semantic similarity measurement methods agree with the human perspective: two GO terms at higher levels of the gene ontology should have larger dissimilarity than two GO terms with the same graph distance at lower levels. We calculate the semantic similarity between GO:0044424 and GO:0005622 (Group 1) and the semantic similarity between GO:0044444 and GO:0005737 (Group 2). The semantic similarity values are shown in Table 2. These two groups of GO terms have similar structure in the GO graph except group 1 is closer to the root of the GO graph. Based on human perception, the semantic similarity of GO terms in group 1 should be less than that in group 2 since GO terms in group 2 are at a lower level of the GO graph. However, only methods A, D and our AIC method satisfy this property. The

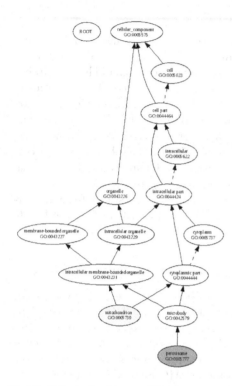

Fig. 3. GO graph of terms GO:0005739 and GO:0005777

Table 2. Semantic similarity values of GO term pairs obtained by different methods

Dataset	Method	Similarity
SW(GO:0005739, GO:0005777)	A	0.135
	B	0.335
	C	0.464
	D	0.797
	AIC	0.915
SW(GO:0044424, GO:0005622)	A	0.049
	B	0.948
	C	0.990
	D	0.845
	AIC	0.902
SW(GO:0044444, GO:0005737)	A	0.104
	B	0.872
	C	0.960
	D	0.879
	AIC	0.942

semantic similarity values obtained by methods B and C are inconsistent with the human perception because these two methods do not consider the specialization level of two terms' LCA in the semantic similarity measure. The "shallow annotation" problem is clearly shown in these experiments.

4.2 Evaluating AIC Using Correlation with Gene Expression Data

In our next set of experiments, we first use Pearson's correlation to compute the gene expression similarity with the Spellman dataset [27]. Then, we calculate the correlation between the functional similarity of these genes obtained from BP ontology and the gene expression similarity. The objective is, as stated in [7], to test the hypothesis that pairs of genes exhibiting similar expression levels which are measured by the absolute correlation values in gene expression data tend to have high functional similarities between each other. The average of correlation coefficients between genes within an expression similarity interval estimates the mean of the statistical distribution of correlations; and it shows the underlying trend that relates expression similarity and functional similarity. We split the gene pairs into groups with equal intervals according to the absolute

Table 3. Pearson's correlation coefficients between gene expression data and gene functional similarities obtained by different semantic similarity measurement methods

Groups	Method B [18]	Method C [19]	Method A [17]	Method D [6]	**Proposed AIC**
4	0.789	0.930	0.614	0.929	**0.966**
5	0.717	**0.889**	0.561	0.802	0.850
6	0.569	0.700	0.413	0.745	**0.774**
7	0.622	**0.761**	0.519	0.725	0.733
8	0.597	0.675	0.496	0.706	**0.714**
9	0.659	0.664	0.417	0.745	**0.778**
10	0.620	0.730	0.403	0.733	**0.772**
11	0.665	0.691	0.419	0.725	**0.761**
12	0.485	0.722	0.246	0.716	**0.782**
13	0.525	0.715	0.321	0.709	**0.791**

Table 4. Computation Efficiency of Methods D and AIC

	Execution Time (seconds)		
# of Gene Pairs	200	500	2000
Method D	173	3506	36123
Method AIC	56	261	7632

gene expression correlation values between gene pairs, as in previous studies [13, 5, 7, 8], and then compute Pearson's correlation coefficient between the mean of gene functional similarities and the mean of gene expression correlation values in each group. We split gene pairs into 4-13 groups respectively. We again compare the results obtained using four existing methods (Methods A, B, C and D) and those obtained using our AIC method, as shown in Table 3. The experimental results show that our AIC method generally outperforms other four methods with higher correlation coefficients between gene functional similarity and gene expression similarity.

4.3 Evaluating the Computation Efficiency of the AIC Method

While methods D and AIC show superiority to other three methods in agreement with human perception and in correlation with gene expression data, Method D is computationally expensive due to the recursive computation of semantic values of GO terms. On the other hand, our proposed AIC method uses the aggregate IC value, which can be precomputed, to represent the semantic value of a GO term. Thus, method AIC should be computationally more effective. We use the execution time of computing the functional similarities of a large number of gene pairs to evaluate the computation efficiency of our proposed AIC method. In this experiment, we use methods D and AIC to compute the functional similarities of three sets of gene pairs. The numbers of genes in these sets are 200, 500 and 2000 respectively. The experiment was conducted on a Linux box with a i7-2600K CPU @ 3.40GHz, 8G memory. The execution time are shown in Table 4. As demonstrated by the experimental results, method AIC is considerably faster than method D.

5   Conclusion

Experimental results in Section 4 demonstrate the superiority of the proposed AIC method over the current ones. Method AIC is characterized with the following unique features:

- It does not suffer from "shallow annotation". Note that, in Equation 14 the denominator is smaller when terms are annotated at the top levels, i.e., the equal difference on the numerator will result in a larger difference in the semantic similarity value. Thus, the semantic similarity value of two terms at top levels is less than that of two terms with the same graph distance at lower levels. This is consistent with human perspectives.
- It exhibits high correlation coefficient between the gene expression similarity and the GO based functional similarity.
- It is computationally much faster than the popular hybrid method [6].

In summary, the proposed method AIC is very promising in that it outperforms all existing state-of-the-art methods in terms of consistency with human perception, correlation with gene expression data and computational efficiency.

Acknowledgement. The work was partially supported by NSF Awards DBI-0960586, DBI-0960443 and CCF 0832582, and NIH award 1 R15 CA131808-01.

References

1. The Gene Ontology Consortium. Gene ontology: tool for the unification of biology. Nature Genetics 25, 25–29 (2000)
2. Stein, L.D., Mungall, C., Shu, S., Caudy, M., Mangone, M., Day, A., Nickerson, E., Stajich, J.E., Harris, T.W., Arva, A., Lewis, S.: The generic genome browser: A building block for a model organism system database. Genome Research 12, 1599–1610 (2002)
3. The UniProt Consortium. The uniprot consortium: The universal protein resource (uniprot). Nucleic Acids Research, pp. 190–195 (2008)
4. Kriventseva, E.V., Fleischmann, W., Zdobnov, E.M., Apweiler, R.: Clustr: a database of clusters of swiss-prot+trembl proteins. Nucleic Acids Research 29, 33–36 (2001)
5. Xu, T., Du, L., Zhou, Y.: Evaluation of go-based functional similarity measures using s.cerevisiae protein interaction and expression profile data. BMC Bioinformatics 9, 472 (2008)

6. Wang, J.Z., Du, Z., Payattakool, R., Yu, P.S., Chen, C.-F.: A new method to measure the semantic similarity of go terms. Bioinformatics 23, 1274–1281 (2007)

7. Wang, H., Azuaje, F., Bodenreider, O., Dopazo, J.: Gene expression correlation and gene ontology-based similarity: An assessment of quantitative relationships. In: Proc. of the 2004 IEEE Symposium on Computational Intelligence in Bioinformatics and Computational Biology, pp. 25–31 (2004)

8. Sevilla, J.L., Segura, V., Podhorski, A., Guruceaga, E., Mato, J.M., Martinez-Cruz, L.A., Corrales, F.J., Rubio, A.: Correlation between gene expression and go semantic similarity. IEEE/ACM Transactions on Computational Biology and Bioinformatics 2, 330–338 (2005)

9. Schlicker, A., Domingues, F.S., Rahnenfuhrer, J., Lengauer, T.: A new measure for functional similarity functional similarity of gene products based on gene ontology. BMC Bioinformatics 7, 302 (2006)

10. Cheng, J., Cline, M., Martin, J., Finkelstein, D., Awad, T., Kulp, D., Siani-Rose, M.A.: A knowledge-based clustering algorithm driven by gene ontology. Journal of Biopharmaceutical Statistics 14(3), 687–700 (2004)

11. Pesquita, C., Faria, D., Falcao, A.O., Lord, P., Couto, F.M.: Semantic similarity in biomedical ontologies. PLoS Computational Biology 5(7), e1000443 (2009)

12. Azuaje, F., Wang, H., Bodenreider, O.: Ontology-driven similarity approaches to supporting gene functional assessment. In: Proc. of the ISMB 2005 SIG Meeting on Bio-ontologies, pp. 9–10 (2005)

13. Li, B., Wang, J.Z., Luo, F., Feltus, F.A., Zhou, J.: Effectively integrating information content and structural relationship to improve the gene ontology similarity measure between proteins. In: The 2010 International Conference on Bioinformatics & Computational Biology (BioComp 2010), pp. 166–172 (2010)

14. Pesquita, C., Faria, D., Bastos, H., Falcao, A.O., Couto, F.M.: Evaluating go-based semantic similarity measures. In: Proc. of the 10th Annual Bio-Ontologies Meeting 2007, pp. 37–40 (2007)

15. Ravasi, T., et al.: An atlas of combinatorial transcriptional regulation in mouse and man. Cell 140(5), 744–752 (2010)

16. Washington, N.L., Haendel, M.A., Mungall, C.J., Ashburner, M., Westerfield, M., Lewis, S.E.: Linking human diseases to animal models using ontology-based phenotype annotation. PLoS Biology 7(11), e1000247 (2009)

17. Resnik, P.: Semantic similarity in taxonomy: An information-based measure and its application to problems of ambiguity in natural language. Journal of Artificial Intelligence Research 11, 95–130 (1999)

18. Lin, D.: An information-theoretic definition of similarity. In: Proc. Int. Conf. on Machine Learning, pp. 296–304 (1998)

19. Jiang, J.J., Conrath, D.W.: Semantic similarity based on corpus statistics and lexical taxonomy. In: Proc. Int. Conf. on Research in Computational Linguistics, pp. 19–33 (1997)

20. Pekar, V., Staab, S.: Taxonomy learning: factoring the structure of a taxonomy into a semantic classification decision. In: Proc. Int. Conf. on Computational Linguistics, vol. 2, pp. 786–792 (2002)

21. Wu, H., Su, Z., Mao, F., Olman, V., Xu, Y.: Prediction of functional modules based on comparative genome analysis and gene ontology application. Nucleic Acids Research 33(9), 2822–2837 (2005)

22. Du, Z., Li, L., Chen, C.-F., Yu, P.S., Wang, J.Z.: G-sesame: web tools for go-term-based gene similarity analysis and knowledge discovery. Nucleic Acids Research 37, W345–W349 (2009)

23. Froehlich, H., Speer, N., Poustka, A., Beissbarth, T.: Gosim - an r-package for computation of information theoretic go similarities between terms and gene products. BMC Bioinformatics 8, 166 (2007)
24. Heyer, L.J., Kruglyak, S., Yooseph, S.: Exploring expression data: Identification and analysis of coexpressed genes. Genome Research 9, 1106–1115 (1999)
25. Jiang, D., Tang, C., Zhang, A.: Cluster analysis for gene expression data: A survey. IEEE Transactions on Knowledge and Data Engineering 16, 1370–1386 (2004)
26. Gibbons, F.D., Roth, F.P.: Judging the quality of gene expression-based clustering methods using gene annotation. Genome Research 12, 1574–1581 (2002)
27. Spellman, P.T., Sherlock, G., Zhang, M.Q., Iyer, V.R., Anders, K., Eisen, M.B., Brown, P.O., Botstein, D., Futcher, B.: Comprehensive identification of cell cycle-regulated genes of the yeast saccharomyces cerevisiae by microarray hybridization. Molecular Biology of the Cell 9, 3273–3297 (1998)

Scalable and Versatile k-mer Indexing for High-Throughput Sequencing Data

Niko Välimäki[1,*] and Eric Rivals[2,**]

[1] Genome-Scale Biology Research Program, and Department of Medical Genetics,
Faculty of Medicine, University of Helsinki, Finland
niko.valimaki@helsinki.fi
[2] LIRMM and Institut de Biologie Computationelle,
CNRS & Université Montpellier 2, France
rivals@lirmm.fr

Abstract. Philippe et al. (2011) proposed a data structure called *Gk arrays* for indexing and querying large collections of high-throughput sequencing data in main-memory. The data structure supports versatile queries for counting, locating, and analysing the coverage profile of k-mers in short-read data. The main drawback of the *Gk* arrays is its space-consumption, which can easily reach tens of gigabytes of main-memory even for moderate size inputs. We propose a compressed variant of *Gk* arrays that supports the same set of queries, but in both near-optimal time and space. In practice, the compressed *Gk* arrays scale up to much larger inputs with highly competitive query times compared to its non-compressed predecessor. The main applications include variant calling, error correction, coverage profiling, and sequence assembly.

1 Introduction

High Throughput Sequencing (HTS) gives access to the whole complement of DNA or RNA sequences present in a biological sample. A single machine yields hundreds of million of short sequencing reads in a short time for a price that is steadily decreasing. Large sequencing centers produce daily tens of terabytes of data, and for instance the Beijing Genome Institute has launched in 2012 a project for sequencing 3 millions of genomes. Applications of HTS go far beyond genome sequencing, and are now used in the medical context for diagnostic and disease follow-up, or in ecology for monitoring biodiversity. In the later context, HTS sequence all DNA/RNA coming from all species present in an environmental sample (*i.e.*, a soil, a sea, or a gut sample). In such *meta-genomics* or *-transcriptomics* experiments, one aims at identifying the species or the genes they expressed in this environment, which is achieved by clustering and mining the reads based on sequence similarity.

* Funded by Academy of Finland CoE in Cancer Genetics Research (No 250345).
** Funded by the ANR Colib'read, MASTODONS, PICS.

Z. Cai et al. (Eds.): ISBRA 2013, LNBI 7875, pp. 237–248, 2013.
© Springer-Verlag Berlin Heidelberg 2013

HTS pushes life sciences in a Big Data era and fosters the development of efficient and scalable algorithms for analyzing huge read sets. A variety of computational questions need to be solved from genome assembly, to read clustering by similarity, going through read mapping (*i.e.* alignment) on a reference genome. Many tasks require indexing data structures that allow querying the reads for an exact or approximate sequence pattern. Most efficient programs for mapping reads onto a reference genome resort to a *FM-index* of the genome [7], which is small enough to fit in memory (*e.g.* [16,20]). However, in many applications, a reference genome is missing and the read set must be mined on its own (de novo genomics, transcriptomics, or in meta-genomics), but the volume of reads is much larger than a reference genome. Let us review shortly existing work on read indexing data structures.

Error Correction and k-mer Counting. Sequencing errors cause important difficulties for read analysis or genome assembly. The correction or elimination of erroneous reads is made possible by the redundancy due to high sequencing coverage. The solution is to monitor the occurrence number of all k-mers within the reads to see if they conform to the expected coverage. For the task of k-mer counting, efficient hashing techniques have been developped using parallel algorithms or Bloom filters [17,18]. However, their lack of scalability hinders indexing all 27-mers of typical Human sequencing data set [22]. A recent paper achieves scalability by partitioning the index between memory and disk [22]. Some assemblers use a parallel k-mer counting index to discard erroneous reads during the deBruijn graph construction [4]. Various error correction methods implement the same strategy using a hash table. For example, Coral [23] identifies sequencing errors by indexing k-mers into a hash table and then computing multiple alignments over reads that share a common k-mer. The hash table requires $\Theta(n \log n)$ bits [23], which can make the approach infeasible for HTS data.

Read Indexing in Mapping. Following the idea of error correction, it has been proposed to compute the *local coverage* of any k-mer in a read, that is the number of reads in which in appears. Inspecting the local coverage profile of k-mers along the read enables the tool CRAC to distinguish erroneous positions from point mutations directly during the mapping [20] (`http://crac.gforge.inria.fr`). For this, CRAC resorts to a data structure called *Gk* arrays which indexes all k-mers occurring within each read of the *collection*[1] using a modified suffix array and complementary tables [21]. It takes advantage from the fact that reads are often compared against themselves and that queried k-mers are taken from a read and can be given by a starting position rather than in extenso. *Gk* arrays offers seven types of locate and counting queries: either for getting the read identifiers in which a k-mer occurs (Q1/Q2), occurs at most once (Q5/Q6), and the occurrence positions with (Q7) or without this restriction (Q3/Q4). Table 1 gives an overview of the queries and theoretical properties of the data structure.

---

[1] In a collection, each read sequence can occur many times, but differ by their identifier, sequence quality, or mate partner. It is a multi-set rather than a set.

Table 1. Theoretical time and space complexities. Here n is the input size, f is the query k-mer, σ is the alphabet size, and $H_h \leq H_0 \leq \log \sigma$ denotes the h-th order entropy. The output size of each query is denoted by $|Q7| = |Q5| \leq |Q1| \leq |Q3|$, where $|Q3|$ is the total number of occurrences and, thus, can be significantly larger than the others. Philippe et al. [21] reported a linear time construction, but omitted their worst-case time of radix-sorting over $\lceil \log n \rceil$-bit integers.

Data structure		Compressed Gk arrays	Gk arrays
Construction	time	$O(n \log n)$	$O(n \log n)$
	space (bits)	$O(n(H_0 + 1))$	$\Theta(n \log n)$
Final index size (bits)		$nH_h \log \log_\sigma n + O(n)$	$\Theta(n \log n)$
Query time for	a k-mer	$O(k \log \sigma + \mathrm{polylog}(n))$	$O(k \log n)$
	a position	$O(\log \log n)$	$O(1)$
Additional query time to answer:			
Q1 In which reads does f occur?		$O(|Q1| \log \log n)$	$O(|Q3|)$
Q2 In how many reads does f occur?		$O(1)$	$O(|Q3|)$
Q3 What are the occurrence positions of f?		$O(|Q3| \log \log n)$	$O(|Q3|)$
Q4 What is the number of occurrences of f?		$O(1)$	$O(1)$
Q5 In which reads does f occur only once?		$O(|Q5| \log \log n)$	$O(|Q3|)$
Q6 In how many reads does f occur only once?		$O(1)$	$O(|Q3|)$
Q7 What are the occurrence positions of f in the reads where it occurs only once?		$O(|Q7| \log \log n)$	$O(|Q3|)$

Gk arrays uses a space proportional to the length of the read collection. Hence, indexing for instance a metagenomics dataset will exhaust the main memory of most computers. Gk arrays are also limited to queries on a single k value.

For Similarity Searching for Assembly and Clustering. When many transcribed RNAs are sequenced in proportion of their abudance, it is useful to reduce the data by clustering reads or ESTs that originate from the same molecule. EST clustering was already critical before the advent of HTS [3]. The effiency and scalabilty of clustering algorithms rely on their indexing strategy. KABOOM implements a modified suffix array for this task [10], but cannot scale up to nowadays huge read sets (as shown in [21]). The detection of similarity between reads is also used to discover overlaps and then build the overlap graph for genome assembly. The sparse representation of the relations between substrings and reads is major issue for scalable assembly programs, as exemplified by [6].

2 Compressed Gk Arrays

We aim at supporting the same set of queries as the original Gk arrays [21]. See Table 1 for the definition of each query Q1–Q7. Notice that Q3 and Q4 are the typical queries found in *full-text indexes* such as suffix trees and suffix arrays,

i	1	2	3	4	5	6	7	8	9	10	11	12	13	14
T	B	A	N	A	N	A	$\$_1$	A	N	A	N	A	S	$\$_2$
SA	7	14	6	4	2	8	10	12	1	5	3	9	11	13
SA^{-1}	8	5	11	4	10	3	1	6	12	7	13	8	14	2
LCP	0	0	0	1	3	5	3	1	0	0	2	4	2	0
B_{lcp}	1	1	1	1	0	0	0	1	1	1	1	0	1	1
B_{last}	0	0	0	1	0	0	1	0	1	0	1	1	1	0
B_{once}	0	0	1	1	1	1								
$T[\mathsf{SA}[i]\mathinner{..}n]$	$\$_1$	$\$_2$	A	A	A	A	A	A	B	N	N	N	N	S
		$\$_1$	N	N	N	N	S	A	A	A	A	A	$\$_2$
			A	A	A	A	$\$_2$	N	$\$_1$	N	N	S	
			$\$_1$	N	N	S		A		A	A	$\$_2$	
				A	A	$\$_2$		N		$\$_1$	S		
				$\$_1$	S			A			$\$_2$		
					$\$_2$			$\$_1$					

Fig. 1. An example of the (inverse) suffix array and LCP array for the input string $T = \mathtt{BANANA\$_1ANANAS\$_2}$ and resulting bit-vectors B_{lcp}, B_{last} and B_{once} for $k = 3$. Notice that the size of B_{once} is equal to $\mathsf{rank}_1(B_{\mathsf{last}}, n)$ and it is queried via the 1-bits in B_{last}.

and in their compressed variants (see [19] for a survey). Q1 and Q2 are also known as the *document listing problem* in the field of information retrieval [13].

At the core of our data structure is a *compressed suffix array* (CSA) [9] built on top of all the input reads. More precisely, the collection of reads is given as a (multi-)set[3] of strings $\mathcal{R} = \{r_1, r_2, \ldots, r_d\}$. We assume that the strings are from ordered alphabet Σ of size $|\Sigma| = \sigma = O(\mathrm{polylog}(n))$. We represent $\mathcal{R}$ as one long concatenated string, say $T = r_1 \$_1 r_2 \$_2 \cdots k_d \$_d$, where each $\$_i$ denotes a special separator-symbol having the lexicographical order $\$_{i-1} < \$_i < c \in \Sigma$ for all $i \in [1, d]$. Let n denote the total length of T. Substrings of T are denoted by $T[i \mathinner{..} j] = T[i]T[i+1] \cdots T[j]$ for any $1 \leq i \leq j \leq n$. The *suffix array* $\mathsf{SA}[1, n]$ of string T stores all suffixes of T in lexicographical order. The lexicographically i-th suffix is given by $T[\mathsf{SA}[i] \mathinner{..} n]$. The *inverse suffix array* is $\mathsf{SA}^{-1}[j] = i$ iff $\mathsf{SA}[i] = j$. The *Longest Common Prefix (lcp)* table, denoted $\mathsf{LCP}[1, n]$, stores in $\mathsf{LCP}[i]$ the length of the lcp of suffixes $T[\mathsf{SA}[i-1] \mathinner{..} n]$ and $T[\mathsf{SA}[i] \mathinner{..} n]$ for any $1 < i \leq n$, and $\mathsf{LCP}[1] = 0$. Fig. 1 gives an example of the SA, SA^{-1}, and LCP values. The compressed suffix array [9] requires $nH_h \log \log_\sigma n + O(n)$ bits[2] and allows $t_{\mathsf{SA}} = O(\log \log_\sigma n + \log \sigma) = O(\log \log n)$ time access to SA and SA^{-1}.

We aim to support two query-types, that is, queries for both a *given k-mer* and for *a given position* p in the set of indexed reads. Let f denote the given k-mer or the k-mer at the given position p, say $f = T[p \mathinner{..} p + k - 1]$ for any $p \in [1, n - k + 1]$. Our first problem is to identify the suffix array range $[s, e]$, which covers all the suffixes of T that have f as a prefix. If the query-type is

---

[2] We denote the *empirical entropy* of a string T with $H_0(T)$ (or simply H_0 if T is clear from the context). The *h-th order entropy* is denoted by $H_h(T)$ (or simply H_h). Notice that $0 \leq H_{h+1}(T) \leq H_h(T) \leq \log \sigma$ for all $h \geq 0$.

a k-mer, we can simply utilize the search functionality built into the CSA to identify the range $[s, e]$ in $O(k \log \sigma + \text{polylog}(n))$ time [9]. In order to support queries for given read positions, we propose the following data structure:

Lemma 1. *Given the CSA of the string $T[1..n]$, a fixed constant k and a query position $p \in [1, n-k+1]$, we can identify the suffix array range $[s, e]$, which covers all the suffixes of T that have $f = T[p..p+k-1]$ as a prefix, using $n + o(n)$ additional bits and $O(\log \log n)$ time.*

Proof. We introduce a bit-vector $B_{\text{lcp}}[1, n]$, which is set to $B_{\text{lcp}}[i] = 1$ if and only if $\text{LCP}[i] < k$. Fig. 1 gives an example of the arrays SA, SA^{-1} and LCP and the resulting B_{lcp}. See the following subsection on details about constructing B_{lcp}. Recall that the CSA can simulate the inverse suffix array in $O(\log \log n)$ time. We compute $j = \text{SA}^{-1}[p]$, which gives us a position j in the suffix array. It follows that $s \leq j$ and $e \geq j$, because f is a prefix of suffix $T[\text{SA}[j]..n] = T[\text{SA}[\text{SA}^{-1}[p]]..n] = T[p..n]$. Now we need to identify the suffixes surrounding j that also have f as a prefix. If such suffixes exists, they are identified by taking the smallest $s \in [1, j]$ and largest $e \in [j, n]$ such that $\text{LCP}[i] \geq k$ holds for all $i \in [s+1, e]$. Notice that the bit-vector B_{lcp} encodes this information, and it is accessible in constant time using rank and select queries: The $\text{rank}_b(B, i)$ query over a bit-vector $B[1..n]$ returns, for any $i \in [1, n]$, the number of times the bit $b \in \{0, 1\}$ occurs in $B[1..i]$. The inverse query, $\text{select}_b(B, j)$, returns the position of the j-th bit b in B (moreover, if $j > \text{rank}_b(B, n)$, we agree that $\text{select}_b(B, j)$ returns $n + 1$). We first compute $r = \text{rank}_1(B_{\text{lcp}}, j)$, which leads us to the final answer $[s, e] = [\text{select}_1(B_{\text{lcp}}, r), \text{select}_1(B_{\text{lcp}}, r+1) - 1]$. The rank and select queries over B_{lcp} can be computed in constant time using $n + o(n)$ bits [14]. $\square$

Notice that the range $[s, e]$ immediately reveals the total number of occurrences f has in the reads, which is $e - s + 1$. The occurrence positions can be enumerated by outputting $\text{SA}[j]$ for each $j \in [s, e]$. That said, the above lemma allows us to reveal the correct range $[s, e]$ and answer queries Q3 and Q4 in additional $O(|\text{Q3}| \log \log n)$ and $O(1)$ time, respectively. We introduce another data structure to answer Q1 and Q2:

Lemma 2. *Given the CSA of the string $T[1..n]$, a fixed constant k and a suffix array range $[s, e]$ covering all the suffixes that have f as a prefix, we can answer the query Q1 (resp. Q2) using $n + o(n)$ additional bits and $O(|\text{Q1}| \log \log n)$ time (resp. $O(1)$ time), where $|\text{Q1}|$ is the number of reads having an occurrence of f.*

Proof. Let $B_{\text{last}}[1, n]$ denote a bit-vector, which is initialized as follows: we set $B_{\text{last}}[j] = 1$ if and only if the k-mer $f = T[\text{SA}[j]..\text{SA}[j] + k - 1]$ starting from text position $p = \text{SA}[j]$ is the last occurrence of f within the corresponding read. That is, we mark the "unique" k-mers for each read and, as an important detail, this marking is stored in the suffix array order. See the following subsection on details about constructing B_{last}. Furthermore, k-mers that span over a separator-symbol are never marked. We can use B_{last} to directly count and enumerate the reads that contain at least one occurrence of f. Recall that $[s, e]$ covers all the suffixes that have f as a prefix. Now Q2 can be answered in constant time simply by computing $\mathsf{rank}_1(B_{\mathsf{last}}, e) - \mathsf{rank}_1(B_{\mathsf{last}}, s-1)$. For Q1, we first compute $r = \mathsf{rank}_1(B_{\mathsf{last}}, s-1)+1$, and then output the values $\mathsf{SA}[i]$ for all $i = \mathsf{select}_1(B_{\mathsf{last}}, r')$ such that $r' \geq r$ and $i \leq e$. This requires in total $O(|\mathsf{Q1}|\log\log n)$ time, where $|\mathsf{Q1}|$ is the number of reads having one or more occurrences of f. Finally, $n+o(n)$ bits are required to compute rank and select over B_{last} in constant time [14]. $\square$

To answer the queries Q5–Q7, we employ yet another data structure:

Lemma 3. *Given the CSA of the string $T[1..n]$, a fixed constant k and a suffix array range $[s, e]$ covering all the suffixes that have f as a prefix, we can answer the queries Q5 and Q7 (resp. Q6) using $n + o(n)$ additional bits and $O(|\mathsf{Q5}|\log\log n)$ time (resp. $O(1)$ time), where $|\mathsf{Q5}| = |\mathsf{Q7}|$ is the number of reads having exactly one occurrence of f.*

Proof. Let B_{once} denote a bit-vector of length $\mathsf{rank}_1(B_{\mathsf{last}}, n)$. We set $B_{\mathsf{once}}[i] = 1$ if and only if the k-mer starting from text position $p = \mathsf{SA}[\mathsf{select}_1(B_{\mathsf{last}}, i)]$ occurs only once within the corresponding read. The following subsection describes how to construct the bit-vector B_{once}. Now, similar to previous lemma, the query Q6 can be answered in constant time by first computing $s' = \mathsf{rank}_1(B_{\mathsf{last}}, s-1) + 1$ and $e' = \mathsf{rank}_1(B_{\mathsf{last}}, e)$. Then the result for Q6 is given by $\mathsf{rank}_1(B_{\mathsf{once}}, e') - \mathsf{rank}_1(B_{\mathsf{once}}, s'-1)$. For Q5 and Q7, we output the values $\mathsf{SA}[\mathsf{select}_1(B_{\mathsf{last}}, i')]$ for all $i' \in [s', e']$ such that $B_{\mathsf{once}}[i'] = 1$. Such positions can be found using one select_1 operation (over B_{once}) per outputted element. Thus, the query is solved in $O(|\mathsf{Q5}|\log\log n)$ time using $\mathsf{rank}_1(B_{\mathsf{last}}, n)(1 + o(1)) \leq n + o(n)$ bits. $\square$

Theorem 1. *Given a set of reads $\mathcal{R} = \{r_1, r_2, \ldots, r_d\}$ of total length n, and a fixed constant k, there exists a data structure that requires $nH_h \log\log_\sigma n + O(n)$ bits of space and supports the queries Q1–Q7 with the time complexities given in Table 1. If the query is a k-mer (resp. a position in $\mathcal{R}$), the queries require additional $O(k\log\sigma + \mathrm{polylog}(n))$ time (resp. $O(\log\log n)$ time).*

Proof. See the above lemmas about supporting each query Q1–Q7. The combined space complexity of the required bit-vectors and their rank and select data structures is $3n + o(n)$ bits. The space complexity is dominated by the compressed suffix array, which requires $nH_h \log\log_\sigma n + O(n)$ bits of space [9]. $\square$

2.1 Construction

We propose a construction algorithm that can build the above data structures in $O(n\log n)$ time and $O(n(H_0 + 1))$ bits of space, assuming that the largest read-length in the collection is limited, that is, $\ell = \max\{|r_i| : r_i \in \mathcal{R}\} = O(n/\log n)$. The theoretical complexities can be achieved by (1) building the CSA, (2) building the LCP array, (3) scanning through the LCP array once to construct B_{lcp}, and finally (4) scanning through the (implicit) suffix array once to construct B_{last} and B_{once}. In practice, B_{lcp} is constructed directly (see Sect. 3).

More precisely, the compressed suffix array can be constructed $O(n\log n)$ time using $O(n(H_0 + 1))$ bits [11]. The final index requires $nH_h \log\log_\sigma n + O(n)$ bits and supports random access to the $\mathsf{SA}[i]$ and $\mathsf{SA}^{-1}[j]$ values in $t_{\mathsf{SA}} = O(\log \log n)$ time for polylog-sized alphabets [9]. The LCP array can then be constructed in $O(n \cdot t_{\mathsf{SA}})$ time and in $4n + o(n)$ bits of space on top of the CSA [12]. The resulting LCP array admits access to values $\mathsf{LCP}[i]$ in $O(t_{\mathsf{SA}})$ time, thus, we can also construct B_{lcp} in $O(n \cdot t_{\mathsf{SA}})$ time. The bit-vectors B_{last} and B_{once} can be attained as follows:

Lemma 4. *Given the CSA of the string $T[1 .. n]$, a fixed constant k and the bit-vector B_{lcp}, we can construct the bit-vectors B_{last} and B_{once} in $O(n \log \log n)$ time and $2n + o(n) + O(\ell \log n)$ additional bits. If the largest read-length is $\ell = O(n/\log n)$, the additional space is $O(\ell \log n) = O(n)$ bits.*

Proof. Let $B_{\mathsf{last}}[1, n]$ and $B'_{\mathsf{once}}[1, n]$ denote two bit-vectors. We initialize B_{last} to all zeros, and B'_{once} to all ones. We traverse over the suffixes of T in backwards order, say $T[n - 1 .. n], T[n - 2 .. n], \ldots, T[1 .. n]$. At each step i of the traversal, we first compute $j = \mathsf{SA}^{-1}[i]$ (in practice, we replace SA^{-1} with LF-mapping; see Sect. 3). Then we compute $r = \mathsf{rank}_1(B_{\mathsf{lcp}}, j)$ and check if the key r exists in a y-fast trie [25]. If it does not yet exists, we set $B_{\mathsf{last}}[j] = 1$, and insert the key-value pair $\langle r, j \rangle$ into the y-fast trie. Moreover, if the key r already exists in the y-fast trie with value j', we set $B'_{\mathsf{once}}[j'] = 0$. Finally, if $T[i]$ is a separator-symbol, we remove all elements in the current y-fast trie, thus, the maximum number of elements in the trie is bounded by $O(\ell)$. Since we traverse T in backwards order, we can easily keep track of the position of the nearest separator-symbol, and avoid marking B_{last} for k-mers that overlap a separator-symbol. All this requires $O(t_{\mathsf{SA}} + \log \log n)$ time per each step (with the exception of the removal of all trie elements, which can be amortized to $O(n \log \log n)$ time over all steps). The trie size is at most $O(\ell \log n)$ bits at any step of the construction. After traversing the whole text, we can construct the final bit-vector B_{once} of length $\mathsf{rank}_1(B_{\mathsf{last}}, n)$. We set $B_{\mathsf{once}}[\mathsf{rank}_1(B_{\mathsf{last}}, j)] = B'_{\mathsf{once}}[j]$ for each j such that $B_{\mathsf{last}}[j] = 1$. $\square$

Corollary 1. *Given a set of reads $\mathcal{R} = \{r_1, r_2, \ldots, r_d\}$ of total length n, and a fixed constant k, the data structure in Theorem 1 can be constructed in $O(n \log n)$ time and $O(n(H_0 + 1))$ bits of space.*

2.2   Query Extensions

Read Coverage Profile. The *coverage profile* of a read r gives, for each position $i \in [1, |r| - k + 1]$ in the read r, the number of reads that share the k-mer $r[i .. i + k - 1]$. The coverage profile can be utilized, for example, to discriminate between sequencing errors and SNVs/SNPs [20]. The read coverage profile can be efficiently computed for any $r \in \mathcal{R}$ by resorting to $|r| - k + 1$ calls to Q2, which requires in total $O(|r| \log \log n)$ time. (In practice, we use LF-mapping and backward search, and answer Q2 at each step via constant time query over B_{lcp} and B_{last}. The resulting time complexity is $O(|r| \cdot t_{\mathsf{LF}})$.)

Queries over Multiple k. We can support queries over multiple $k_1, k_2, \ldots, k_z$ by building separate bit-vectors for each k_i. Now, the final index consists of one CSA built for the input reads, and z sets of bitvectors requiring in total $3nz + o(nz)$ bits of space. For any $z = O(\log \sigma)$, the total index size becomes $O(n \log \sigma)$ bits, which is still less than the original Gk arrays require for one fixed k. For large z, another time–space tradeoff is to replace all the LCP bit-vectors with the full LCP array, which requires just $4n + o(n)$ bits [12], and resort to *Previous Smaller Value* and *Next Smaller Value* queries over the LCP table similar to [8]. PSV and NSV can solved in sublogarithmic time with $o(n)$ extra bits of space.

3   Experiments

We implement the compressed Gk arrays (CGkA) using the *FM-index* concept [7] and Huffman-shaped wavelet trees [19]. We use Heng Li's implementation of the BCR algorithm [15,1] to construct the Burrows–Wheeler Transform (BWT) for the input reads. The resulting FM-index requires $nH_0(T) + o(n \log \sigma)$ bits of space and supports *LF-mapping* in average $t_{\mathsf{LF}} = O(H_0(T)) = O(\log \sigma)$ time. We build the B_{lcp} bit-vector directly from the wavelet tree in $O(n\sigma)$ time by adapting the algorithm of [2]. For Lemma 4, we use LF-mapping instead of explicitly computing $j = \mathsf{SA}^{-1}[i]$ for each step. It allows us to construct bit-vectors B_{last} and B_{once} simultaneously with the (inverse) suffix array samples, using one pass over the text. We store (inverse) suffix array samples for every s text positions, which allows an $t_{\mathsf{SA}} = O(s \cdot t_{\mathsf{LF}})$ time access to SA (SA^{-1}). We test the sampling rates $s \in \{2, 4, 8, 16, 32\}$.

We compare the compressed Gk arrays against a performant hash table Jellyfish 1.1.6 [17], a *Run-Length Compressed Suffix Array* (RLCSA[3] Jan. 2013 version) [24], and the original Gk arrays (GkA[4] version 1.0.1) [21]. GkA offer a native support for queries Q1–Q7, the RLCSA supports only queries Q3–Q4, and Jellyfish the counting query Q4. We run the RLCSA using sampling rates $s \in \{3, 4, 8, 16\}$ (the construction ran out of memory for $s = 2$) and *nibble encoded* bit-vectors, which are faster and, in our experiments, require only around 2% more space. We use block size 16 for the internal bit-vectors.

Remark 1. Claude et al. [5] proposed a *compressed k-mer index* for indexing highly-repetitive biological sequences. However, their experimental results show that RLCSA is faster and uses less space for any $k \geq 6$ [5]. Also, the construction space of Claude et al. is about twice larger than RLCSA. We omit the compressed k-mer index of Claude et al. from our experiments for these reasons.

The input reads are taken from a set of 151bp Illumina reads sequenced from an E. Coli strain MG1655[5]. We truncate the low-quality tails, using a Phred threshold of 10, and include only the full-length reads in the final set. This

---

[3] http://www.cs.helsinki.fi/group/suds/rlcsa/ The latest RLCSA package includes an unpublished data structure to solve Q1, however, its construction time and space do not yet scale up gracefully (J. Sirén, Personal communication, 2013).

[4] http://crac.gforge.inria.fr/gkarrays/

[5] http://www.illumina.com/systems/miseq/scientific_data.ilmn

filtering leaves a total of 8.5 million reads. All experiments are ran using a single core of an Intel Xeon E5540 2.53GHz processor equipped with 32GB of main memory, Linux 3.2.0 (Ubuntu x86_64) and gcc 4.6.3. We report the final index size, average query times for Q1–Q4, and the construction time and space for each data structure.

The final index size of CGkA represents 10% to 60% of the size of the original Gk arrays depending on the sampling rate. CGkA require 5.6 GB for $s = 2$ and 1.3 GB for $s = 32$, while the non-compressed GkA require between 9.0–9.2 GB depending on k. Jellyfish and RLCSA have the smallest index sizes at the cost of supporting only Q4 and Q3–Q4, respectively.

Remark 2. The RLCSA implementation could be extended to support Q1–Q7 by adding the bit-vectors B_{lcp}, B_{last} and B_{once} over the RLCSA. It would then give yet another time–space trade-off for the compressed Gk arrays: a smaller index size, but slightly slower query times as the results in Fig. 2 suggest.

Fig. 2 gives an overview of the average query times for Q1–Q4, when querying a set of 1–100 million randomly chosen k-mers (depending on size of the k-mer). Jellyfish is the most space-efficient and also the fastest, since its hash table is tailored for simple counting queries (Q4). The compressed data structures are still competitive regarding both query time and space, while providing a more versatile set of queries. The differences for Q2 are more significant with $k = 11$ due to the $O(|\mathbf{Q3}|)$ worst-case time of GkA. Regarding the locate queries (Q1 and Q3), the sampling rates $s \leq 8$ are competitive against the non-compressed GkA for $k = 22$. This is mostly due to small numbers of occurrences (i.e. large k) and faster backward search. For smaller k, the numbers of occurrences are significantly higher, and the time to locate the suffix array interval has a smaller impact on the average query times. Fig. 3 gives the query times for Q1–Q3, when the query is given as a randomly chosen position from the indexed read set. The query times are averaged over 1–100 million randomly chosen positions (depending on query). The compressed representation is slower for all queries but Q2, mostly because the (inverse) suffix array values must be computed via the sampled array. CGkA have $O(t_{\mathsf{SA}}) = O(s \cdot t_{\mathsf{LF}})$ time access to inverse SA, which is notably slower than the constant time access within GkA. However, for Q2 the compressed representation can be faster due to the worst-case $O(|\mathbf{Q3}|)$ query time of GkA. RLCSA and Jellyfish do not support these types of queries.

Construction Time and Space. We also measure the construction time and space for all data structures. For RLCSA, we use the fastest construction method in the Jan. 2013 package [24]. As a second hash table approach, we include Coral version 1.4 [23], which use the GNU C++ hash_map for storing a list of occurrence positions for each k-mer. Table 2 reports the construction times and maximum memory usages for the 8.5 million 151bp Illumina reads, including the figures for each construction step of the compressed Gk arrays.

Here, CGkA take roughly the same construction time as RLCSA, but use less memory. CGkA require only twice the construction time of non-compressed data structures, and achieves the most space-efficient construction (Jellyfish could use

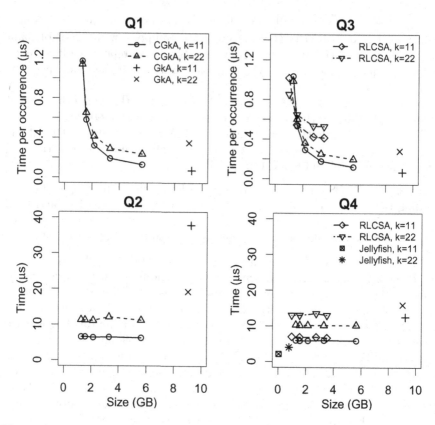

Fig. 2. Average query times and the index size when the query is given as a k-mer. RLCSA supports queries Q3 and Q4, and Jellyfish only a counting query (Q4).

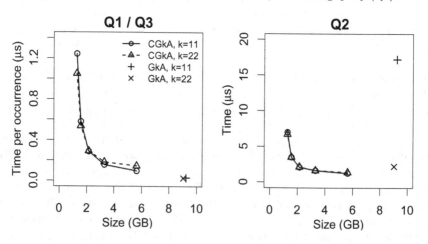

Fig. 3. Average query times and the index size when the query is given as a read position. RLCSA and Jellyfish do not support these types of queries.

Table 2. Construction time and space for 8.5 million 151bp Illumina reads

	k	Time (s)	Memory usage (MB)
Gk arrays [21]	11	611	9,452
Gk arrays	22	605	9,251
RLCSA [24] ($s = 16$)	n/a	1,095	16,446
Coral [23] (GNU C++ hash_map)	11–22	861	16,016
Jellyfish [17] (counting only, $M = 2^{24}$)	11	88	2,911
Jellyfish [17] (counting only, $M = 2^{24}$)	22	405	2,965
Compressed Gk arrays ($s = 16$)	11	957	2,881
Compressed Gk arrays ($s = 16$)	22	1,086	2,881
CGkA construction steps:			
ropebwt+BCR [15,1]	n/a	288	506
Wavelet tree (Huffman)	n/a	44	1,471
B_{lcp}	11	10	1,471
B_{lcp}	22	139	1,471
B_{last}, B_{once}, sampling SA, SA^{-1}	n/a	615	2,881

less memory by balancing between the hash table size and merge cost). Hence, the CGkA achieve a much better scalability in term of memory requirements than its uncompressed version, while still offering competitive query times.

4 Discussion

We presented a space-efficient data structure for indexing all k-mers in HTS data. The data structure supports a comprehensive set of locate and count queries with competitive query times. It is also more scalable than its non-compressed predecessor, the Gk arrays [21], due to the time–space trade-off we achieve: for a fixed amount of main memory, the compressed representation can index up to seven times more data (regarding the final index size). Both the construction and query algorithms are completely different from those of Gk arrays. It also allows queries over multiple $k_1, k_2, \ldots, k_z$ with an overhead of 3.19 bits per input character per k_i. A parallized, secondary memory [1] construction, as well as enhancements to allow navigation in a de Bruijn graph belong to future work.

References

1. Bauer, M.J., Cox, A.J., Rosone, G.: Lightweight BWT construction for very large string collections. In: Giancarlo, R., Manzini, G. (eds.) CPM 2011. LNCS, vol. 6661, pp. 219–231. Springer, Heidelberg (2011)
2. Beller, T., Gog, S., Ohlebusch, E., Schnattinger, T.: Computing the longest common prefix array based on the burrows-wheeler transform. In: Grossi, R., Sebastiani, F., Silvestri, F. (eds.) SPIRE 2011. LNCS, vol. 7024, pp. 197–208. Springer, Heidelberg (2011)

3. Burkhardt, S., Crauser, A., Ferragina, P., Lenhof, H.-P., Rivals, E., Vingron, M.: q-gram Based Database Searching Using a Suffix Array (QUASAR). In: 3rd Int. Conf. on Computational Molecular Biology, pp. 77–83. ACM Press (1999)
4. Chikhi, R., Lavenier, D.: Localized genome assembly from reads to scaffolds: Practical traversal of the paired string graph. In: Przytycka, T.M., Sagot, M.-F. (eds.) WABI 2011. LNCS, vol. 6833, pp. 39–48. Springer, Heidelberg (2011)
5. Claude, F., Fariña, A., Martínez-Prieto, M.A., Navarro, G.: Compressed q-gram indexing for highly repetitive biological sequences. In: Proc. 10th IEEE Intl. Conf. on Bioinformatics and Bioengineering, pp. 86–91 (2010)
6. Conway, T.C., Bromage, A.J.: Succinct Data Structures for Assembling Large Genomes. Bioinformatics 27(4), 479–486 (2011)
7. Ferragina, P., Manzini, G.: Opportunistic data structures with applications. In: Proc. 41st Annual Symposium on Foundations of Computer Science (FOCS), pp. 390–398. IEEE Computer Society (2000)
8. Fischer, J., Mäkinen, V., Navarro, G.: Faster entropy-bounded compressed suffix trees. Theor. Comput. Sci. 410(51), 5354–5364 (2009)
9. Grossi, R., Gupta, A., Vitter, J.S.: High-order entropy-compressed text indexes. In: 14th Ann. ACM-SIAM Symp. on Discrete Algorithms, pp. 841–850 (2003)
10. Hazelhurst, S., Lipták, Z.: Kaboom! a new suffix array based algorithm for clustering expression data. Bioinformatics 27(24), 3348–3355 (2011)
11. Hon, W.-K., Lam, T.-W., Sadakane, K., Sung, W.-K., Yiu, S.-M.: A space and time efficient algorithm for constructing compressed suffix arrays. Algorithmica 48(1), 23–36 (2007)
12. Hon, W.-K., Sadakane, K.: Space-economical algorithms for finding maximal unique matches. In: Apostolico, A., Takeda, M. (eds.) CPM 2002. LNCS, vol. 2373, pp. 144–152. Springer, Heidelberg (2002)
13. Hon, W.-K., Shah, R., Vitter, J.S.: Space-efficient framework for top-k string retrieval problems. In: FOCS, pp. 713–722. IEEE Computer Society (2009)
14. Jacobson, G.: Succinct Static Data Structures. PhD thesis, Carnegie–Mellon (1989)
15. Li, H.: Implementation of BCR, https://github.com/lh3/ropebwt
16. Li, H., Durbin, R.: Fast and accurate short read alignment with Burrows-Wheeler transform. Bioinformatics 25(14), 1754–1760 (2009)
17. Marçais, G., Kingsford, C.: A fast, lock-free approach for efficient parallel counting of occurrences of k-mers. Bioinformatics 27(6), 764–770 (2011)
18. Melsted, P., Pritchard, J.: Efficient counting of k-mers in dna sequences using a bloom filter. BMC Bioinformatics 12(1), 333 (2011)
19. Navarro, G., Mäkinen, V.: Compressed full-text indexes. ACM Comput. Surv. 39(1) (2007)
20. Philippe, N., Salson, M., Commes, T., Rivals, E.: CRAC: an integrated approach to read analysis. Genome Biology (in press, 2013)
21. Philippe, N., Salson, M., Lecroq, T., Léonard, M., Commes, T., Rivals, E.: Querying large read collections in main memory: a versatile data structure. BMC Bioinformatics 12, 242 (2011)
22. Rizk, G., Lavenier, D., Chikhi, R.: DSK: k-mer counting with very low memory usage. Bioinformatics, page Advance access (January 2013)
23. Salmela, L., Schröder, J.: Correcting errors in short reads by multiple alignments. Bioinformatics 27(11), 1455–1461 (2011)
24. Sirén, J.: Compressed Full-Text Indexes for Highly Repetitive Collections. PhD thesis, Dept. of Computer Science, Report A-2012-5, University of Helsinki (2012)
25. Willard, D.E.: Log-logarithmic worst-case range queries are possible in space Theta(N). Inf. Process. Lett. 17(2), 81–84 (1983)

POMAGO: Multiple Genome-Wide Alignment Tool for Bacteria

Nicolas Wieseke[1], Marcus Lechner[2], Marcus Ludwig[3], and Manja Marz[3]

[1] University of Leipzig, Faculty of Mathematics and Computer Science,
Augustusplatz 10, 04109 Leipzig, Germany
`wieseke@informatik.uni-leipzig.de`

[2] Philipps-Universität Marburg, Institut für Pharmazeutische Chemie, Marbacher
Weg 6, 35032 Marburg, Germany
`lechner@staff.uni-marburg.de`

[3] Friedrich-Schiller-University Jena, Faculty of Mathematics and Computer Science,
Leutragraben 1, 07743 Jena, Germany
`{m.ludwig,manja}@uni-jena.de`

Abstract. Multiple Genome-wide Alignments are a first crucial step to compare genomes. Gain and loss of genes, duplications and genomic rearrangements are challenging problems that aggravate with increasing phylogenetic distances. We describe a multiple genome-wide alignment tool for bacteria, called POMAGO, which is based on orthologous genes and their syntenic information determined by `Proteinortho`. This strategy enables POMAGO to efficiently define anchor points even across wide phylogenetic distances and outperform existing approaches in this field of application. The given set of orthologous genes is enhanced by several cleaning and completion steps, including the addition of previously undetected orthologous genes. Protein-coding genes are aligned on nucleotide and protein level, whereas intergenic regions are aligned on nucleotide level only. We tested and compared our program at three very different sets of bacteria that exhibit different degrees of phylogenetic distances: 1) 15 closely related, well examined and described *E. coli* species, 2) six more divergent Aquificales, as putative basal bacteria, and 3) a set of eight extreme divergent species, distributed among the whole phylogenetic tree of bacteria. POMAGO is written in a modular way which allows extending or even exchanging algorithms in different stages of the alignment process. Intergenic regions might for instance be aligned using an RNA secondary structure aware algorithm rather than to rely on sequence data alone. The software is freely available from `http://www.rna.uni-jena.de/supplements/pomago`

Keywords: Multiple Genome Alignment, Synteny, Annotation.

1 Introduction

Multiple Genome-wide Alignments (MGAs) are a first crucial step for the comparison of genomes, and subsequent analysis, such as annotation of non-coding

Z. Cai et al. (Eds.): ISBRA 2013, LNBI 7875, pp. 249–260, 2013.
© Springer-Verlag Berlin Heidelberg 2013

genes [20], discovery of functional elements [13] or structure conservation [16]. Many MGA generating programs exist already, such as Mugsy [2], NcDNAlign [20], TBA [3], progressiveMauve [11], CHAOS [8], or Mavid [7]. Most of the software packages are written for a specific class of organisms or problems and therewith have their specific advantages and disadvantages. However, there is still an urgent need for more practicability and accuracy [20,10,15].

Nowadays, common algorithms use some kind of anchored alignment. Most of them use pairwise alignment scores of homologous subsequences as anchors, e.g. calculated by Blast [1], on nucleotide or amino acid level. TBA is an accurate state of the art multiple genome alignment tool which is based on threaded blocksets that are aligned. The approach was designed to align as many regions as possible. progressiveMauve on the other hand is a very different approach that uses pairwise alignment scores and iteratively determines homologous single-copy subsequences that can be aligned and merged. While we are not aware of any other approach that directly facilitates orthologous genes, this concept is closest to the strategy we propose here.

POMAGO is a multiple genome-wide alignment approach for bacteria, which is based on orthologous genes and their syntenic information determined by Proteinortho [18]. Obtained orthologous genes are enhanced by several cleaning and completion steps, including the addition of previously undetected orthologs. We align our protein anchors with a modified version of CAUSA [21] and intergenic regions, as well as 5' and 3' UTRs with ClustalW [17]. While less advanced algorithms for determination of anchor points are sufficient for closely related species, this approach is efficient and robust irrespective of species relatedness. Improved in this way, MGAs can enhance the discovery and analysis of genes, functional elements and structure conservation beyond the level of related species. They might also enable deeper insights into the evolution of species.

To illustrate the improvement regarding more diverged species sets, we tested and compared our program with progressiveMauve and TBA at three very different sets of bacteria that exhibit different degrees of phylogenetic distances: 1) 15 closely related, well examined and described *E. coli* species, 2) six more diverged Aquificales, as putative basal bacteria, and 3) a set of eight extreme divergent species, distributed among the whole phylogenetic tree of bacteria.

2   Methods

The approach proposed here is based on preannotated subsequences of whole genomes to compute a genome wide alignment. Therefore, e.g. annotated protein sequences from the NCBI can be used. In a first step Proteinortho is applied, which predicts co-orthologous groups by performing a pairwise all-against-all Blast followed by spectral partitioning. To complement for potentially incomplete annotations, all groups are reviewed. For every species where no member was found, the genome is scanned for a member that was not present in the given annotation using tblastn. The highest scoring alignment to an ORF above a fairly high E-value threshold of $1e^{-20}$ is used to complement the initial annotation. Proteinortho is applied a second time using the extended annotation.

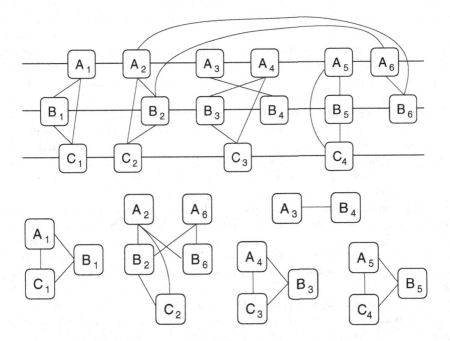

Fig. 1. Example of input data and its graph representation. **Top:** Each horizontal line represents a genome of species A, B or C. Accordingly, A_i, B_i, and C_i represent genes encoded in these genomes. Non-horizontal lines show the estimated orthologous relationships predicted by `Proteinortho`. **Bottom:** Graph G of the above given example.

After this preprocessing step, a set of coding sequences is obtained together with an estimation of the pairwise orthologies. This data can be interpreted as a colored graph $G = (V, E)$ with V being the set of coding sequences and E the estimated orthology relation, see Fig. 1. Nodes are colored according to the species they belong to. Nodes of the same color are ordered with respect to the position of the respective gene within its genome. The node A_i therefore refers to the i-th gene in species A. This graph is called the orthology graph.

2.1 Orthology Cleaning

As the initial orthology graph is just an estimate of the orthology relations in the dataset, it will be noisy. Due to sequence similarities, orthologs and paralogs might be clustered together or some relations are missing. Therefore, several cleaning steps are performed to prepare the data before it can be used to align the genomes. These cleaning steps are described in the following subsections.

Gene Duplication. Whenever a gene duplication occurs in one species the two copies A_i and A_j might be orthologous to a gene B_k from another species and therefore both copies will be connected to B_k within the orthology graph. This is consistent with the definition of orthology by Fitch [14], but when it comes to an alignment it is unclear which of the sequences A_i or A_j has to be aligned against B_k. Therefore, we refer to Braga *et al.* [6] who assume the sequence which remains in its genomic context to be the 'more' correct ortholog. We first extend the orthology graph by transitivity to obtain all candidates of orthologs, i.e., in each connected component of the graph new edges are inserted such that all pairs of nodes from different species are connected, see Fig. 2 Left.

To select groups of genes which have to be aligned against each other, a decomposition of the connected components has to be done, such that each connected component contains only nodes with different colors, i.e., it will be a clique in the resulting orthology graph with one gene per species at maximum.

The decomposition of the connected components is done with a greedy strategy taking into account the similarity of each two orthologous sequences A_i and B_k as well as their synteny regarding neighboring genes. For the sequence similarities each edge is weighted by the E-value taken from the pairwise all-against-all `Blast` computed in the preprocessing step. For synteny, each edge is weighted by the number of orthologous pairs of genes $A_{i\pm x}$ and $B_{k\pm x}$ in a certain neighborhood x around A_i and B_k. Such pairs are counted whenever the nucleotide distance of A_i and $A_{i\pm x}$ is similar to the distance of B_k and $B_{k\pm x}$. As a default, the neighborhood was defined to be ten genes upstream and downstream of a gene ($x = 10$). Nucleotide distances were assumed to be similar if they do not differ by a factor of more than two.

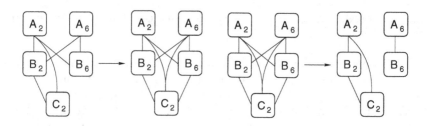

Fig. 2. Left: Extension of the orthology graph by transitivity. Whenever two nodes of the same connected component are colored differently, they get connected. **Right:** Decomposition of connected components. Connected components are decomposed into cliques according to synteny and pairwise similarity of sequences. Each subset contains no more than one node of each color.

The decomposition $\mathcal{D}_i$ of a connected component $\mathcal{C}_i$ into cliques is constructed as follows. Starting with an empty graph $\mathcal{D}_i = (V_{\mathcal{C}_i}, \emptyset)$ on nodeset $V_{\mathcal{C}_i}$ the best edge according to its synteny weight is chosen. For two edges with the same synteny value the one with the better E-value is selected. This edge is added to the new graph whenever it results in a graph where all connected components have no more than one node of each color. Otherwise the edge is rejected. This step is repeated until all edges were selected, see e.g. Fig. 2 Right. Finally, each connected component in the orthology graph is replaced by its decomposition. The pseudocode for this step is given in Algorithm 1 in the supplement.

Synteny-Based Orthology Insertion. Whenever two related coding sequences differ by more than the allowed threshold of `Proteinortho`, they are not detected as direct orthologs in the preprocessing step. However, if these sequences are located within syntenic regions a different threshold should be used. We account for that by considering the following two cases:

In the first case a pair of orthologous genes A_i, B_k exists and therefore an edge $e = (A_i, B_k)$ in the orthology graph. Furthermore, there are two nodes $A_{i\pm1}$ and $B_{k\pm1}$ which refer to neighboring genes, which are not connected and belong to connected components with node sets of distinct colors. For the sequences of $A_{i\pm1}$ and $B_{k\pm1}$ a pairwise alignment is computed. If the score of this alignment is better than a user defined threshold, the respective connected components are merged by graph join.

In the second case two pairs of orthologous genes and therefore the two edges $e_1 = (A_i, B_k)$ and $e_2 = (A_j, B_l)$ exist. Let $|i - j| < x$ and $|k - l| < x$ with x being the size of the assumed syntenic region. Furthermore, let the intermediate nodes A_n, B_m with $n \in (i, j)$ and $m \in (k, l)$ be pairwise disconnected in the orthology graph. For all pairs of nodes A_n and B_m alignments are computed. Whenever the score of one of these alignments is better than the threshold and the node sets of the respective connected components have distinct colors, these components are merged by graph join, see Fig. 3 Top. The pseudocode for synteny-based insertion is given in Algorithm 2 in the supplement.

Pairwise alignments are computed on a combined nucleotide-amino acid level using a modified version of `CAUSA` [21] and `ClustalW` [17] by default. An alignment is rejected if the `ClustalW` score is below 20. This is approximately one half of the average alignment score of orthologous coding sequences. The alignment approach is described in detail in section 2.2.

In some cases this approach might be to strict, as it only allows the insertion of an edge between two nodes A_n and B_m whenever the respective connected components are color distinct. Alternatively, one can add such an edge for connected components which are not color distinct. This will result in a new connected component which does not fulfill the property of having no more than one node of each color/species. However, this can be curated with the same decomposition approach described in the previous section. It is not guaranteed that the edge between A_n and B_m will be contained in the decomposition, but the edge will be added if there is sufficient syntenic support among the genes of that component.

Synteny-Based Sequence Annotation. Synteny is also used to annotate new coding sequences which are orthologous to already annotated sequences.

Assume there are two pairs of orthologous genes resulting in the two edges $e_1 = (A_i, B_k)$ and $e_2 = (A_{i+1}, B_l)$ with $1 < |k - l| < x$ in the orthology graph, i.e., the two genes of species A are next to each other while there are 1 to $x - 1$ other genes between those of species B. If the nucleotide distance of A_i and A_{i+1} is similar to the distance of B_k and B_l, then all sequences B_m with

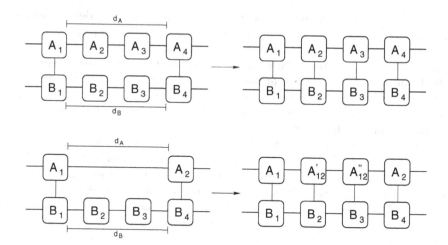

Fig. 3. Top: Synteny-based orthology insertion. If distances d_A and d_B (in nucleotides or number of genes) are sufficiently similar, each gene within d_A of species A is compared to each gene within d_B of species B. New edges are added to the orthology graph. **Bottom:** Synteny-based sequence annotation. If d_A and d_B (in nucleotides or number of genes) are similar, the genomic subsequence of d_A of species A is compared to each gene within d_B of species B. New nodes and edges are added to the orthology graph.

$m \in (k, l)$ are aligned against the intermediate sequence between A_i and A_{i+1} considering all three possible reading frames. For each alignment between B_m and the intermediate sequence of A_i and A_{i+1} with a score above a user defined threshold, a new node $A_{i'}$ is added to the orthology graph. The final graph includes the edge $e_n = (A_{i'}, B_m)$, see Fig. 3 Bottom. Pairwise alignments are computed in the same way as in the orthology insertion step. The pseudocode for synteny-based annotation is given in Algorithm 3 in the supplement.

2.2 Anchored Multiple Genome Alignment

After the cleaning and completion steps mentioned above, the orthology graph consists of several cliques, each representing a set of orthologous genes. These genes have to be aligned in a sophisticated way to each other in order to achieve a convincing genome wide alignment.

The final alignment is calculated with respect to a chosen reference species. Therefore, cliques containing nodes from this genome are ordered according to the position of the respective genes therein. For the gene sequences of each clique a multiple sequence alignment is computed with a modified version of CAUSA [21]. CAUSA computes combined nucleotide-amino acid sequences such that after three nucleotides the respective amino acid is inserted. Furthermore, these sequences are aligned with ClustalW using a customized substitution matrix to ensure that nucleotides are not mistakenly aligned against amino acids. We modified CAUSA to deal with genomic sequences containing characters $c \notin \{A, C, G, T, U, N\}$.

These are converted into N. Each character U is converted into T. If a character of a nucleotide is detected to be an N and the assignment of the amino acid is not affected by this letter, the appropriate amino acid is chosen.

For each group of orthologous genes, i.e., each clique in the orthology graph, a multiple sequence alignment is computed and truncated such that there is no overlap with consecutive genes and no gaps at the borders of the alignment. Each of these alignments is used as an anchor to additionally align the intergenic regions between the respective genes. Whenever possible, two consecutive gene alignments are connected by extending the alignments across the intergenic regions between them. Therefore, a pairwise similarity of each non-reference intergenic region with the reference sequence is computed using ClustalW. The regions with a score above a certain threshold (by default 40) are aligned and added between both single gene alignments. Note that two gene alignments can only be merged if the genes of all species are consecutively within their genome and all intergenic regions could be aligned. Otherwise the two alignment blocks remain separated and the intergenic regions are appended in the same manner, either at the end of the previous block or at the beginning of the next block, see Fig. 4. The final output is returned in multiple alignment format (MAF).

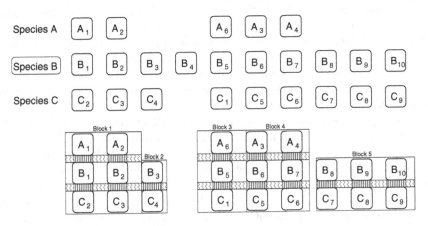

Fig. 4. The final anchored multiple genome alignment is calculated with a modified version of CAUSA for protein-coding genes (vertical lines between genes) and ClustalW for intergenic regions (zigzag lines between genes). In this example species B is chosen as reference organism, containing consecutive genes from B_1 to B_n. The final alignment consists of five blocks as outlined at the bottom.

3   Results

To illustrate the performance and improvement of POMAGO regarding more diverged species sets, we used three datasets with largely varying diversity: 1) A set of *E. coli* stains, being very closely related and well examined. 2) A set of Aquificales species, discussed to be phylogenetically located at the root of bacteria [12,9]. 3) A set of extremely diverged bacteria distributed among the whole phylogenetic tree.

The first dataset was taken from [19] and contained very closely related sequences consisting of 15 *Escherichia coli* genomes: *E. coli APEC O1*, *E. coli BW2952*, *E. coli CFT073*, *E. coli IAI1*, *E. coli K12 DH10B*, *E. coli K12 MG1655*, *E. coli K12 W3110*, *E. coli O103 H2 12009*, *E. coli O111 H 11128*, *E. coli O157 H7 EDL933*, *E. coli O157 H7 Sakai*, *E. coli O157 H7 TW14359*, *E. coli S88*, *E. coli SMS* and *E. coli UTI89*. *E. coli K12 MG1655*, as well examined and least genetically manipulated strain [4], was selected as reference. Nearly the complete reference genome (99.9%) was aligned to one or more non-reference sequences. Moreover, 95.2% of the reference sequence could be aligned to at least half the sequences and 77.3% to all sequences of the dataset. On average each of the 1,554 alignment blocks contained 13.91 out of 15 species with a mean length of 2,982 nt. The alignment covered 18 continuous intervals of the reference genome with an average length of 257,486 nt and a median of 209,015 nt.

The second dataset contained six hyperthermophile bacteria from the Aquificales group, which are believed to contain many archeal genes obtained by horizontal gene transfer [5]: *Aquifex aeolicus*, *Hydrogenobaculum sp.*, *Persephonella marina*, *Sulfurihydrogenibium azorense*, *Sulfurihydrogenibium YO3AOP1* and *Thermocrinis albus*. This set exhibits a number of interesting criteria making a multiple genome alignment worthwhile but harder than for the *E. coli* set. In comparison, the genomes are poorly examined and annotated, they have a high number of invading genes (due to horizontal gene transfer) and possibly form a root group. We selected *Aquifex aeolicus* as reference genome for this group. *A. aeolicus* appeared to be the most rearranged genome and asks therewith for a critical inspection of POMAGO.

The computed alignment consisted of 1,202 alignment blocks covering 187 continuous intervals with an average length of 7,097 nt and a median of 4,957 nt. The longest sequence of consecutive genes conserved in the same order and present in all the six genomes consists of 13 genes, respectively 9,000 nt. Taking any of the other species as reference resulted in a much longer maximal alignment block consisting of 23 (with *T. albus* as reference), respectively 45 (with any of the other four species), consecutive genes and around 22,000 nt, respectively 35,000 nt. Nevertheless, more than 83% of the *Aquifex aeolicus* genome was alignable to at least one of the other species. Furthermore, the alignment blocks contained 4.97 out of the 6 species on average.

For the third dataset the following very distant bacterial genomes were selected: *Aquifex aeolicus*, *Borrelia burgdorferi B31*, *Burkholderia gladioli BSR3*, *Escherichia coli K12 W3110*, *Geobacter sulfurreducens PCA*, *Helicobacter pylori PeCan4*, *Methylobacterium extorquens DM4* and *Streptococcus mutans GS5*. This group is interesting due to diversity. Only a low number of genes conserved in a syntenic range is expected. As in the second dataset *Aquifex aeolicus* was selected as reference.

Despite the diversity of sequences more than half of the *Aquifex aeolicus* genome (56.1%) could be aligned to at least one of the other sequences and

11.4% could be aligned to all other genomes. On average an alignment block contains 5.61 out of 8 genomes. The alignment consisted of 808 blocks with an average length of 1,104 nt spanning over 346 continuous intervals with an average length of 2,578 nt and a median of 1,844 nt. As expected, most of the blocks covered only a single gene and the surrounding intergenic regions (98%). None of the alignment blocks spanned more than four genes in a row.

4 Discussion

We compared our results with `progressiveMauve` and TBA, see Table 1. We achieved the highest coverage of the reference genome within the alignment for nearly all sets. The only exception was the closely related *E. coli* dataset where POMAGO aligned 0.1% (4,933 nt) less than `progressiveMauve`. The advantage of our new approach increased the more diverged the species sets became. POMAGO obtained a coverage of 99.9% for the *E. coli* group, 83.4% for Aquificales group and 56.1% for the 'Divergent Group', compared to 100%, 37.0% and 29.4%, resp. 97.2%, 79.6% and 32.8% for `progressiveMauve` and TBA, respectively.

Table 1. Comparison of genome-wide alignments of closely related *E. coli* strains, more distant Aquificales, and a set of extreme divergent species ('Divergent Group') calculated by `progressiveMauve`, TBA and POMAGO.

	E. coli (15)			Aquificales (6)			Divergent Group (8)		
	pMauve	TBA	POMAGO	pMauve	TBA	POMAGO	pMauve	TBA	POMAGO
No. of blocks	86	1693	1554	343	2475	1202	49	1372	808
Coverage (%)	**100**	97.2	99.9	37.0	79.6	**83.4**	29.4	32.8	**56.1**
Block length (nt)	53949	2663	2982	1715	512	1104	9534	381	1104
Mean no. of seq.	14.83	13.62	13.91	5.05	4.70	4.97	3.96	4.92	5.61
Gap rate	0.43	0.01	0.04	0.50	0.07	0.09	0.71	0.10	0.14
Weighted SoP	12.21	13.19	**13.20**	2.70	2.78	**2.81**	1.92	2.68	**2.71**
CPU time (min)	183	597	1098	46	12	41	128	41	42

The three tools perform apparently different in terms of number and length of produced alignment blocks. While `progressiveMauve` produces much less but longer alignment blocks than POMAGO, the number of blocks from TBA is generally higher but the blocks are shorter. Although POMAGO achieved a moderate number of alignment blocks, the average block length was relatively high.

POMAGO was able to align on average 13.91 out of 15 organisms, 4.97 out of 6 and 5.61 out of 8, for the *E. coli*, Aquificales and 'Divergent Group', respectively. This is slightly more compared to TBA. For the *E. coli* and Aquificales dataset more organisms are alignable using `progressiveMauve`. This is due to the fact that `progressiveMauve` generates lots of gaps within the alignment. 43%, respectively 50% and 71%, of the `progressiveMauve` alignment consisted of gaps while TBA and POMAGO had only 7% to 14% gaps. POMAGO always had a slightly higher gap rate than TBA. The huge number of gaps in the progressiveMauve alignment on the other hand results in much less but longer alignment blocks.

In addition to coverage and fraction of gaps, we computed the length-weighted Sum-of-Pairs score (WSoP) [20] to compare the quality of alignments:

$$\sigma = \frac{1}{n-1} \sum_{x,y \in A} \frac{1}{l(A)} \sum_{i=1}^{l(A)} \delta(x_i, y_i),$$

where $\delta(p, q)$ is Kronecker's delta and $l(A)$ denotes the length of alignment A.

For all datasets POMAGO outperformed the other tools in terms of the WSoP score, although there is only a small difference between TBA and POMAGO.

While progressiveMauve tends to fill huge parts of the alignment with gaps the alignments of TBA and POMAGO are quite similar. Especially, when aligning very closely related sequences the differences are small, except that TBA is much faster. On the other hand if the sequences are only distantly related, POMAGO was able to align a much higher percentage of nucleotides (56.1%) with more sequences (5.61) on average and with a better WSoP score (2.71), compared to both other tools. In this case POMAGO clearly benefits from selecting orthologous genes as anchor points and obtains a run time similar to TBA (42 minutes).

In general POMAGO consumed more CPU time in comparison to both other tools. However, less than 5% of the runtime was used for the cleaning and completion steps. Faster methods than ClustalW may speed up the alignment step and compensate this difference.

To evaluate the influence of the selected reference sequence on the alignment quality we computed the alignments with each of the six genomes from the Aquificales dataset as reference. The results are given in Table S1 in the supplement. While the alignments are quite similar in terms of number of blocks (ranging from 1,101 to 1,330), average block lengths (1,045 to 1,186), mean number of sequences per alignment block (4.83 to 5.16), gap rate (0.07 to 0.12) and WSoP score (2.77 to 2.95), there is a significant difference in the coverage (70.3% to 88.9%). This difference comes with no surprise, as the coverage of the reference genome within the alignment mainly depends on the most closely related species in the dataset. Therefore, selecting one of the closely related species *S.azorense* and *S.YO3AOP1* as reference results in the highest coverage.

Although different reference genomes are chosen, the same (cleaned) orthology graph and hence the same anchors are used for the alignment. In turn, the alignment of the coding sequences does not depend on the reference. When selecting a different genome. However, alignment blocks can be divided or merged, depending on the similarity of intergenic regions to the reference sequence.

5   Conclusion

With POMAGO we present a multiple genome-wide alignment program for bacteria. We used an orthology based graph obtained by Proteinortho, which was extended and decomposed in order to obtain a biological meaningful alignment.

Protein-coding regions are aligned on nucleotide and amino acid level, whereas intergenic regions are aligned on nucleotide level only. Additionally, we provide the possibility of exchangeable methods. Therewith, intergenic alignments might be performed with tools based on secondary RNA structure.

In terms of weighted SoP score POMAGO outperforms progressiveMauve and even TBA, which we consider the reference alignment tool so far. We could show that our approach delivers outstanding coverage while maintaining good block lengths. This advantage scales with increasing phylogenetic distances. It appears that with the use of synteny information POMAGO is most suitable for datasets with distantly related species where orthologous sequences can be detected even if the sequence similarity is quite small.

Further improvements in runtime are desirable when an application to eukaryotic species is intended. This is currently possible as well, but would take rather long. Improvements can be conducted, e.g. by choosing different alignment methods. Also, several processing steps do not yet use multiple CPUs. Moreover, an extension to handle introns correctly should improve the results for these genomes. Additional extensions for repeats and pseudogenes are not necessary as these are already covered by our method.

We plan to extend the orthology graph to not only contain nodes for coding sequences but also for the intergenic regions. The additional nodes will be divided into groups to differentiate between several types of sequences, including different classes of ncRNAs. Each group should be aligned with a different method, e.g. structure-based alignment using LocaRNA [22].

The POMAGO software and supplementary material is available at http://www. rna.uni-jena.de/supplements/pomago.

Acknowledgements. This work was funded by the Carl-Zeiss-Stiftung, DFG GRK-1384, MA5082/1-1 and MI439/14-1.

References

1. Altschul, S.F., Gish, W., Miller, W., Myers, E.W., Lipman, D.J.: Basic local alignment search tool. J. Mol. Biol. 215(3), 403–410 (1990)
2. Angiuoli, S.V., Salzberg, S.L.: Mugsy: fast multiple alignment of closely related whole genomes. Bioinformatics 27(3), 334–342 (2011)
3. Blanchette, M., Kent, W.J., Riemer, C., Elnitski, L., Smit, A.F., Roskin, K.M., Baertsch, R., Rosenbloom, K., Clawson, H., Green, E.D., Haussler, D., Miller, W.: Aligning multiple genomic sequences with the threaded blockset aligner. Genome Res. 14(4), 708–715 (2004)
4. Blattner, F.R., Plunkett, G., Bloch, C.A., Perna, N.T., Burland, V., Riley, M., Collado-Vides, J., Glasner, J.D., Rode, C.K., Mayhew, G.F., Gregor, J., Davis, N.W., Kirkpatrick, H.A., Goeden, M.A., Rose, D.J., Mau, B., Shao, Y.: The complete genome sequence of Escherichia coli K-12. Science 277(5331), 1453–1462 (1997)
5. Boussau, B., Guéguen, L., Gouy, M.: Accounting for horizontal gene transfers explains conflicting hypotheses regarding the position of aquificales in the phylogeny of bacteria. BMC Evol. Biol. 8, 272–272 (2008)

6. Braga, M.D., Machado, R., Ribeiro, L.C., Stoye, J.: Genomic distance under gene substitutions. BMC Bioinformatics 12(suppl. 9) (2011)
7. Bray, N., Pachter, L.: MAVID: constrained ancestral alignment of multiple sequences. Genome Res. 14(4), 693–699 (2004)
8. Brudno, M., Chapman, M., Göttgens, B., Batzoglou, S., Morgenstern, B.: Fast and sensitive multiple alignment of large genomic sequences. BMC Bioinformatics 4, 66–66 (2003)
9. Burggraf, S., Olsen, G.J., Stetter, K.O., Woese, C.R.: A phylogenetic analysis of Aquifex pyrophilus. Syst. Appl. Microbiol. 15(3), 352–356 (1992)
10. Chen, X., Tompa, M.: Comparative assessment of methods for aligning multiple genome sequences. Nat. Biotechnol. 28(6), 567–572 (2010)
11. Darling, A.E., Mau, B., Perna, N.T.: progressiveMauve: multiple genome alignment with gene gain, loss and rearrangement. PLoS One 5(6) (2010)
12. Deckert, G., Warren, P.V., Gaasterland, T., Young, W.G., Lenox, A.L., Graham, D.E., Overbeek, R., Snead, M.A., Keller, M., Aujay, M., Huber, R., Feldman, R.A., Short, J.M., Olsen, G.J., Swanson, R.V.: The complete genome of the hyperthermophilic bacterium Aquifex aeolicus. Nature 392(6674), 353–358 (1998)
13. Dieterich, C., Wang, H., Rateitschak, K., Luz, H., Vingron, M.: CORG: a database for COmparative Regulatory Genomics. Nucleic Acids Res. 31(1), 55–57 (2003)
14. Fitch, W.M.: Distinguishing homologous from analogous proteins. Syst. Zool. 19, 99–113 (1970)
15. Frith, M.C., Hamada, M., Horton, P.: Parameters for accurate genome alignment. BMC Bioinformatics 11, 80–80 (2010)
16. Gruber, A.R., Findeiß, S., Washietl, S., Hofacker, I.L., Stadler, P.F.: RNAz 2.0: improved noncoding RNA detection. Pac. Symp. Biocomput. 15, 69–79 (2010)
17. Larkin, M.A., Blackshields, G., Brown, N.P., Chenna, R., McGettigan, P.A., McWilliam, H., Valentin, F., Wallace, I.M., Wilm, A., Lopez, R., Thompson, J.D., Gibson, T.J., Higgins, D.G.: Clustal W and Clustal X version 2.0. Bioinformatics 23(21), 2947–2948 (2007)
18. Lechner, M., Findeiss, S., Steiner, L., Marz, M., Stadler, P.F., Prohaska, S.J.: Proteinortho: detection of (co-)orthologs in large-scale analysis. BMC Bioinformatics 12, 124–124 (2011)
19. Qi, Z.-H., Du, M.-H., Qi, X.-Q., Zheng, L.-J.: Gene comparison based on the repetition of single-nucleotide structure patterns. Computers in Biology and Medicine 42, 975–981 (2012)
20. Rose, D., Hertel, J., Reiche, K., Stadler, P.F., Hackermüller, J.: NcDNAlign: plausible multiple alignments of non-protein-coding genomic sequences. Genomics 92(1), 65–74 (2008)
21. Wang, X., Fu, Y., Zhao, Y., Wang, Q., Pedamallu, C.S., Xu, S.Y., Niu, Y.: Accurate reconstruction of molecular phylogenies for proteins using codon and amino acid unified sequence alignments (CAUSA). Nature Proceedings (2001)
22. Will, A., Joshi, T., Hofacker, I.L., Stadler, P.F., Backofen, R.: LocARNA-P: accurate boundary prediction and improved detection of structural RNAs. RNA 18(5), 900–914 (2012)

Effect of Incomplete Lineage Sorting on Tree-Reconciliation-Based Inference of Gene Duplication

Yu Zheng and Louxin Zhang

Department of Mathematics, National University of Singapore,
10 Lower Kent Ridge, Singapore 119076

Abstract. Incomplete lineage sorting (ILS) gives rise to stochastic variation in the topology of a gene tree and hence introduces false duplication events when gene tree and species tree reconciliation method is used for inferring the duplication history of a gene family. We quantify the effect of ILS on inference of gene duplication by examining the expected number of false duplication events inferred from reconciling a random gene tree, which occurs with a probability predicted in coalescent theory, and the given species tree. We computationally analyze the relationships between the number of false duplication events inferred on a branch and its length in a species tree, and the relationships between the expected number of false duplication events in a species tree and its topological parameters. This study provides evidence that inference of gene duplication based on tree reconciliation was affected by ILS to a greater extent on an asymmetric species tree than on a symmetric one. Our findings also suggest that the bias caused by ILS in reconciliation-based inference of gene duplication might not be negligible. Hence, when gene duplication is inferred via tree reconciliation or any other method that takes gene tree topology into account, the ILS-induced bias should be examined cautiously.

1 Background

A gene tree is the phylogenetic tree of a family of homologous genes. A species tree is the phylogenetic tree of a collection of species. In population genomics and phylogenetics, it is important to distinguish gene trees and species trees, as a gene tree reconstructed from the DNA sequences of the given gene family is sometimes discordant with the species tree that contains it [31,33]. The incongruence of gene trees and species trees can be caused by gene duplication and loss, horizontal gene transfer, hybridization, or incomplete lineage sorting (ILS) [17,21,29]. Accordingly, the relationships between gene trees and species trees have been the focus of many studies over the past two decades [10,18]. Gene trees have been used to estimate the species trees [8,12,19,20,22], to estimate species divergence time [4] and ancestral population size [11,16,34], and to infer the history of gene duplication [1,3,5,6,13,23,32].

Z. Cai et al. (Eds.): ISBRA 2013, LNBI 7875, pp. 261–272, 2013.
© Springer-Verlag Berlin Heidelberg 2013

One popular approach for gene duplication inference is gene tree and species tree reconciliation. It is formalized from the following fact: If the descendants of a node in a gene tree are distributed in the same set of species as one of its children, then the node corresponds to a gene duplication event [12,23]. Clearly, this approach takes gene tree topology into account. If incorrect gene trees are used, duplication events are often mis-inferred [14].

In a species tree, each internodal branch represents an ancestral population; each internal node represents a time point at which the ancestral population split into two subpopulations. It is assumed that there was no gene flow between the subpopulations after split. When the population of each species is large, the DNA sequences sampled from two species are more unlikely to have their common sequence ancestor living at the moment that the most recent common ancestor (MRCA) of the two species split; instead, the time back to the common sequence ancestor is uncertain and typically longer than the time back to the MRCA of the species. This evolutionary phenomenon is ILS or deep coalescence. Clearly, ILS gives rise to considerable stochastic variations in gene tree topology [27,31], implying that different unlinked loci might have different genealogical histories, and different samplings might also lead to different gene tree topologies for the same gene. Consider the two different gene tree topologies in Fig. 1. The gene topology in red is concordant to the species tree; reconciling this gene tree and the species tree does not infer any gene duplication events, whereas reconciliation with the gene tree in green gives one (false) duplication event. Hence, ILS affects gene duplication inference. To the best of our knowledge, the effects of ILS on gene duplication inference has not been examined quantitatively although they have been noticed for long time (see [21] for example).

The present paper examines quantitatively the effect of ILS on inference of gene duplication. Here, we assume that no genetic exchange has occurred between unrelated species and there is no sequence error to facilitate our quantitative study. Notice that the effects of horizontal gene transfer and hybridization events on gene duplication inference have been studied by proposing general evolutionary models to coordinate these events or by computational simulation [2,9,36].

2   Results and Discussion

In our study, we shall consider only gene trees over single-gene families. In other words, we assume only one gene is sampled from each species. Under such an assumption, any inferred gene duplication event is a false one, and the gene tree distribution can be computed using coalescent theory [7,27]. Accordingly, the assumption greatly simplifies our discussion and allows us to find out crucial connections between the effect of ILS and species tree topologies.

When calculating the probability that a gene tree is seen in the corresponding species tree, we consider a simple coalescent model each species has a constant diploid effective population size N during its entire existence and evolutionary time of t generations equals $T = t/(2N)$ coalescent time units [31].

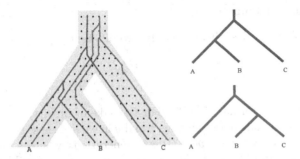

Fig. 1. Schematic view of two different coalescent histories in a species tree. The species tree (light blue) of three species is given in the left panel. DNA sequences sampled from different individuals within a species may give different collapsed gene trees (red and green, right panel) for a gene family. If the green gene tree is used, a gene duplication event is inferred in the branch entering the species tree root and a gene loss event is inferred in the lineage leading to C.

2.1 Measuring the Effect of ILS on Gene Duplication Inference

Consider a single-gene family $\mathcal{F}$ sampled from a set X of species. Let S be the phylogeny over X. If no gene duplication occurred to the gene family during the evolution of the species, the tree of the gene family has the same topology as S. If ILS events have occurred, however, the gene tree reconstructed from the gene sequences might be different from S. To quantify the effect of ILS on inference of gene duplication for a gene family on S, we use the expected number $D(S)$ (or $L(S)$) of false gene duplication (or loss) events output from the lca reconciliation of a random gene tree and S. For a gene tree G, we use $c_{\mathrm{dup}}(G, S)$ (or $c_{\mathrm{loss}}(G, S)$) to denote the gene duplication (or loss) cost of the lca reconciliation between G and S (Materials and Methods). Since $c_{\mathrm{dup}}(G, S) = c_{\mathrm{loss}}(G, S) = 0$ if $G = S$. $D(S)$ and $L(S)$ are simply:

$$D(S) = \sum_{G \in \mathcal{G}} c_{\mathrm{dup}}(G, S) \Pr[G \mid S], \tag{1}$$

$$L(S) = \sum_{G \in \mathcal{G}} c_{\mathrm{loss}}(G, S) \Pr[G \mid S], \tag{2}$$

where $\mathcal{G}$ is the set of all possible gene tree topologies, and $\Pr[G \mid S]$ the probability that G is the collapsed gene tree of a coalescent history of the sampled genes from the species belonging to X (see Fig. 1 for an illustration of a coalescent history and its collapsed gene tree).

Let $\mathcal{H}(G)$ be the set of all possible coalescent histories that give the gene tree G. For each $H \in \mathcal{H}(G)$, we use $\Pr[H \mid S]$ to denote the probability that H occurs in S. By definition, we compute $\Pr[G \mid S]$ by:

$$\Pr[G \mid S] = \sum_{H \in \mathcal{H}(G)} \Pr[H \mid S], \tag{3}$$

where $\Pr[H \mid S]$ can be computed efficiently given H and S [7,35].

2.2   The Case of Four Species

In the case of four species, there are only two different topologies

$$S_1 = ((A, B) : t_1, (C, D) : t_2),$$
$$S_2 = (((A, B) : \tau_1, C) : \tau_2, D),$$

in Newick phylogeny format (Fig. 2). For the sake of brevity, we use $p(x, y)$ to denote the parental node of two siblings x and y in each of these two species trees. In S_1, the evolutionary time of $p(A, B)$ is t_1 generations, whereas that of $p(C, D)$ is t_2 generations. Let G be an arbitrary gene tree of a gene family. Consider the lca reconciliation between G and S_1. Any gene tree node is mapped to $p(A, B)$ if and only if its two children are mapped to A and B respectively (Materials and Methods). This fact also holds for $p(C, D)$. Therefore, false duplication events can only be inferred on the branch entering the root in S_1. Set $T_i = t_i/(2N)$ for $i = 1, 2$. By calculating the distribution of the gene trees [24,27], we obtain:

$$D(S_1) = \frac{2}{3}(e^{-T_1} + e^{-T_2}), \tag{4}$$

$$L(S_1) = 2(e^{-T_1} + e^{-T_2}) + \frac{2}{9}e^{-(T_1+T_2)} \tag{5}$$

from Eqn. (1) and (2).

Now, we switch to consider S_2. Setting $\bar{T}_i = \tau_i/(2N)$ for $i = 1, 2$, we have:

$$D(S_2) = \frac{2}{3}\left(e^{-\bar{T}_1} + e^{-\bar{T}_2}\right) - \frac{1}{3}e^{-(\bar{T}_1+\bar{T}_2)} + \frac{5}{18}e^{-(\bar{T}_1+3\bar{T}_2)}, \tag{6}$$

$$L(S_2) = 2\left(e^{-\bar{T}_1} + e^{-\bar{T}_2}\right) - \frac{1}{3}e^{-(\bar{T}_1+\bar{T}_2)} + \frac{5}{6}e^{-(\bar{T}_1+3\bar{T}_2)}. \tag{7}$$

Since $e^{-x} < 1$ for any $x > 0$, we have:

$$D(S_1) < 1\frac{1}{3} \text{ and } L(S_1) < 4\frac{2}{9};$$
$$D(S_2) < 1\frac{5}{18} \text{ and } L(S_2) < 4\frac{1}{2}.$$

If all branches have equal length (i.e. $T_1 = T_2 = \bar{T}_1 = \bar{T}_2 = T$), $D(S_1) \geq D(S_2)$ for any T. However, $L(S_1) \geq L(S_2)$ only if $T \geq \frac{1}{2}\ln(3/2)$.

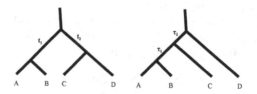

Fig. 2. Two topologies S_1 (left) and S_2 (right) of the species trees of 4 species

Assume S_1 and S_2 are ultrametric and have the same height. We further assume that $T_1 = T_2 = 2T$ and $\bar{T}_1 = \bar{T}_2 = T$, implying that $\tau_1 + \tau_2 = t_1 = t_2$ and that A and B diverged at the same time in both trees. Then,

$$D(S_1) = \frac{4}{3}e^{-2\mathrm{T}} \text{ and } L(S_1) = 4e^{-2\mathrm{T}} + \frac{2}{9}e^{-4\mathrm{T}};$$

$$D(S_2) = \frac{4}{3}e^{-\mathrm{T}} - \frac{1}{3}e^{-2\mathrm{T}} + \frac{5}{18}e^{-4\mathrm{T}} \text{ and } L(S_2) = 4e^{-\mathrm{T}} - \frac{1}{3}e^{-2\mathrm{T}} + \frac{5}{6}e^{-4\mathrm{T}}.$$

Using numerical computation, we obtained $D(S_1) < D(S_2)$ only if $T > 0.0649$, but $L(S_1) < L(S_2)$ for any T. This analysis suggests that the effect of ILS is closely related to species tree structure.

2.3 Effect Analysis on a *Drosophila* Species Tree

Genome-wide analysis provides strong evidence for the prevalence of ILS events in *Drosophila* evolution [25]. Here, we examined the expected number of false duplication events caused by ILS in the phylogeny of 12 *Drosophila* species [15], in which evolutionary time is dated for all branches. Since the effective population size N for the *Drosophila* species is unknown, we considered four different effective population sizes ($2 \times 10^6, 6 \times 10^6, 10 \times 10^6$, and 14×10^6) and set the generation time to be $1/10$ years [25]. The expected numbers of false duplication events caused by ILS for different effective population sizes are plotted (Fig. 3). Here, we point out that our conclusion does not depend on the specific effective population sizes we used.

Since only one gene is sampled from the population of each species, no gene duplication is inferred on branches connecting to the leaves. In other words,

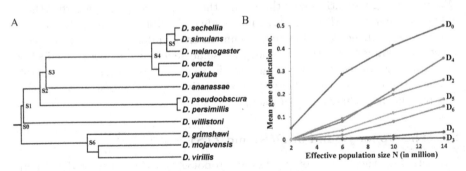

Fig. 3. Effect analysis for a *Drosophila* species tree. (A). A tree of 12 *Drosophila* species given in [15]. All the branches are drawn in proportion to evolutionary time. False duplication events caused by ILS can only be inferred on seven branches that enter S0–S6 respectively for single gene families. (B). The expected number D_i of false gene duplication events on the branch entering Si is plotted against the effective population size N, with the generation time being set to $\frac{1}{10}$ years. Four different effective population sizes (2×10^6, 6×10^6, 10×10^6 and 14×10^6) were examined. It shows that the number of false gene duplication events on a branch correlates largely with its evolutionary time.

false gene duplication events can only be inferred on the seven branches that are denoted by their end nodes Si ($0 \leq i \leq 6$) (Fig. 3A). The expected total number of false gene duplication events in the tree can range from 0.0534 to 1.7663 for each of the selected effective population sizes. Let D_i be the expected number of false gene duplication events on the branch Si for each i. Although the exact values of these D_i are different for the different effective population sizes, their relative ranks remain almost the same, correlating well with the branches' evolutionary time. For instance, D_0 has the largest value for each effective population size. This is because the branch entering the root is assumed to be long enough that all the lineages coexisting at the moment that the MRCA of all the extant species split will coalesce on it. For the longest branch entering S4, D_4 is the second largest for effective population sizes of 10 and 14 million, and the third largest for other sizes.

Another finding is that on the branches close to the root, the expected number of false gene duplication events is relatively large. For example, for the shortest branch S1, D_1 is not the smallest; instead, it is larger than D_3, probably due to the closeness of S_1 to the tree root. Similarly, branch S6 is longer than S2, but D_6 is smaller than D_2 for each effective population size because S6 is closer to the tree root.

We now switch to 6698 gene trees in the *Drosophila* species tree [14]. We inferred gene duplication events for the corresponding gene families by reconciling the gene trees and the species tree (Fig. 3A). In total, we inferred 10,264 gene duplication events that are distributed on the seven branches as: 1.8% (S3), 6.5% (S0), 7.4% (S2), 8.0% (S5), 15.1% (S1), 20.5% (S6) and 40.6% (S4). Such a distribution is not quite consistent with the computational analysis presented above. The proportion of inferred duplication events on the branches entering S0 and S2 is significantly lower than what the analysis suggests, whereas those on branches entering S1 and S6 are much higher. Possible reasons for this are either because sequence sampling and alignment errors influenced gene tree reconstruction, leading to incorrect topology for some gene trees, or because effective population size varies for different ancestral species. At this stage, we are unable to assess the effect of these factors, as the estimation of ancestral population sizes remains as a challenging problem.

2.4   The Upper Bound of $D(S)$ and $L(S)$

To interrogate the impact of species tree topology on the effect of ILS for inference of gene duplication, we considered 10 ultrametric tree topologies over 10 species (Fig. 4). In each of the 5 asymmetric species trees, the two subtrees rooted at the children of the root are linear trees. In each of the 5 symmetric trees, the subtrees are balanced binary trees instead.

We define the height of a ultrametric species tree to be the coalescent time of a path from the root to a leaf, measured in coalescent time units. $D(S)$ and $L(S)$ for these 10 topologies with heights of 2 and 10 units are respectively presented in two panels in Fig. 4. Although each path from the root to a leaf has the same evolutionary time, the number of branches contained in each path

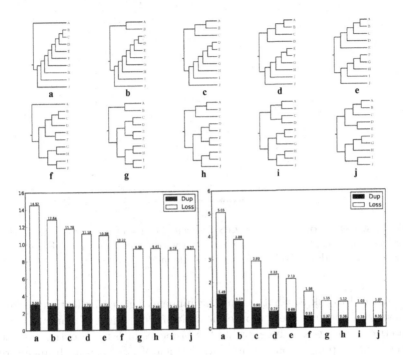

Fig. 4. $D(S)$ **and** $L(S)$ **for asymmetric topologies (first row) and symmetric topologies (second row).** Branches in each of 10 ultrametric topologies are drawn in proportion to their length. In the bottom row, the left and right plots are drawn to different scales for the topologies of heights 2 and 10, respectively. In each plot, the white and black bars represent $L(S)$ and $D(S)$, respectively.

varies. For each leaf, we define its depth to be the number of branches in the unique path from it to the root. Although each path from the root to a leaf has the same evolutionary time, different leaves may have different depths. The Sackin index of a species tree is defined as the average depth of a leaf in the tree [28]. The ten tree topologies listed in the figure have the following Sackin indexes:

Tree	a	b	c	d	e	f	g	h	i	j
Sackin index	5.4	4.7	4.2	3.9	3.8	3.9	3.8	3.5	3.4	3.4

Hence, our experiments suggest that:

- $D(S)$ and $L(S)$ increase with the Sackin index of a species tree S;
- Asymmetric trees have a larger $D(S)$ and $L(S)$ than symmetric ones of the same height.

In [7], the authors studied the probability distribution of all the gene trees in a species tree over 5 species. Since our study focuses the mean duplication and gene loss costs of a gene tree defined in (1) and (2), the facts reported here are not direct consequences of those reported in [7].

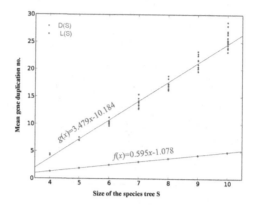

Fig. 5. Regression of $D(S)$ and $L(S)$. Given the size of S, $D(S)$ varies with the topology of S in a narrow range, whereas $L(S)$ varies in a wide range. For each size, we generated 20 random species trees in the Yule model.

It is natural to ask to what extent ILS influences gene duplication inference. To answer this question, we compute the limit of $D(S)$ and $L(S)$ for an arbitrary ultrametric species tree S by allowing the branches of S to be extremely short. Fix the effective population size for each branch of S. When all branches of S become very short, two lineages are unlikely to coalesce in any branch below the tree root; in other words, there is a high probability that any pair of lineages will coalesce in the branch entering the root. Therefore, in the limit case, for each gene tree G:

$$\Pr[G \mid S] \sim \sum_{H \in \mathcal{H}'(G)} \Pr[H \mid S],$$

where $\mathcal{H}'(G)$ is the set of the coalescent histories of n lineages whose collapsed gene tree is G in the root branch. Based on this fact, we computed the limit of $D(S)$ and $L(S)$ for 20 random species tree for each size (i.e. the number of species) from 4 to 10 (Fig. 5). We found that $D(S)$ varies in a narrow range for each tree size and linearly increases with species tree size. However, $L(S)$ changes in a different manner. First, $L(S)$ varies in a wide range for a fixed species tree size. Secondly, although $L(S)$ also fits a linear function for the tree size in the range of 4 to 10, it remains unclear if it grows linearly or not because of its wide range for a fixed tree size.

3   Conclusion

ILS introduces stochastic variation into the topology of the gene tree of a gene family. For the first time, we have quantified the effect of ILS on gene duplication inference by examining the expected number of false gene duplication events inferred from reconciling a random gene tree and the species tree that contains it. In this preliminary study, we have also analyzed the connection between the topological parameters of the corresponding species tree and the effect of ILS on gene duplication inference.

One of our findings is that inference of gene duplication based on tree reconciliation was affected by ILS to a greater extent on an asymmetric species tree than on a symmetric one. Considering gene duplication events arising from ILS on different species tree branches separately, we also found that the longer an internodal branch is, the more likely gene duplication events are to be mis-inferred on it. Additionally, gene duplication events are more likely to be mis-inferred on a branch close to the species tree root.

In analyzing the limit of $D(S)$ and $L(S)$ for a species tree S when its branches are extremely short, we found that $D(S)$ increases linearly with the species tree size in the range of 4 to 10. This fact indicates that $D(S)$ increases with $|S|$, the size of S in general and hence it is not bounded above. It also raises a theoretical problem: $D(S) \leq 0.6|S|$? Since $L(S) \geq 3D(S)$ for a species tree [37], $L(S)$ is not bounded above by a constant if $D(S)$ is not. Our findings imply that the bias caused by ILS in reconciliation-based gene duplication inference is not negligible. Therefore, when gene duplication is inferred via tree reconciliation or any other method that takes gene tree topology into account, the ILS-induced bias should be examined cautiously. Alternatively, one may use a unified reconciliation approach that considers gene duplication, loss and ILS simultaneously [26,30].

Finally, we remark that ILS also affect the gene trees for genes which are from different genera. How much ILS is expected to affect gene duplication inference for gene families cross different genera is definitely a research topic for future study.

4   Material and Methods

4.1   Computing the Gene Tree Distribution in a Species Tree

The probability that a gene tree occurs in a given species tree is computed by Eqn. (3). For the purpose of computing the gene tree distribution in a species tree, COAL is too slow to be used, although it has many useful features [7]. Our analysis used a home-made computer program implemented in C. It speeds up computation via the dynamic programming technique, which had also been used by Wu in STELLS [35]. Presently, it allows us to examine the effect of ILS on gene duplication inference for species trees of up to 12 species. For the case of 12 species, one needs to consider about 13.7 billion gene trees for the analysis.

4.2   Gene Duplication Inference

Consider a collection X of extant species. The species tree of the given species is a rooted tree in which each leaf uniquely represents (and hence is labeled by) an extant species. Here, we further assume species trees are fully binary and branch-weighted. Therefore, in a species tree, each non-leaf node has exactly two children; each internodal branch represents an ancestral species and has the evolutionary time of the ancestral species as its length.

For a gene family sampled from X, its gene tree is a rooted tree in which each leaf represents a gene and is labelled by the species where the gene is found. Since the gene family is assumed to have one gene sampled from each species, the gene tree is uniquely leaf-labelled in our study.

Let S be the species tree of X and let G be a binary gene tree for a gene family $\mathcal{F}$ over X. For any two nodes x, y of S, we use $\mathrm{lca}(x, y)$ to denote the MRCA of x and y in S. The lca reconciliation $\mathcal{R}$ between S and G is a node-to-node mapping from $V(G)$ to $V(S)$ defined as:

$$\mathcal{R}(g) = \begin{cases} \text{the unique leaf in } S \text{ that has the same label as } g, & \text{if } g \text{ is a leaf;} \\ \mathrm{lca}\left(\mathcal{R}(g_1), \mathcal{R}(g_2)\right), & \text{otherwise,} \end{cases}$$

for any gene tree node g, where g_1 and g_2 are the children of g.

The duplication history of $\mathcal{F}$ can be inferred through the lca reconciliation $\mathcal{R}$ [12,23]. For a non-leaf node g of G, if $\mathcal{R}(c(g)) = \mathcal{R}(g)$ for some child $c(g)$ of g, then a duplication event is inferred in the branch entering $\mathcal{R}(g)$ in S.

The number of the gene duplication events inferred by using the lca reconciliation is denoted by $c_{\mathrm{dup}}(G, S)$. All the inferred gene duplication events form a putative duplication history of $\mathcal{F}$ in which some genes might become lost. The number of gene loss events assumed in the gene duplication history is computed as follows.

For any two nodes s and t such that s is below t in S, we write $s \subset h \subset t$ to denote that h is a node in the path from t to s for a node h. We define:

$$l(s, t) = |\{h \in S \mid s \subset h \subset t\}|.$$

Note that $l(s, t)$ is equal to the number of lineages off the evolutionary path from t to s. For a non-leaf node g with children g_1 and g_2 of G, define:

$$l(g) = \begin{cases} 0, & \text{if } \mathcal{R}(g) = \mathcal{R}(g_1) = \mathcal{R}(g_2), \\ l(\mathcal{R}(g_1), \mathcal{R}(g_2)) + 1, & \text{if } \mathcal{R}(l_g) \subset \mathcal{R}(g) = \mathcal{R}(r_g), \\ l(\mathcal{R}(g_1), \mathcal{R}(g)) + l(\mathcal{R}(g_2), \mathcal{R}(g)), & \text{if } \mathcal{R}(g_1) \subset \mathcal{R}(g) \supset \mathcal{R}(g_2). \end{cases}$$

The number of genes that have to be assumed to be lost in the inferred duplication history is equal to $\sum_{g \in G} l(g)$, denoted by $c_{\mathrm{loss}}(G, S)$ and called the *gene loss cost* of the lca reconciliation between G and S.

In this work, we used our computer program to compute the gene duplication and loss costs for a gene tree and a species tree [38].

Acknowledgements. The work was supported by Singapore AcRF Tier-2 Grant R-146-000-134-112.

References

1. Åkerborg, Ö., Sennblad, B., Arvestad, L., Lagergren, J.: Simultaneous bayesian gene tree reconstruction and reconciliation analysis. Proc. Natl. Acad. Sci. U. S. A. 106(14), 5714–5719 (2009)
2. Bansal, M.S., Alm, E.J., Kellis, M.: Efficient algorithms for the reconciliation problem with gene duplication, horizontal transfer and loss. Bioinformatics 28(12), i283–i291 (2012)
3. Berglund-Sonnhammer, A.C., Steffansson, P., Betts, M.J., Liberles, D.A.: Optimal gene trees from sequences and species trees using a soft interpretation of parsimony. J. Mol. Evol. 63(2), 240–250 (2006)
4. Cann, R.L., Stoneking, M., Wilson, A.C.: Mitochondrial DNA and human evolution. Nature 325(6099), 31–36 (1987)
5. Chauve, C., El-Mabrouk, N.: New perspectives on gene family evolution: Losses in reconciliation and a link with supertrees. In: Batzoglou, S. (ed.) RECOMB 2009. LNCS, vol. 5541, pp. 46–58. Springer, Heidelberg (2009)
6. Chen, K., Durand, D., Farach-Colton, M.: Notung: a program for dating gene duplications and optimizing gene family trees. J. Comput. Biol. 7(3-4), 429–447 (2000)
7. Degnan, J.H., Salter, L.A.: Gene tree distributions under the coalescent process. Evolution 59(1), 24–37 (2005)
8. Doyle, J.J.: Gene trees and species trees: molecular systematics as one-character taxonomy. Syst. Botany 17, 144–163 (1992)
9. Doyon, J.-P., Scornavacca, C., Gorbunov, K.Y., Szöllősi, G.J., Ranwez, V., Berry, V.: An Efficient Algorithm for Gene/Species Trees Parsimonious Reconciliation with Losses, Duplications and Transfers. In: Tannier, E. (ed.) RECOMB-CG 2010. LNCS, vol. 6398, pp. 93–108. Springer, Heidelberg (2010)
10. Edwards, S.V.: Is a new and general theory of molecular systematics emerging? Evolution 63(1), 1–19 (2008)
11. Edwards, S.V., Beerli, P.: Perspective: gene divergence, population divergence, and the variance in coalescence time in phylogeographic studies. Evolution 54(6), 1839–1854 (2000)
12. Goodman, M., Czelusniak, J., Moore, G.W., Romero-Herrera, A.E., Matsuda, G.: Fitting the gene lineage into its species lineage, a parsimony strategy illustrated by cladograms constructed from globin sequences. Syst. Biol. 28(2), 132–163 (1979)
13. Górecki, P., Tiuryn, J.: DLS-trees: a model of evolutionary scenarios. Theor. Comput. Sci. 359(1), 378–399 (2006)
14. Hahn, M.W.: Bias in phylogenetic tree reconciliation methods: implications for vertebrate genome evolution. Genome Biol. 8(7), R141 (2007)
15. Hahn, M.W., Han, M.V., Han, S.G.: Gene family evolution across 12 *Drosophila* genomes. PLoS Genetics 3(11), e197 (2007)
16. Hey, J., Nielsen, R.: Multilocus methods for estimating population sizes, migration rates and divergence time, with applications to the divergence of *Drosophila pseudoobscura* and *D. persimilis*. Genetics 167(2), 747–760 (2004)
17. Keeling, P.J., Palmer, J.D.: Horizontal gene transfer in eukaryotic evolution. Nat. Rev. Genet. 9(8), 605–618 (2008)
18. Knowles, L.L., Kubatko, L.S.: Estimating Species Trees: Practical and Theoretical Aspects. Wiley-Blackwel, New Jersey (2010)
19. Liu, L., Yu, L., Kubatko, L., Pearl, D.K., Edwards, S.V.: Coalescent methods for estimating phylogenetic trees. Mol. Phylogenet. Evol. 53(1), 320–328 (2009)

20. Ma, B., Li, M., Zhang, L.X.: From gene trees to species trees. SIAM J. Comput. 30(3), 729–752 (2000)
21. Maddison, W.P.: Gene trees in species trees. Syst. Biol. 46(3), 523–536 (1997)
22. Maddison, W.P., Knowles, L.L.: Inferring phylogeny despite incomplete lineage sorting. Syst. Biol. 55(1), 21–30 (2006)
23. Page, R.D.M.: Maps between trees and cladistic analysis of historical associations among genes, organisms, and areas. Syst. Biol. 43(1), 58–77 (1994)
24. Pamilo, P., Nei, M.: Relationships between gene trees and species trees. Mol. Biol. Evol. 5(5), 568–583 (1988)
25. Pollard, D.A., Iyer, V.N., Moses, A.M., Eisen, M.B.: Widespread discordance of gene trees with species tree in *Drosophila*: evidence for incomplete lineage sorting. PLoS Genet. 2(10), e173 (2006)
26. Rasmussen, M.D., Kellis, M.: Unified modeling of gene duplication, loss, and coalescence using a locus tree. Genome Research 22(4), 755–765 (2012)
27. Rosenberg, N.A.: The probability of topological concordance of gene trees and species trees. Theor. Popul. Biol. 61(2), 225–247 (2002)
28. Sackin, M.J.: Good and bad phenograms. Syst. Zool. 21, 225–226 (1972)
29. Sang, T., Zhong, Y.: Testing hybridization hypotheses based on incongruent gene trees. Syst. Biol. 49(3), 422–434 (2000)
30. Stolzer, M., Lai, H., Xu, M., Sathaye, D., Vernot, B., Durand, D.: Inferring duplications, losses, transfers and incomplete lineage sorting with nonbinary species trees. Bioinformatics 28(18), i409–i415 (2012)
31. Takahata, N.: Gene genealogy in three related populations: consistency probability between gene and population trees. Genetics 122(4), 957–966 (1989)
32. Wehe, A., Bansal, M.S., Burleigh, J.G., Eulenstein, O.: Duptree: a program for large-scale phylogenetic analyses using gene tree parsimony. Bioinformatics 24(13), 1540–1541 (2008)
33. Wong, K.M., Suchard, M.A., Huelsenbeck, J.P.: Alignment uncertainty and genomic analysis. Science 319(5862), 473–476 (2008)
34. Wu, C.I.: Inferences of species phylogeny in relation to segregation of ancient polymorphisms. Genetics 127(2), 429–435 (1991)
35. Wu, Y.: Coalescent-based species tree inference from gene tree topologies under incomplete lineage sorting by maximum likelihood. Evolution 66, 763–775 (2012)
36. Yu, Y., Than, C., Degnan, J.H., Nakhleh, L.: Coalescent histories on phylogenetic networks and detection of hybridization despite incomplete lineage sorting. Syst. Biol. 60(2), 138–149 (2011)
37. Zhang, L.X.: From gene trees to species trees ii: Species tree inference by minimizing deep coalescence events. IEEE-ACM Trans. Comput. Biol. Bioinform. 8(6), 1685–1691 (2011)
38. Zheng, Y., Wu, T., Zhang, L.X.: Reconciliation of gene and species trees with polytomies. arXiv preprint, arXiv:1201.3995 (2012)

Ellipsoid-Weighted Protein Conformation Alignment

Hyuntae Na and Guang Song

Department of Computer Science, Iowa State University,
226 Atanasoff Hall, Ames, IA 50011, USA
{htna,gsong}@iastate.edu

Abstract. Conformation alignment is a critical step for properly interpreting protein motions and conformational changes. The most widely
used approach for superposing two conformations is by minimizing their
root mean square distance (RMSD). In this work, we treat the alignment problem from a novel energy-minimization perspective. To this end
we associate each atom in the protein with a mean-field potential well,
whose shape, ellipsoidal in general, is to be inferred from the observed
or computed fluctuations of that atom around its mean position. The
scales and directions of the fluctuations can be obtained experimentally
from anisotropic B-factors for crystal structures or computationally. We
then show that this "ellipsoid-weighted" RMSD alignment can be reformulated nicely as a point-to-plane matching problem studied in computational geometry. This new alignment method is a generalization of
standard RMSD and Gaussian-weighted RMSD alignment. It is heavily
weighted by immobile regions and immobile directions of the protein and
hence highlights the directional motions of the flexible parts. It has an
additional advantage of aligning conformations of proteins along their
preferred directions of motions and could be applied to order protein
conformations along its trajectory.

Keywords: conformation alignment, conformation change, root mean
square distance, ellipsoid-weighted alignment, ensemble alignment, point-
to-plane matching, computational geometry.

1 Introduction

Proteins are one of the fundamental units of living organisms. Besides experimental structure determination using methods such as X-ray crystallography
or NMR, many computational methods have been developed with the aim to
predict protein structure from sequence, using homology modeling, threading,
or ab initio methods. Proteins are not static and protein dynamics plays an
important role in the realization of protein function. Therefore much effort has
been devoted also to study and understand protein dynamics and conformation
changes [1–6].

In the process of studying different conformations of a protein and understanding its dynamic behavior and conformation changes, one inevitably runs

Z. Cai et al. (Eds.): ISBRA 2013, LNBI 7875, pp. 273–285, 2013.
© Springer-Verlag Berlin Heidelberg 2013

into a fundamental question – what is the best way to align a set of conformations? The question is important since the interpretation of many conformation changes depends to a large extent on the alignment method used.

The most widely-used method for aligning a pair of conformations or structures is based on the root mean-square distance (RMSD) [7, 8]. This purely geometry-based approach is widely-used since its problem definition is the simplest and efficient analytical solutions exist [9, 10]. Its major drawback is that it treats all the atoms in a structure equally. To overcome this, a number of methods have been developed over the years that identify only an appropriate subset of the structures for alignment and structure comparison [11–13]. One problem with many of these approaches is that the choices of the subsets are somewhat arbitrary. To address this, other methods were developed that include all the atoms in the alignment but different weights are assigned to the atoms according to their mobility, such as in Gaussian-weighted RMSD alignment [14] and alignment by the maximization of statistical likelihood [15–17].

In this work, we treat the alignment problem from a new energy-minimization perspective and provide a set of alignment methods that are a generalization of RMSD alignment. To this end we associate each atom in the protein with a mean-field potential well, whose shape is ellipsoidal in general and is inferred from the observed or computed fluctuations of the atom around its mean position. The magnitudes and directions of the fluctuations can be obtained from experimental anisotropic/isotropic B-factors for crystal structures or computationally. When viewed from this perspective, the optimal alignment is achieved by minimizing the energy cost to transit from one conformation to the other. This new set of alignment methods remove one major constraint of the RMSD alignment – that the centers of mass of the conformations being aligned have to be overlapped. When provided with right potential fields, these new energy-minimization based methods have the flexibility to align conformations at immobile regions (such as hinge regions) or other places that are more appropriate for understanding the motion.

The paper is organized as follows. Section 2 describes a set of conformation alignment methods, all of which are developed under the idea of energy minimization. These methods can be used to align a pair of conformations or an ensemble of many conformations. In Section 3, the proposed methods are applied to a number of protein systems. The paper is concluded in Section 4.

2   Methods

RMSD alignment treats atoms in a protein as points and assumes that all points contribute to the alignment equally. It is purely geometrical and the goal of alignment is to minimize root mean squared distances between the points in the two aligned conformations. It is the geometry-driven dissimilarity minimization method.

However, proteins and atoms are physical objects and their displacements are influenced by physical laws. If we associate each atom in the protein with a mean-field potential well and assume the fluctuation of each atom is dictated by its potential, then the alignment problem can be rephrased as finding the optimal alignment that minimizes the energy cost to transit from one conformation to the other. It is noted that in such a mean-field potential model the correlated motions of atoms are not considered, and each atom moves independently under the influence of the potential field surrounding it. Under this formulation, RMSD alignment is a special case where the mean field potential for each atom is identical and spherical. Improvement over RMSD alignment can thus be achieved by finding a better mean-field potential model that is more suited for a given protein system. To this end, we explore several possibilities for the potential model, and investigate their advantages in understanding protein conformation changes and motions.

2.1   Alignment by Energy Minimization

Let X be a conformation of a given protein with the coordinates of its atoms denoted by $x_1, ..., x_n$. Similarly, denote another conformation of the same protein as Y. In this work, the mean-field potential well around each atom i is treated to be ellipsoid-shaped whose spring constant is represented by a spring tensor K_i, a 3×3 matrix. The relative potential energy $V(x, y, K)$, when an atom is positioned at x at conformation X and y at conformation Y, can be computed by

$$V(x, y, K) = \frac{1}{2}(x - y)^t K (x - y) , \tag{1}$$

where v^t denotes the transpose of a vector v. Under this formulation, the best alignment between conformations X and Y is defined as finding the optimal rotation R and translation b that transform X so that the energy cost to transit from X to Y is minimized:

$$\text{AlignPair}(X, Y, K) = \underset{R,b}{\text{argmin}} \frac{1}{N} \sum_{i=1}^{n} V_i(Rx_i + b, y_i, K_i) , \tag{2}$$

where K represents a set of spring tensors, $\{K_1, ..., K_n\}$.

Now to solve the alignment problem, One needs to: (i) provide appropriate potential wells for all the atoms, in the form of K_i, $1 \leq i \leq n$, (ii) solve the minimization problem Eq. (2) once K_i are known. Depending on the choice of K_i and thus the shape of the potential it represents, different algorithms may be used to solve the minimization problem.

Fig. 1 illustrates the idea of viewing conformation alignment as an energy minimization problem. The atoms in one conformation (conformation Y) are represented by the cross marks, the mean-field potentials surrounding which are depicted by the gray ellipses. The atoms in the other conformation (conformation X, blue filled circles) are overlaid onto the mean-field potentials of conformation Y in such a way that the total potential energy of X relative to Y is minimized. The red open circles are another superposition of X onto Y, using RMSD alignment, which minimizes the root mean-square distance between atoms in X and Y. The blue filled circles, though having a larger root mean square distance to the cross marks than the read open circles, fall into places where the total potential energy is the lowest.

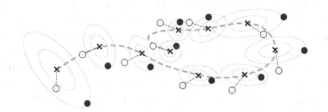

Fig. 1. Comparison of the alignment by energy minimization with the RMSD alignment. Under the perspective of energy minimization, atoms at conformation X (represented by the blue filled circles) are overlaid onto the mean-field potentials surrounding atoms at conformation Y (the black 'x' marks) in such a way that the total potential energy of X relative to Y is minimized. The contours of the mean-field potentials are shown in gray ellipses. The red open circles are another superposition of X onto Y, using the standard RMSD alignment.

In this work, we have explored a few different models for the potential wells. Common to all of these models, the magnitude of the potential surrounding each atom is chosen to reflect, to some degree or other, that atom's scale of flexibility. The intuition is that atoms that have larger scale of fluctuations should have a mean-field potential that is more flat, while atoms with small scale of fluctuations should have a potential that increases sharply as they deviate from their mean positions. Apparently, choices of potential models are not limited to what we explore here. For some alignment cases, other potential models may be more appropriate.

In the following section, we present the details of several potential models K_i. Their implementations are presented in Section 2.3.

2.2   Potential Models for Alignment

Spherical Potentials. As aforementioned, RMSD alignment is a special case under the current formulation, where the potential for every atom is identical and its spring tensor is an identity matrix, i.e., $K_i = I$. Indeed, under such a potential, Eq. (2) becomes

$$\text{AlignPair}(X, Y, K) = \underset{R, b}{\text{argmin}} \frac{1}{n} \sum_{i=1}^{n} (R\boldsymbol{x}_i + \boldsymbol{b} - \boldsymbol{y}_i)^2 , \tag{3}$$

which indeed is the same as the standard RMSD alignment.

Alignment by Temperature B-factor. In reality each atom in a protein has different scale of flexibility, as commonly observed in X-ray crystallography B-factors. Taking this into consideration, we could use a spherical mean-field potential whose K_i is inversely proportional to B_i, the temperature factor of ith atom. The temperature factor, or the mean-square fluctuations, of each atom can be obtained directly from the PDB files for proteins determined by X-ray crystallography, or computationally, from the hessian matrix [3]. Each isotropic temperature factor represents a spherical fluctuation and is proportional to the mean-square fluctuations of the atom ($\langle u_i^2 \rangle = B_i/8\pi^2$). Thus, define a spherical potential whose spring constant $K_{i,\mathrm{iso}}$ is

$$K_{i,\mathrm{iso}} = \frac{1}{B_i} \cdot I , \tag{4}$$

where I is the 3×3 identity matrix. Under such a potential, Eq. (2) becomes

$$\mathrm{AlignPair}(X, Y, K) = \underset{R,b}{\mathrm{argmin}} \frac{1}{n} \sum_{i=1}^{n} \frac{(Rx_i + b - y_i)^2}{B_i} , \tag{5}$$

which is a weighed RMSD alignment and whose solution can be easily obtained by using the same Kabsch's algorithm [7].

Ellipsoidal Potentials. There are increasingly more protein structures that are determined at atomic or near-atomic resolutions by X-ray crystallography. And for many of these structures, anisotropic B-factors have been reported in a 3×3 symmetric matrix form, which represents an anisotropic, ellipsoidal fluctuation. They thus include not only the magnitudes of atomic fluctuations but also their directional preference.

Let U_i be the 3×3 symmetric positive definite matrix that represents the anisotropic mean-square fluctuation of the ith atom. Now define a mean-field potential whose K_i is inversely proportional to the scale of the anisotropic fluctuations. Specifically, let K_i be the inverse matrix of U_i:

$$K_{i,\mathrm{ani}} = U_i^{-1} . \tag{6}$$

Note that the isotropic thermal fluctuation is a special case of this. Since K_i is ellipsoid-shaped in general, the alignment problem can no longer be solved by Kabsch's algorithm. Fortunately, the problem can be reformulated nicely as a point-to-plane matching problem studied in computational geometry. The full algorithm is given in Section 2.3.

Ensemble Alignment. For problems where an ensemble of conformations need to be aligned, such as for an NMR ensemble, there exists another way to construct the potential wells – that is by estimating the scale of fluctuations directly from the structure variations existing in the ensemble. This is an iterative process. First, the ensemble of conformations are initially aligned, for example, using the RMSD alignment. This can be achieved by picking one reference structure from the ensemble (which could be any of the conformations or the conformation that is the closest to the geometrically-averaged mean structure of the ensemble), and align the rest of the conformations to the reference structure using pairwise RMSD alignment. Second, pseudo anisotropic B-factors are computed from the aligned ensemble [18]. Then, the ensemble is realigned using the pseudo anisotropic B-factors just computed and the anisotropy-weighted alignment method described above. The whole process is repeated until it converges to a final alignment. The detailed implementation of this process is given in the next section.

2.3   Implementations

Alignment Using Spherical Potentials. As aforementioned, the minimization problem (Eq. (2)) with spherical potentials is the same to the weighted RMSD alignment problem, with the weight factor of ith atom being $1/B_i$ (see Eq. (5)). The optimal translation b can be determined using the weighted centroids of X and Y: $b = \sum_{i=1}^{n} (y_i - x_i)/(nB_i)$. The optimal rotation matrix R can be obtained by extending methods that use SVD [7, 9, 17] or the quaternion [19, 20].

Alignment Using Ellipsoidal Potentials. The minimization problem (Eq. (2)) with ellipsoidal potentials (see Eq. (6)) seems a little intimidating at the first glance. However, it turns out that this problem can be nicely reformulated into a point-to-plane alignment problem studied in computational geometry. To do so, we first denote $u_{i,j}$ and $\lambda_{i,j}$ as the jth eigenvector and eigenvalue of $K_{i,\text{ani}}$, respectively, where $1 \leq i \leq n$ and $1 \leq j \leq 3$. Using $u_{i,j}$ and $\lambda_{i,j}$, Eq. (2) can be rewritten as follows:

$$\text{AlignEllipsoid}(X, Y, K_{\text{ani}}) = \underset{R,b}{\text{argmin}} \sum_{i=1}^{n} \sum_{j=1}^{3} \lambda_{i,j}(u_{i,j} \cdot (Rx_i + b - y_i))^2 , \quad (7)$$

where $K_{\text{ani}} = \{K_{1,\text{ani}}, ..., K_{n,\text{ani}}\}$, and $u \cdot v$ is the dot product of two vectors u and v.

Now let $\mathcal{P}(y, u)$ be the plane that contains a point y and whose unit normal vector is u. Then the value $(u_{i,j} \cdot (Rx_i + b - y_i))^2$ in Eq. (7) can be thought of as the squared distance from a transformed point $Rx_i + b$ to a plane $\mathcal{P}(y_i, u_{i,j})$. Thus, the minimization problem in Eq. (7) is the same as the point-to-plane alignment problem from $3n$ points to $3n$ planes, with weight factors $\lambda_{i,j}$. This problem can be solved using the well-known ICP (iterative closest point) [21] algorithm.

Algorithm 1 lists the steps to solve the minimization problem in Eq. (7) using the ICP algorithm. In step 11, we use $\text{AlignSphere}(P, Q, \Lambda)$ to find the best superposition between two point sets P and Q with weight factors Λ (i.e., by applying Kabsch's algorithm). If we let k be the number of iterations needed to update the transformation $\langle R, b \rangle$, the algorithm takes $O(nk)$ computation time.

Ensemble Alignment. Denote by $U_{i,\text{psu}}$ the pseudo anisotropic B-factor [18] of ith atom in the ensemble, where $1 \leq i \leq n$. The potential well $K_{i,\text{ens}}$ is determined in the same way as in Eq. (6), i.e., $K_{i,\text{ens}} = U_{i,\text{ens}}^{-1}$. Let $\lambda_{i,j}$ and $u_{i,j}$ be the jth eigenvalue and eigenvector of the potential well K_i, respectively, where $1 \leq j \leq 3$. The minimization problem in Eq. (2) can be rewritten as follows:

$$\text{AlignEnsemble}(X_1, ..., X_m) = \underset{R,b,Y}{\text{argmin}} \sum_{i=1}^{m} \sum_{j=1}^{n} \sum_{k=1}^{3} \lambda_{j,k}(u_{j,k} \cdot (R_i x_{i,j} + b_i - y_j))^2, \quad (8)$$

Algorithm 1. AlignEllipsoid(X, Y, K_{ani})

1: Determine eigenvalues $\lambda_{i,1}, \lambda_{i,2}, \lambda_{i,3}$ and eigenvectors $\boldsymbol{u}_{i,1}, \boldsymbol{u}_{i,2}, \boldsymbol{u}_{i,3}$ of $K_{i,\text{ani}}$ for all $i \in \{1, 2, ..., n\}$
2: $\langle R, \boldsymbol{b} \rangle \leftarrow \langle I, \ (0, 0, 0)^t \rangle$
3: $\Lambda \leftarrow (\lambda_{1,1}, \lambda_{1,2}, \lambda_{1,3}, \lambda_{2,1}, ..., \lambda_{n,3})$
4: **repeat**
5:    **for all** $(i, j) \in \{1, 2, ..., n\} \times \{1, 2, 3\}$ **do**
6:       $\boldsymbol{p}_{i,j} \leftarrow R\boldsymbol{x}_i + \boldsymbol{b}$
7:       $\boldsymbol{q}_{i,j} \leftarrow$ point in $\mathcal{P}(\boldsymbol{y}_i, \boldsymbol{u}_{i,j})$ closest to $\boldsymbol{p}_{i,j}$
8:    **end for**
9:    $P \leftarrow (\boldsymbol{p}_{1,1}, \boldsymbol{p}_{1,2}, \boldsymbol{p}_{1,3}, \boldsymbol{p}_{2,1}, ..., \boldsymbol{p}_{n,3})$
10:   $Q \leftarrow (\boldsymbol{q}_{1,1}, \boldsymbol{q}_{1,2}, \boldsymbol{q}_{1,3}, \boldsymbol{q}_{2,1}, ..., \boldsymbol{q}_{n,3})$
11:   $\langle R', \boldsymbol{b}' \rangle \leftarrow$ AlignSphere(P, Q, Λ)
12:   $\langle R, \boldsymbol{b} \rangle \leftarrow \langle R'R, \ R'\boldsymbol{b} + \boldsymbol{b}' \rangle$
13: **until** $\langle R, \boldsymbol{b} \rangle$ no longer changes

---

where $\boldsymbol{x}_{i,j}$ is the coordinate of jth atom at conformation X_i, $\boldsymbol{y}_j$ is the coordinate of jth atom at the mean conformation Y, and R_i (and $\boldsymbol{b}_i$) are the rotation (and translation) needed to align X_i to Y.

To solve Eq. (8), we iteratively update rotation R, translation $\boldsymbol{b}$, and the mean conformation Y. Denote by $R_i^{(k)}$ the rotation for the ith conformation X_i at the kth iteration. Similarly are $\boldsymbol{b}_i^{(k)}$ and $Y^{(k)}$ defined. Given the values at kth iteration, the new mean value $\boldsymbol{y}_i^{(k+1)}$ that minimizes the summations in Eq. (8) can be determined as follows:

$$y_i^{(k+1)} = \frac{1}{m} \sum_{j=1}^{m} (R_i^{(k)} \boldsymbol{x}_{j,i} + \boldsymbol{b}_i^{(k)}) . \tag{9}$$

Subsequently, the rotation $R_i^{(k+1)}$ and translation $\boldsymbol{b}_i^{(k+1)}$ can be determined, using the ellipsoid-weighted alignment, as follows:

$$\langle R_i^{(k+1)}, \boldsymbol{b}_i^{(k+1)} \rangle = \text{AlignEllipsoid}(X_i, Y^{(k+1)}, K_{i,\text{ens}}) , \tag{10}$$

where $K_{i,\text{ens}}$ is the potential well determined from the pseudo anisotropic B-factors [18]. The iteration continues until it converges to the final solution.

This iterative optimization requires an initial value for the mean conformation $Y^{(0)}$, which is set to be the first conformation X_1 of the ensemble. The iteration is repeated until $Y^{(k)}$ converges. Algorithm 2 outlines the procedure for aligning an ensemble of conformations, where AlignRMSD(X, Y) stands for the standard RMSD alignment.

3 Results

The ensemble alignment described in Algorithm 2 is tested over 196 NMR ensembles that contain more than 5 conformations. Among 196 ensembles, about

---

Algorithm 2. AlignEnsemble($X_1, X_2, ..., X_m$)

1: $Y \leftarrow X_1$
2: $\langle R_i, b_i \rangle \leftarrow$ AlignRMSD(X_i, Y) $\forall 1 \leq i \leq m$

3: **repeat**
4:     determine Y according to (9)
5:     determine pseudo anisotropic B-factor and their inverse $K_{i,\text{ens}}$ $\forall 1 \leq i \leq m$

6:     $\langle R_i, b_i \rangle \leftarrow$ AlignEllipsoid($X_i, Y, K_{i,\text{ens}}$) $\forall 1 \leq i \leq m$
7: **until** Y converges

---

half of them show distinct alignment differences between our method and the standard RMSD alignment. We present the results for a few selected ensembles here, while the rest of the results are given at the following website: http://www.cs.iastate.edu/~gsong/CSB/Alignment/list.html.

3.1   Identify Immobile Regions in Structures

Fig. 2 shows that the ellipsoid-based ensemble alignment is able to identify the immobile regions of the structure and renders the motions of the flexible parts more pronounced. Fig. 2(a) shows the 10 conformations of an NMR ensemble as they are initially superimposed using the standard RMSD alignment. The initial alignment is iteratively refined using the proposed ensemble alignment method, until it converges to the final alignment displayed in (b).

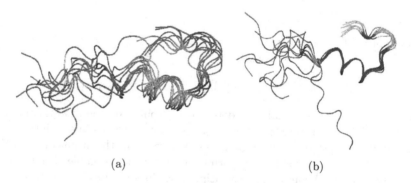

(a)   (b)

Fig. 2. Immobile regions determined by the ensemble alignment. The 10 conformations in an NMR ensemble of a long-sarafotoxin protein (pdbid: 2LDE) are aligned using the proposed ensemble alignment and all of its 384 atoms. In (a), the cartoon image of the 10 conformations that are initially superimposed using the RMSD alignment. The colors of the cartoon image represent the mean square fluctuations of backbone atoms, ranging from blue (small fluctuation, $\sim 0.40\mathring{A}^2$) to red (large fluctuations, $\sim 12\mathring{A}^2$). The final superposition by the ensemble alignment is displayed in (b) using the same color scheme, with a pronounced rigid region in blue.

The result from RMSD alignment in Fig. 2(a) are less informative, as the alignment is greatly skewed by the protein's flexible tails. The result from iterative ensemble alignment, as shown in Fig. 2(b), on the other hand, clearly identifies a stable secondary structure (a helix) in the middle. With a consistent core as the reference frame, the tail motions become more pronounced and are more accurately portrayed. A similar phenomenon has been observed by Damm and Clarson [14] in aligning conformation pairs using Gaussian-weighted RMSD alignment.

3.2   Mine Directional Motion Tendency of a Protein from Its Ensemble

Besides being able to align properly pairs of protein conformations and protein ensembles by assigning more weights to the immobile regions, our ellipsoid-weighted alignment has the additional advantage of being capable of identifying the directional motions of a protein.

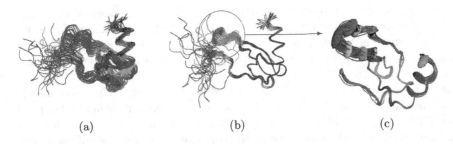

(a)   (b)   (c)

Fig. 3. The loop motions of stromal cell-derived factor-1 protein (pdbid: 2SDF, 30 conformations) as portrayed by the RMSD alignment (a), where no clear motion pattern of the protein is visible, and by ellipsoid-weighted alignment (b), where it is clearly seen that most of the protein motion takes place at the loops or the tails. The motion tendency of the loop is then portrayed using ellipsoids in (c).

Fig. 3 shows the alignment results of stromal cell-derived factor-1 protein using (a) standard RMSD and (b) our ellipsoid-weighted ensemble alignment method. The alignment in (a) portrays the protein with a near-uniform scale of fluctuations along its chain and with little visible motion pattern. On the contrary, ellipsoid-weighted alignment shown in (b) recognizes that the protein chain is mostly immobile except at the tails and the loop region. The anisotropic mean-square fluctuations of the C_α in the loop region are drawn using ellipsoids and shown in Fig. 3(c). The directions of the axes of these ellipsoids point towards the probable motion directions of these residues. And in the case of this protein, the ellipsoids clearly display the loop's motion tendency: swinging up and down, and to a lesser extent, in and out of the paper, with little tendency of stretching away or pressing towards the body of the protein.

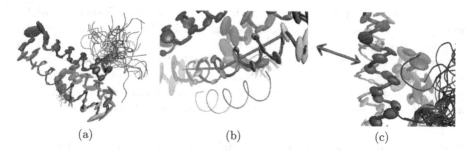

(a)   (b)   (c)

Fig. 4. Six intrinsic motions of the kDa receptor associated protein (pdbid: 1OV2, 39 conformations). (a) displays the backbones in gray lines, which are superimposed using the ellipsoid-weighted alignment, and the anisotropic mean-square fluctuation of C_α atoms in ellipsoids. The backbone is roughly divided into six segments due to the difference in their motion patterns, which are colored with orange, blue, cyan, green, red, or magenta. (b) and (c) show in an enlarged view two of these segments.

The ellipsoids of the anisotropic fluctuations present themselves as a way to visualize the directional motions of a protein. In the following example, we show how these ellipsoids can help display the motion tendency of secondary structures also. Some of these motions appear to be screw-like, while others sliding motions [22].

In Fig. 4(a), after applying ellipsoid-based alignment, the anisotropic fluctuations of C_α atoms of the kDa receptor associated protein (pdbid: 1OV2) are drawn using ellipsoids along its backbone. The backbone is roughly divided into six segments due to the difference in their motion patterns. Two of these segments are further shown in (b) and (c) in an enlarged view. For the orange segment shown in Fig. 4(b), it is seen that the longest axes of most of the ellipsoids lie along the direction of the backbone trace, thus displaying a screw-like motion pattern. The blue portion in Fig. 4(c), on the other hand, displays a back and forth sliding motion tendency along the direction marked by the arrows, probably due to the fact that it is being dragged along by the large fluctuations of the nearby tail.

It is worth pointing out that eigenvectors-based decomposition methods such as NMA (normal mode analysis) or PCA (principal component analysis) have also been used to analyze collective motion tendencies existing in a structure or in a structure ensemble. But usually it is difficult for them to capture, with a single mode or a principal component, a localized screw-like motion or sliding motion as displayed here.

4   Discussion and Conclusion

Protein conformation alignment is an essential step in understanding protein motion and conformational changes. An appropriate alignment can help correctly identify the nature of protein motion.

In this work, we treat the alignment problem from an energy-minimization perspective. To this end we associate each atom in the protein with a mean-field potential well, whose shape, ellipsoidal in general, is inferred from the observed or computed fluctuations of that atom around its mean position. We show that this "ellipsoid-weighted" RMSD alignment can be reformulated nicely as a point-to-plane matching problem studied in computational geometry [23, 24]. This new alignment method is a generalization of standard RMSD and Gaussian-weighted RMSD alignments. It is able to identify and distinguish the immobile regions of a protein from the flexible components. One end result of such an alignment is a description of the anisotropic movements of the atoms in protein, which are represented by ellipsoids. These ellipsoids present themselves as a way to visualize the directional motions of the protein. We have shown that these ellipsoids can help display not only the motion tendency of the loops or the tails, but also that of the secondary structures, some of whose motions appear to be screw-like, while others are sliding motions.

The functional processes of many proteins involve little overall structural change but localized motions of a small piece of the structure, such as that of a loop. The rest of the structure thus provides a reference frame for describing the motions of the flexible part. Standard RMSD alignment aligns conformations at the universal frame, using the center of mass as the origin, unwittingly assuming the motion of the protein is always a motion of the whole protein. While this may be the right choice for proteins that go through global conformation changes as whole, for proteins with localized motions, RMSD alignment is not able to provide the best reference frame for describing the conformation changes.

Our proposed alignment method, however, is able to correctly identify the immobile portions of a protein and use them as the reference frame. As a result, the motions of the flexible part(s) can be more clearly portrayed. Such an alignment has a few advantages. First, by isolating the motions mostly to the tails or loops, it identifies what are the key movable parts of the protein in its dynamic process. Second, when the motions are isolated to a small part of a protein instead of being expressed through the whole body, their behavior may be describable in a lower dimension of space and may thus become more tractable to elastic network model studies, using fewer modes [3, 25]. Lastly, since the ellipsoids provide a description of the motion tendency, it is perceivable that such a method may be used to order conformations along the directions of motions, or even to reconstruct a trajectory from a number of isolated conformations. Another advantage of formulating the alignment problem as an energy-minimization problem is that this weighted RMSD measure is linearly proportional to the effective mean field potential. Consequently, this alignment has the nice property that the effective energy of a conformation is perfectly correlated with its weighted RMSD distance to the native state conformation. Such a weighted RMSD measure thus may be useful for evaluating the quality of predicted structures.

Acknowledgments. The authors would like to thank Santhosh Vammi for valuable discussions. Funding from National Science Foundation (CAREER award, CCF-0953517) is gratefully acknowledged.

References

1. Karplus, M., McCammon, J.A.: Molecular dynamics simulations of biomolecules. Nat. Struct. Biol. 9, 646–652 (2002)
2. Tama, F., Sanejouand, Y.H.: Conformational change of proteins arising from normal mode calculations. Protein Eng. 14, 1–6 (2001)
3. Atilgan, A.R., Durell, S.R., Jernigan, R.L., Demirel, M.C., Keskin, O., Bahar, I.: Anisotropy of fluctuation dynamics of proteins with an elastic network model. Biophys J. 80, 505–515 (2001)
4. Song, G., Jernigan, R.L.: An enhanced elastic network model to represent the motions of domain-swapped proteins. Proteins 63, 197–209 (2006)
5. Henzler-Wildman, K., Kern, D.: Dynamic personalities of proteins. Nature 450, 964–972 (2007)
6. Dror, R.O., Dirks, R.M., Grossman, J.P., Xu, H., Shaw, D.E.: Biomolecular simulation: a computational microscope for molecular biology. Annu. Rev. Biophys 41, 429–452 (2012)
7. Kabsch, W.: Solution for Best Rotation to Relate 2 Sets of Vectors. Acta Crystallogr. A 32, 922–923 (1976)
8. Mclachlan, A.D.: Rapid Comparison of Protein Structures. Acta Crystallogr. A 38, 871–873 (1982)
9. Kabsch, W.: Discussion of Solution for Best Rotation to Relate 2 Sets of Vectors. Acta Crystallogr. A 34, 827–828 (1978)
10. Coutsias, E.A., Seok, C., Dill, K.A.: Using quaternions to calculate RMSD. J. Comput. Chem. 25, 1849–1857 (2004)
11. Khazanov, N.A., Damm-Ganamet, K.L., Quang, D.X., Carlson, H.A.: Overcoming sequence misalignments with weighted structural superposition. Proteins 80, 2523–2535 (2012)
12. Wriggers, W., Schulten, K.: Protein domain movements: Detection of rigid domains and visualization of hinges in comparisons of atomic coordinates. Proteins-Structure Function and Genetics 29, 1–14 (1997)
13. Irving, J.A., Whisstock, J.C., Lesk, A.M.: Protein structural alignments and functional genomics. Proteins 42, 378–382 (2001)
14. Damm, K.L., Carlson, H.A.: Gaussian-weighted RMSD superposition of proteins: A structural comparison for flexible proteins and predicted protein structures. Biophysical Journal 90, 4558–4573 (2006)
15. Theobald, D.L., Wuttke, D.S.: Empirical Bayes hierarchical models for regularizing maximum likelihood estimation in the matrix Gaussian Procrustes problem. Proc. Natl. Acad. Sci. U S A 103, 18521–18527 (2006)
16. Theobald, D.L., Wuttke, D.S.: THESEUS: maximum likelihood superpositioning and analysis of macromolecular structures. Bioinformatics 22, 2171–2172 (2006)
17. Liu, Y.S., Fang, Y., Ramani, K.: Using least median of squares for structural superposition of flexible proteins. BMC Bioinformatics 10, 29 (2009)
18. Yang, L., Song, G., Jernigan, R.L.: Comparisons of experimental and computed protein anisotropic temperature factors. Proteins 76, 164–175 (2009)

19. Liu, P., Agrafiotis, D.K., Theobald, D.L.: Fast determination of the optimal rotational matrix for macromolecular superpositions. J. Comput. Chem. 31, 1561–1563 (2010)
20. Horn, B.K.P.: Closed-Form Solution of Absolute Orientation Using Unit Quaternions. J. Opt. Soc. Am A 4, 629–642 (1987)
21. Besl, P.J., Mckay, N.D.: A Method for Registration of 3-D Shapes. IEEE T Pattern Anal. 14, 239–256 (1992)
22. Gerstein, M., Krebs, W.: A database of macromolecular motions. Nucleic Acids Res. 26, 4280–4290 (1998)
23. Chen, Y., Medioni, G.: Object Modeling by Registration of Multiple Range Images. Image Vision Comput. 10, 145–155 (1992)
24. Low, K.-L.: Linear Least-Squares Optimization for Point-to-Place ICP Surface Registration. Chapel Hill, University of North Carolina (2004)
25. Yang, L., Song, G., Jernigan, R.L.: How well can we understand large-scale protein motions using normal modes of elastic network models? Biophys J. 93, 920–929 (2007)

Construction of Uncertain Protein-Protein Interaction Networks and Its Applications

Bihai Zhao[1,2], Jianxin Wang[1], Fang-Xiang Wu [1,3], and Yi Pan[4]

[1] School of Information Science and Engineering, Central South University,
Changsha, 410083, China
[2] Department of Information and Computing Science, Changsha University,
Changsha, 410003, China
[3] Department of Mechanical Engineering and Division of Biomedical Engineering,
University of Saskatchewan, Saskatoon, SK S7N 5A9, Canada
[4] Department of Computer Science, Georgia State University, Atlanta, GA 30302-4110, USA

Abstract. Recent developments in experiments have resulted in the publication of many high-quality, large-scale protein-protein interaction (PPI) data. Unfortunately, a significant proportion of PPI networks have been found to contain false positives, which have negative effects on the further research of PPI networks. We construct an uncertain protein-protein interaction (UPPI) network, in which each protein-protein interaction is assigned with an existence probability using the topology of the PPI network solely. Based on the uncertainty theory, we propose the concept of expected density to assess the density degree of a subgraph, the concept of the relative degree to describe the relationship between a protein and a subgraph in a UPPI network. To verify the effectiveness of the UPPI network, we propose a novel complex prediction method named CPUT (Complex Prediction based on Uncertainty Theory). In CPUT, the expected density combined with the absolute degree is used to determine whether a mined subgraph from the UPPI network can be represented as a core component with high cohesion and low coupling while the relative degree is the criterion of binding an attachment protein to a core component to form a complex. We employ CPUT and the existing competitive algorithms on two yeast PPI networks. Experimental results indicate that CPUT performs significantly better than the state-of-the-art methods.

Keywords: Protein-protein interaction, Uncertainty, Protein complex, Core-attachment.

1 Introduction

Recent developments in experiments such as yeast two-hybrid [1], tandem affinity purification [2] and mass spectrometry [3] have resulted in the publication of many high-quality, large-scale PPI data. These data provide a stepping stone for finding protein complexes, which are very important for understanding the cell's functional organization, to carry out their biological functions.

Unfortunately, a significant proportion of PPI networks obtained from these high-throughput biological experiments have been found to contain false positives [4], due

Z. Cai et al. (Eds.): ISBRA 2013, LNBI 7875, pp. 286–297, 2013.
© Springer-Verlag Berlin Heidelberg 2013

to the limitations of the associated experimental techniques and the dynamic nature of protein interaction maps. The research [5] shows that for a filtered yeast two-hybrid data set, the fraction of false positives has also been predicted to be of the order of 50%. These errors in the experimental PPI data will have negative effects on the further study of PPI networks. Several computational approaches have been proposed to predict protein interactions for false positive reduction in PPI networks. Numerous approaches have been proposed using a variety of biological information [6, 7], some others rely on statistical scoring functions [8, 9] to calculate the extent of similarity of protein phylogenetic profiles, employ machine learning techniques [10, 11] to predict PPIs or use support vector machines method [12] to construct supervised classifiers for identifying interacting proteins.

To assess the reliability of high-throughput protein interactions, many computational approaches have been proposed. Some methods are designed to estimate the overall error rate of PPIs identified in yeast [13, 14]. Comparing interaction data is difficult, because they are often derived under different conditions and are presented in different format. Therefore, some more complicated approaches try to assess the reliability of individual interactions [15, 16]. Several pieces of genomic information such as gene annotations, gene expression etc. have been used in these methods, while others use solely the topology of PPI networks.

In spite of the advances in these computational approaches and experimental techniques, it is still impossible to construct an absolutely reliable PPI network. So, for applications on PPI networks, such as functional module identifications [17, 18], protein function predictions [19, 20], essential protein discoveries [21], protein complexes detections [22, 23, 24, 25] etc, the tolerance of false positives is more meaningful and important than the assessment of reliability or the reduction of false positives. To improve the prediction accuracy, some algorithms take into account the reliability of PPI networks and weight interactions in PPI networks. For example, CDdistance [26] and FSWeight [27] are two measures calculated based on the number of common neighbors of two proteins. They have been shown to perform well. Even for these algorithms running on weighted PPI networks, the weight of interactions usually represents the priority of proteins or interactions and the criterion of describing a subgraph or a protein is still similar to the un-weighted algorithms. Weighted degree is introduced into some methods [28, 29]. However, these methods are not sufficient to deduce satisfactory conclusions when a large amount of protein interaction data appears.

In this paper, we take into account the reliability of PPIs and construct an uncertain protein-protein interaction (UPPI) network, in which the reliability of each interaction is represented as a probability. Great contribution of our work is the proposed concepts for assessing the density of a subgraph and the relationship between a protein and a subgraph in a UPPI network based on the uncertainty theory. To test the effectiveness of the UPPI network, we propose a novel complex prediction method named CPUT (Complex Prediction based on Uncertainty Theory). We compare our CPUT approach to a representative set of state-of-the-art complex prediction algorithms: MCODE [22], MCL [23], CFinder [25], CMC [26], SPICi [28], HC-PIN [29], COACH [30] and Cluste-rONE [31]. The experimental results show that our CPUT approach outperforms these algorithms in terms of accuracy and statistical significance.

2   Constructing UPPI Networks

Based on the uncertainty theory, which is the cornerstone of our work, we construct an UPPI network. Some definitions are described firstly.

Definition 1. *The uncertain PPI network* Consider a PPI network $IG= (V, E)$, where $V = \{v_1, v_2,..., v_n\}$ is a set of proteins; $E = \{e_1, e_2,..., e_m\}$ is a set of interactions. An uncertain PPI network is defined as $UG= (V, E, P)$, where $P(E= e_i) =p_i$, $i=1, 2...m$, P is a probability function denoting the existence of an interaction in E and defined as follows:

$p_i =N_{ic}/N_{imax}$, where N_{ic} is the number of common neighbors of the two proteins of interaction e_i and N_{imax} is the maximum possible number of common neighbors of the two proteins. N_{imax} equals to the minimum degree of the two proteins minus 1.

In this paper, we assume that the existence probabilities of different interactions in an uncertain PPI network are independent to each other.

Definition 2. *The possible PPI network* $G= (V_G, E_G)$ is an instantiation of an uncertain PPI network $UG= (V, E, P)$, where $V_G = V$, $E_G \subseteq E$. We denote the relationship between G and UG as $UG \rightarrow G$, and then the sampling probability of G is given by:

$$Pr(UG \rightarrow G)\prod_{e\in E_G} p(e) \prod_{e\in (E\backslash E_G)} (1-p(e)) \qquad (1)$$

According to the uncertainty theory, $\sum_{i=1}^{n} Pr(UG \rightarrow g_i)=1$ (g_i is a possible PPI network of the UPPI network UG, $n= 2^{|E|}$). From the definition 2, we can see that a possible PPI network is a certain network with certain interactions and a whole sampling probability actually. In other words, an uncertain PPI network consists of a large amount of certain PPI networks with sampling probabilities. So, prediction from an UPPI network is transformed to the prediction from a lot of certain PPI networks, but it is different from the current handling of PPI networks, due to the sampling probabilities of possible PPI networks.

Based on the uncertainty theory, the relationship between a protein and an UPPI network is described by the concept of relative degree.

Definition 3. *The relative degree* Given an uncertain PPI network $UG= (V, E, P)$ and a protein vertex $v_a \in V$, $PG= \{g_1, g_2,..., g_n\}$ ($n=2^{|E|}$) is a set of possible PPI networks, which are instantiations of UG. Pr (g_i) is the sampling probability associated with instance g_i ($g_i \in PG$). The relative degree of vertex v_a in UG is defined as:

$$RD(v_a, UG) = \sum_{i=1}^{n} (Pr(g_i) \times m_i /(|V|-1)) \qquad (2)$$

Where m_i is the number of PPIs between v_a and other protein vertices in possible PPI network g_i.

According to Definition 2, with the increase of the number of interactions in PPI networks, the number of possible PPI networks would grow exponentially. To get the relative degree of a protein vertex from the above definition is becoming more and more computationally demanding. The following theorem gives a simple formula to compute the relative degree. Let $p(v_i, v_j)$ denote the probability of the interaction between v_i and v_j. If there is not an interaction between v_i and v_j, $p(v_i, v_j)=0$.

Theorem 1. Given an uncertain PPI network $UG= (V, E, P)$ and a protein vertex $v_a \in V$, the relative degree of vertex v_a in UG can be represented as:

$$RD(v_a, UG) = 1/(|V|-1) \sum_{i=1 \wedge i \neq a}^{|V|} p(v_a, v_i) \qquad (3)$$

Proof. Assume that the protein vertices of UG that connect to v_a is $\{v_1, v_2, ..., v_k\}$, probability values of these interactions are $p(v_a, v_1), p(v_a, v_2), ..., p(v_a, v_k)$ ($1 \leq k \leq |V|$).

$$RD(v_a, UG) = 1/(|V|-1) \sum_{i=1}^{k} [1-p(v_a, v_1)] [1-p(v_a, v_2)]...p(v_a, v_i)...[1-p(v_a, v_k)]+$$

$$2/(|V|-1) \sum_{i=1}^{k} \sum_{j>i}^{k} [1-p(v_a, v_1)][1-p(v_a, v_2)]...p(v_a, v_i)...p(v_a, v_j)...[1-p(v_a, v_k)]+...$$

$$+k/(|V|-1)p(v_a, v_1)p(v_a, v_2)...p(v_a, v_i)...p(v_a, v_{k-1})p(v_a, v_k)$$

$$=1/(|V|-1) \sum_{i=1}^{k-1} [1-p(v_a, v_1)] ...p(v_a, v_i)...[1-p(v_a, v_{k-1})]+2/(|V|-1) \sum_{i=1}^{k-1} \sum_{j>i}^{k-1} [1-p(v_a, v_1)]...$$

$$p(v_a, v_i)...p(v_a, v_j)...[1-p(v_a, v_{k-1})]+...+(k-1)/(|V|-1)p(v_a, v_1)...p(v_a, v_i)...$$

$$p(v_a, v_{k-1})+p(v_a, v_k)/(|V|-1)\{[1-p(v_a, v_1)]...[1-p(v_a, v_i)]...[1-p(v_a, v_k)]+$$

$$\sum_{i=1}^{k} [1-p(v_a, v_1)]...p(v_a, v_i)...[1-p(v_a, v_k)]+\sum_{i=1}^{k} \sum_{j>i}^{k} [1-p(v_a, v_1)]...p(v_a, v_i)...$$

$$p(v_a, v_j)...[1-p(v_a, v_k)]+...+p(v_a, v_1)...p(v_a, v_k)\}$$

$$=1/(|V|-1) \sum_{i=1}^{k-1} [1-p(v_a, v_1)]...p(v_a, v_i)...[1-p(v_a, v_{k-1})]+2/(|V|-1) \sum_{i=1}^{k-1} \sum_{j>i}^{k-1} [1-p(v_a, v_1)]...$$

$$p(v_a, v_i)...p(v_a, v_j)...[1-p(v_a, v_{k-1})]+...+(k-1)/(|V|-1)p(v_a, v_1)...p(v_a, v_i)...$$

$$p(v_a, v_{k-1})+p(v_a, v_k)/(|V|-1)$$

$$=1/(|V|-1) \sum_{i=1}^{k-2} [1-p(v_a, v_1)] ...p(v_a, v_i)...[1-p(v_a, v_{k-2})]+2/(|V|-1) \sum_{i=1}^{k-2} \sum_{j>i}^{k-2} [1-p(v_a, v_1)]...$$

$$p(v_a, v_i)...p(v_a, v_j)...[1-p(v_a, v_{k-2})]+...+(k-2)/(|V|-1)p(v_a, v_1)...p(v_a, v_i)...$$

$$p(v_a, v_{k-2})+[p(v_a, v_{k-1})+p(v_a, v_k)]/(|V|-1) = ...$$

$$=1/(|V|-1)\{p(v_a, v_1)[1-p(v_a, v_2)]+p(v_a, v_2)[1-p(v_a, v_1)]\}+2/(|V|-1)p(v_a, v_1)p(v_a, v_2)$$

$$+\sum_{i=3}^{k} p(v_a, v_i)/(|V|-1)=[p(v_a, v_1)+p(v_a, v_2)]/(|V|-1)+\sum_{i=3}^{k} p(v_a, v_i)/(|V|-1)=\sum_{i=1}^{k} p(v_a, v_i)/(|V|-1) \quad \square$$

Algorithm 1 illustrates the procedure of constructing UPPI network.

Algorithm 1. Constructing UPPI network

Input: the PPI network $IG=(V, E)$;

Output: the uncertain network $UG=(V, E, P)$, attachment threshold AVG_RD

1: for each edge $(u, v) \in E$ compute its probability value $p(u, v)$;

2: generate UG and the set of possible PPI network $PG= \{g_1, g_2, ..., g_n\}$

3: for each vertex $v \in V$

  compute $RD(v, NG)$;// NG is the neighborhood graph of v

4. $AVG\_RD = \overline{RD(v_i, NG_i)}, v_i \in V$ and $RD(v_i, NG_i) > 0$;

In this process, protein vertices are inserted into queue ordered by their relative degree within neighborhood graph descendant and Avg_RD as the average value of all protein vertices' relative degree is computed, which is a threshold value used in the subsequent section.

3   The CPUT Method

Taking into account the inherent organization for extracting dense subgraphs, recent analysis [32, 33] of experimentally detected protein complexes has revealed that a complex consists of a core component and attachments. Core proteins are highly co-expressed and share high functional similarity, each attachment protein binds to a subset of core proteins to form a biological complex. Because of the demonstrated significance of the structure in predicting protein complexes, our CPUT method is based on the core-attachment concept.

Different from the current core-attachment based approaches, we use the concept of expected density to measure whether a subgraph can be selected as a core, instead of the concept of density. To describe the relationship between an attachment protein and a core, we use the concept of relative degree, firstly. We believe that a subgraph representing a core component should contain many reliable PPIs between its subunits and be well separated from their neighbor subgraphs. In other words, a core should be high cohesion and low coupling.

Definition 4. *The expected density* Given an uncertain PPI network $UG= (V, E, P)$, $PG= \{g_1, g_2, ..., g_n\}$ $(g_i = (V, E_i), n=2^{|E|})$ is a set of possible PPI networks, which are instantiations of UG, $\Pr(g_i)$ is the probability associated with instance $g_i \in PG$. The expected density of UG is defined as follow:

$$ED(UG) = \sum_{i=1}^{n} \Pr(g_i) \times 2 \times |E_i| / (|V| \times (|V|-1)) \tag{4}$$

Theorem2 is a simple formula to compute the expected density.

Theorem 2. Given an uncertain PPI network $UG= (V, E, P)$, $V = \{v_1, v_2, ..., v_n\}$, $E = \{e_1, e_2, ..., e_m\}$, $P = \{p(e_1), p(e_2), ..., p(e_m)\}$. The expected density of UG can be computed by $ED(UG) = \sum_{i=1}^{m} p(e_i) \times 2 / (|V| \times (|V|-1))$

Proof. It is similar to the proof of theorem 1.

Our CPUT method consists of three major steps.

(1) Starting from the first protein vertex, a greedy procedure adds protein vertices to form a candidate core with high cohesion and low coupling. The growth process is repeated from all vertices to form non-redundant core sets.

(2) Add attachment vertices to core sets to form complexes, where the relative degree of an attachment vertex within the core is above a specified threshold.

(3) Quantify the extent overlap between each pair of complexes and discard the complex with lower density or smaller size, and whose overlap value is above a specified threshold.

3.1   Core Detection

The first stage of our method is protein core detection, which takes as input the UPPI network *UG* and an expected density threshold *CT*.

In our core detection algorithm, every protein has the same probability of being drawn as a seed. The algorithm consists of three steps as shown in Algorithm 2 below.

Algorithm 2. Core detection

Input: UPPI network $UG = (V, E, P)$;expected density threshold CT

Output: SC: the set of protein cores;

1: for each protein $v \in V$

2.   Insert v into CS;// Candidate core set

3:      $Q = \{ v_i \mid v_i \in V \wedge dis(v, v_i) = 1 \}$

4.   for each element $q \in Q$ insert q into CS; If $ED(CS) < CT$ remove q from CS;

5. $NS = \{ v_i \mid (v_i, v_j) \in E \wedge v_j \in CS \wedge v_i \notin CS \}$;

6.   for each vertex $vc \in CS$

7.   if $RD(vc, CS) * Size(CS) <= RD(vc, NS) * Size(NS)$ then

8.   remove vc from CS;   label vc with $DISCARDED$;

9.   if $CS \not\subset s_i \wedge s_i \not\subset CS$, where $s_i \subset SC$ then insert CS into SC;

Step 1: Inserted the seed into the candidate set CS in lines 3, and then all neighbors of the seed are put into the queue. Finally, we get a protein from the queue in order and insert into CS, if the expected density less than the threshold CT, remove the protein vertex from CS. A high expected density of a subgraph indicates that the subgraph can be represented as a high cohesion core. After the queue is empty, a high cohesion candidate set is formed.

Step 2: The neighbor set NS is consist of neighbors of protein vertices in the candidate set CS. For each protein vertex of CS, we compute an internal absolute degree of the vertex with CS and an external absolute degree of the vertex with NS. The absolute degree of a vertex in a subgraph equals to the relative degree of the vertex in the subgraph multiplied by the total number of vertices in the subgraph. If the internal absolute degree less than the external absolute degree, we believe that the protein vertex has high coupling and should be removed from CS. At the same time, the protein vertex is labeled with *DISCARDED*.

Step 3: If a candidate set *CS* is not a subset of elements in the set protein cores *SC*, *CS* can be represented as a core and inserted into *SC*.

The procedure repeats Step 1-3, until all protein vertexes are handled.

3.2   Attachment Detection

The second stage is the attachment detection. Algorithm 3 gives the detail of attachment detection.

Algorithm 3. Attachment detection

Input: UPPI network $UG= (V, E, P)$;*SC*: the set of protein cores;

Output: *SC*: the set of protein complexes;

1: for each vertex $v \in V$

2.   if v labeled with *DISCARDED*

3:      for each core $S \in SC$ if $RD(v, S) > Avg\_RD$ insert v into S ;

4.   compute $ED(S)$;

For each candidate attachment protein v, if the relative degree between v and a core is greater than the threshold Avg_RD, insert v into the core and compute the expected density of the core again. The threshold Avg_RD is self-adjustable with the UPPI network according to the average value of all protein vertices' relative degree, which obtained in the Algorithm 1 of CPUT.

3.3   Redundancy-Filter

The last stage is redundancy-filter. Although some redundancy may have biological importance, complexes overlapping to a very high extent in comparison to their expected density and size should be discarded. With quantifying the extent of overlap between each pair of complexes, complex with smaller expected density or size is discarded for which overlap score of the pair is above the threshold. In our CPUT method, the overlap threshold is typically set as 0.8 [31], where the overlap score of two complexes A and B is defined as follows [34].

$$NA(A, B) = | A \cap B |^2 / (| A \| B |) \tag{5}$$

4   Results and Discussion

We have applied our CPUT method and other eight competing algorithms on two yeast PPI networks, including DIP data [35] and Krogan data [36]. The DIP dataset consists of 5023 proteins and 22,570 interactions among the proteins, and the Korgan dataset consists of 3672 proteins and 14317 interactions. To evaluate the protein complexes predicted by our method, a benchmark set is adopted from CYC2008 [37], which consist of 408 complexes. For comprehensive comparisons, we employ several evaluation criteria, such as F-measure and functional enrichment of GO terms. For all those competitive algorithms, the optimal parameters are set as recommended by their authors.

4.1   Results on DIP Data

To assess the quality of the produced complexes, we match the generated complexes with the benchmark complex set. Specificity (*Sp*) and sensitivity (*Sn*) are the commonly used measures to evaluate the performance of protein complex prediction methods. Specificity is the fraction of predicted complexes that are true complexes while sensitivity is the fraction of benchmark complexes that are retrieved.

Given the predicted complex set $PC= \{pc_1, pc_2,...,pc_n\}$ and the benchmark complex set $BC= \{bc_1, bc_2,...,bc_m\}$.

$$TP = |\{pc_i \in PC \mid \exists bc_j \in BC, NA(pc_i, bc_j) \geq T\}| \qquad (6)$$

$$FP = |\{pc_i \in PC \mid \forall bc_j \in BC, NA(pc_i, bc_j) < T\}| \qquad (7)$$

$$FN = |\{bc_i \in BC \mid \forall pc_j \in PC, NA(pc_i, bc_j) < T\}| \qquad (9)$$

From (6)-(9), *TP* is the number of correctly predicted complexes and *FP* is the number of incorrectly predicted complexes, while *TN* is the number of predicted benchmark complexes and *FN* is the number of unpredicted benchmark complexes.

$$Sp = TP/(TP+FP), \quad Sn = TP/(TP+FN) \qquad (10)$$

Generally *T* is set as 0.2 [34], which is also used in this paper. F-measure is another measure to evaluate the performance of a method synthetically.

$$F-measure = 2 \times Sn \times Sp/(Sn+Sp) \qquad (11)$$

The basic information about predicted complexes by various algorithms running on DIP data is presented in Table 1. In Table 1, *#PC* is the total number of predicted complexes, while *AS* is the average size of the complexes detected by each algorithm and *MS* is the maximum size of predicted complexes. *PMC* is the number of complexes perfectly matching the known complexes. From Table 1, we can see that our CPUT method contains the second-biggest number of correctly predicted complexes and predicted benchmark complexes after the COACH method, while *#PC* of CPUT is far less than COACH's. Table 1 show that CPUT achieves the largest value of F-measure and *Sp*, the second-largest value of *Sn* after the COACH. The F-measure of CPUT is 375%, 119%, 128%, 54%, 26.7%, 128%, 58.3% and 46.2% higher than MCODE, MCL, CFinder, CMC, COACH, SPICi, HC-PIN and ClusterONE.

To evaluate the statistical and biological significance of the predicted complexes, functional enrichment of GO terms is employed. GO annotation is a useful information resource to measure the reliability of protein interaction pairs. GO::TermFinder [38] is a set of software modules to determine statistically

Table 1. Basic information of predicted complexes by various algorithms

Algorithms	#PC	AS	MS	TN	TP	PMC	Sn	Sp	F-measure
CPUT	581	8.5	52	214	306	16	0.61	0.53	0.57
MCODE	59	13.59	82	30	28	2	0.07	0.47	0.12
MCL	928	5.15	122	195	174	12	0.45	0.19	0.26
CFinder	197	13.31	1821	83	75	12	0.19	0.38	0.25
CMC	235	6.13	32	124	119	8	0.30	0.51	0.37
COACH	902	9.18	59	219	319	15	0.63	0.35	0.45
SPICi	574	4.7	48	143	118	7	0.31	0.21	0.25
HC-PIN	277	5.67	118	149	119	20	0.31	0.43	0.36
ClusterONE	371	4.9	24	136	155	6	0.36	0.42	0.39

significant GO terms shared by a set of genes and to access GO information and annotation information. To determine whether any GO terms annotate a specified list of genes at a frequency greater than that expected by chance, GO::TermFinder calculates a P-value using the hyper geometric distribution:

A low P-value of a predicted complex indicates that those proteins in the complex do not happen merely by chance, so the complex has high statistical significance. Generally a complex is considered to be significant with corrected P-value<0.01 [39].

Research shows that the proportion of significant complexes over all predicted ones can be used to evaluate the overall performance of various algorithms. In addition, P-score is also used as an effective evaluation measure, which is defined as:

$$P-score = \frac{1}{n}\sum_{i=1}^{n}-\lg(p-value_i) \mid p-value_i < Y \qquad (12)$$

Y is set as 0.01 mentioned above. Table 2 shows the comparison results based on these measures on the whole by various algorithms on the DIP data. In Table 2, *#PC* is the number of predicted complexes, and *#SC* is the number of significant complexes. Our method achieves the largest value of P-score and the second-largest value of proportion after MCODE. P-score of our method is 31.61%, 80.21%, 41.86%, 16.31%, 27.66%, 66.4%, 12.83% and 24.7% higher than MCODE, MCL, CFinder, CMC, COACH, SPICi, HC-PIN and ClusterONE, respectively.

Table 2. Statistical significance of predicted complexes by various algorithms

Algorithms	#PC	#SC	Proportion(%)	P-score
CPUT	581	501	86.24%	10.2
MCODE	59	55	93.22%	7.75
MCL	928	414	44.41%	5.66
CFinder	197	122	61.93%	7.19
CMC	235	196	83.4%	8.77
COACH	902	736	81.6%	7.99
SPICi	574	297	51.74%	6.13
HC-PIN	277	176	63.54%	9.04
ClusterONE	371	253	68.19%	8.18

4.2   Results on Krogan Data

To further investigate the results obtained by our method, we also perform our CPUT algorithm on Krogan data mentioned above. The results of each algorithm using Krogan dataset are shown in Table 3, including *Sn*, *Sp* and F-measure. In Table 3, CPUT still performs the best and the F-measure of CPUT is 260%, 100%, 157.14%, 68.75%, 10.2%, 74.19%, 184.21% and 42.11% higher than MCODE, MCL, CFinder, CMC, COACH, SPICi, HC-PIN and ClusterONE, respectively.

Table 3. Results of various algorithms on Krogan data

Algorithms	*Sn*	*Sp*	F-measure
CPUT	0.47	0.63	0.54
MCODE	0.08	0.67	0.15
MCL	0.39	0.21	0.27
CFinder	0.14	0.45	0.21
CMC	0.23	0.55	0.32
COACH	0.51	0.47	0.49
SPICi	0.3	0.31	0.31
HC-PIN	0.11	0.51	0.19
ClusterONE	0.3	0.53	0.38

4.3   Effect of Parameter *CT*

In Algorithm 2, in order to evaluate the expected density of cores detected by our CPUT method, we employ a user-defined parameter *CT*. As *CT* is used to describe the expected density of a subgraph, according to Definition 4, *CT* ∈ [0,1]. Figure 1 show how the F-measure of CPUT method fluctuates under various value of *CT*. From the figure, we can easily see that the F-measure reach the maximum value when *CT* is assigned to 0.1. In our experiment, we set *CT*=0.1.

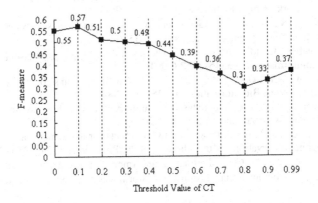

Fig. 1. The effect of threshold *CT*

5   Conclusion

We have constructed an uncertain protein-protein interaction (UPPI) network and introduced some new concepts to assess subgraphs or proteins based on the uncertainty theory. To test the effectiveness, we have proposed a protein complex detection method named CPUT based on the UPPI network. CPUT first mines cores from neighbor graphs of all protein vertices with high cohesion and low coupling, and then binds attachment proteins to form complexes with cores.

Comprehensive comparisons among the state-of-the-art methods and CPUT method have been made on two yeast PPI networks. Experimental results have shown higher accuracy and more significant biological meaning of our CPUT than others.

Acknowledgements. This work is supported in part by the National Natural Science Foundation of China under Grant No. 61232001, No. 61128006, No. 61073036 the Program for New Century Excellent Talents in University (NCET-10-0798).

References

1. Ito, T., et al.: A comprehensive two-hybrid analysis to explore the yeast protein interactome. PNAS 98, 4569–4574 (2001)
2. Rigaut, G., et al.: A generic protein purification method for protein complex characterization and proteome exploration. Nature Biotechnology 17, 1030–1032 (1999)
3. Ho, Y., et al.: Systematic identification of protein complexes in Saccharomyces cerevisiae by mass spectrometry. Nature 405, 180–183 (2002)
4. Mrowka, R., Patzak, A., Herzel, H.: Is There a Bias in Proteome Research? Genome Research 11, 1971–1973 (2001)
5. Mering, C.V., et al.: Comparative assessment of large-scale data sets of protein-protein interactions. Nature 417, 399–403 (2002)
6. Tsoka, S., Ouzounis, C.A.: Prediction of protein interactions: metabolic enzymes are frequently involved in gene fusion. Nature Genetics 26, 141–142 (2000)
7. Wojcik, J., Schächter, V.: Protein–protein interaction map inference using interacting domain profile pairs. Bioinformatics 17, 296–305 (2001)
8. Yamada, T., Kanehisa, M., Goto, S.: Extraction of phylogenetic network modules from the metabolic network. BMC Bioinformatics 7, 130 (2006)
9. Wu, J., Kasif, S., DeLisi, C.: Identification of functional links between genes using phylogenetic profiles. Bioinformatics 19, 1524–1530 (2003)
10. Albert, I., Albert, R.: Identification of functional links between genes using phylogenetic profiles. Bioinformatics 20, 3346–3352 (2004)
11. Bock, J.R., Gough, D.A.: Identification of functional links between genes using phylogenetic profiles. Bioinformatics 19, 125–135 (2003)
12. Lo, S.L., et al.: Effect of training datasets on support vector machine prediction of protein-protein interactions. Proteomics 5, 876–884 (2005)
13. Deane, C.M., et al.: Protein interactions: two methods for assessment of the reliability of high throughput observations. Molecular & Cellular Proteomics 1, 349–356 (2002)
14. D'haeseleer, P., Church, G.M.: Estimating and improving protein interaction error rates. In: Proc. IEEE Computational Systems Bioinformatics Conference, USA, pp. 216–223 (2004)
15. Gilchrist, M.A., et al.: A statistical framework for combining and interpreting proteomic datasets. Bioinformatics 20, 689–700 (2004)

16. Mering, V.C., et al.: Comparative assessment of large-scale data sets of proteinprotein interactions. Nature 417, 399–403 (2002)
17. Hwang, W., et al.: A novel functional module detection algorithm for protein-protein interaction networks. Algorithms for Molecular Biology 1, 24 (2006)
18. Ulitsky, I., Shamir, R.: Identifying functional modules using expression profiles and confidence-scored protein interactions. Bioinformatics 25, 1158–1164 (2009)
19. Gabow, A.P., et al.: Improving protein function prediction methods with integrated literature data. BMC Bioinformatics 9, 198 (2008)
20. Hu, L., et al.: Predicting Functions of Proteins in Mouse Based on Weighted Protein-Protein Interaction Network and Protein Hybrid Properties. PLoS ONE 6, e14556 (2011)
21. Peng, W., et al.: Iteration method for predicting essential proteins based on orthology and protein-protein interaction networks. BMC Systems Biology 6, 87 (2012)
22. Bader, G.D., Hogue, C.W.: An automated method for finding molecular complexes in large protein interaction networks. BMC Bioinformatics 4, 2 (2003)
23. Enright, A.J., Dongen, S.V., Ouzounis, C.A.: An efficient algorithm for large-scale detection of protein families. Nucleic Acids Research 30, 1575–1584 (2002)
24. Palla, G., et al.: Uncovering the Overlapping Community Structure of Complex Networks in Nature and Society. Nature 435, 814–818 (2005)
25. Adamcsek, B., et al.: CFinder: locating cliques and overlapping modules in biological networks. Bioinformatics 22, 1021–1023 (2006)
26. Liu, G., Wong, L., Chua, H.N.: Complex discovery from weighted PPI networks. Bioinformatics 25, 1891–1897 (2009)
27. Chua, H.N., Sung, W.K., Wong, L.: Exploiting indirect neighbours and topological weight to predict protein function from protein–protein interactions. Bioinformatics 22, 1623–1630 (2006)
28. Jiang, P., Singh, M.: SPICi: a fast clustering algorithm for large biological networks. Bioinformatics 26, 1105–1111 (2010)
29. Wang, J.X., et al.: A Fast Hierarchical Clustering Algorithm for Functional Modules Discovery in Protein Interaction Networks. IEEE/ACM Transactions on Computational Biology and Bioinformatics 8, 607–620 (2011)
30. Wu, M., et al.: A core-attachment based method to detect protein complexes in ppi networks. BMC Bioinformatics 10, 169 (2009)
31. Nepusz, T., Yu, H., Paccanaro, A.: Detecting overlapping protein complexes in protein-protein interaction networks. Nature Methods 9, 471–475 (2012)
32. Dezso, Z., Oltvai, Z.N., Barabási, A.L.: Analysis of Experimentally Determined Protein Complexes in the Yeast Saccharomyces cerevisiae. Genome Research 13, 2450–2454 (2003)
33. Gavin, A.C., et al.: Proteome survey reveals modularity of the yeast cell machinery. Genome Research 440, 631–636 (2006)
34. Li, X.L., Foo, C.S., Ng, S.K.: Discovering protein complexes in dense reliable neighborhoods of protein interaction networks. In: Proc. CSB, pp. 157–168 (2007)
35. Xenarios, X., et al.: DIP: the database of interacting proteins. Nucleic Acids Research 28, 289–291 (2000)
36. Krogan, N.J., et al.: Global landscape of protein complexes in the yeast Saccharomyces cerevisiae. Nature 440, 637–643 (2006)
37. Pu, S., et al.: Up-to-date catalogues of yeast protein complexes. Nucleic Acids Research 37, 825–831 (2009)
38. Boyle, E.I., et al.: GO:TermFinder-open source software for accessing GeneOntology information and finding significantly enriched Gene Ontology terms associated with a list of genes. Bioinformatics 20, 3710–3715 (2004)
39. Hu, H., et al.: Mining coherent dense subgraphs across massive biological networks for functional discovery. Bioinformatics 25, 213–221 (2005)

Does Accurate Scoring of Ligands against Protein Targets Mean Accurate Ranking?

Hossam M. Ashtawy and Nihar R. Mahapatra*

Department of Electrical and Computer Engineering, Michigan State University
East Lansing, Michigan 48824, USA
{ashtawy,nrm}@egr.msu.edu

Abstract. Accurately predicting the binding affinities of large sets of protein-ligand complexes efficiently is a key challenge in computational biomolecular science, with applications in drug discovery, chemical biology, and structural biology. Since a scoring function (SF) is used to score, rank, and identify potential drug leads, the fidelity with which it predicts the affinity of a ligand candidate for a protein's binding site has a significant bearing on the accuracy of virtual screening. Despite intense efforts in developing conventional SFs, which are either force-field based, knowledge-based, or empirical, their limited scoring and ranking accuracies have been a major roadblock toward cost-effective drug discovery. Therefore, in this work, we examine a range of SFs employing different machine-learning (ML) approaches in conjunction with a variety of physicochemical and geometrical features characterizing protein-ligand complexes. We compare the scoring and ranking accuracies of these ML SFs as well as those of conventional SFs in the context of the diverse test sets of the 2007 and 2010 PDBbind benchmarks. We also investigate the influence of the size of the training dataset and the number of features used on scoring and ranking accuracies. We find that the best performing ML SF has a scoring power of 0.807 in terms of Pearson correlation coefficient between predicted and measured binding affinities compared to 0.644 achieved by a state-of-the-art conventional SF. Despite this substantial improvement (25%) in binding affinity prediction, the ranking power improvement is only 6% from a success rate of 58.5% achieved by the best conventional SF to 62.2% obtained by the best ML approach when ligands were ranked for 65 unique proteins.

1 Introduction

1.1 Background

Protein-ligand binding affinity (BA) is the principal determinant of many vital processes, such as cellular signaling, gene regulation, metabolism, and immunity, that depend upon proteins binding to some substrate molecule. Consequently, it has a central role in drug design, which involves two main steps: first, the enzyme,

---

* Corresponding author.

Z. Cai et al. (Eds.): ISBRA 2013, LNBI 7875, pp. 298–310, 2013.
© Springer-Verlag Berlin Heidelberg 2013

receptor, or other protein responsible for a disease of interest is identified; second, a small molecule or *ligand* is found or designed that will bind to the target protein, modulate its behavior, and provide therapeutic benefit to the patient. Typically, *high-throughput screening* (HTS) facilities with automated devices and robots are used to synthesize and screen ligands against a target protein. However, due to the large number of ligands that need to be screened, HTS is not fast and cost-effective enough as an *in silico* hit identification method in the initial phases of drug discovery. Therefore, computational methods referred to as *virtual screening* are employed to complement HTS by narrowing down the number of ligands to be physically screened. In virtual screening, information such as structure and physicochemical properties of a ligand, a protein, or both, are used to estimate *binding affinity* (or *binding free energy*), which represents the strength of association between the ligand and its receptor protein. The most popular approach to predicting BA in virtual screening is *structure-based* in which physicochemical interactions between a ligand and receptor are deduced from the 3D structures of both molecules. This *in silico* method is also known as *protein-based* as opposed to the alternative approach, *ligand-based*, in which only ligands that are biochemically similar to the ones known to bind to the target are screened.

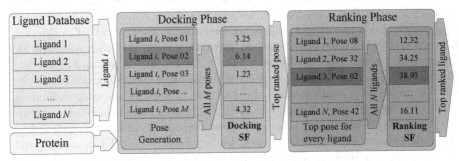

Fig. 1. Protein-ligand docking and ranking workflow

In this work, our focus will be on protein-based drug design, wherein ligands are placed into the active site of the receptor. The 3D structure of a ligand, when bound to a protein, is known as *ligand active conformation*. *Binding mode* refers to the orientation of a ligand relative to the target and the protein-ligand conformation in the bound state. A *binding pose* is simply a candidate binding mode. In molecular *docking*, a large number of binding poses are evaluated using a *scoring function (SF)*, which is a mathematical or predictive model that produces a score representing the binding free energy of a binding pose. The outcome of the docking run, therefore, is a ligand's top pose ranked according to its predicted binding score as shown in Figure 1. Typically, this docking and scoring step is performed iteratively over a database containing thousands to millions of ligand candidates. After predicting their binding modes, another scoring round is performed to rank ligands according to their predicted binding free energies. The top-ranked ligand, considered the most promising drug candidate, is synthesized and physically screened using HTS.

Most commercial molecular docking tools use by default one SF for both the docking and ranking phases as shown in Figure 1. Even though in some cases users may be provided with other scoring options for both phases, almost all the provided SFs are generally similar to the ones we develop and compare in this study. This work focuses on the ranking phase shown in the figure wherein ligands are scored first before being ranked. This approach, however, is extensible to the docking stage since it involves scoring binding poses and then ranking them for each ligand.

1.2   Related Work

Most SFs in use today can be categorized as either force-field-based [1], empirical [2], or knowledge-based [3] SFs. Despite intense efforts into these conventional scoring schemes, several recent studies report that the predictive power of existing SFs is quite limited. Cheng and co-workers recently conducted an extensive test of sixteen SFs from these three categories that are employed in mainstream docking tools and researched in academia [4]. The team analyzed the performance of each SF in predicting the binding energy of protein-ligand complexes whose high resolution structures and binding constants are experimentally known. The main test set used in this study consisted of 195 diverse protein-ligand complexes and four other protein-specific test sets. Considering different evaluation criteria and test datasets, they concluded that no single SF was superior to others in every aspect. In fact, the best SF in terms of predicting binding constants that are most correlated to the experimentally calculated ones was not even in the top five when the goal was to identify the correct binding pose of the ligand. SFs examined in their study were force-field-based, empirical, or knowledge-based, but none were based on sophisticated machine learning (ML) algorithms. Based on this work, we separately studied the scoring and ranking problems by comparing ML and conventional SFs [5,6]. In this work, however, our focus is on comparing the scoring performance of SFs, both conventional ones and those based on ML approaches, to their ranking performance in various test scenarios.

1.3   Key Contributions

One of the main challenges of scoring and ranking is modeling the unknown relationship correlating BA to interactions between a ligand and its receptor. Conventional empirical SFs rest on the hypothesis that a linear regression model is capable of capturing the BA. Such an assumption fails to explain intrinsic nonlinearities latent in the data on which such models are calibrated. Instead of assuming a predetermined theoretical function that governs the unknown relationship between different energetic terms and BA, one can employ more accurate methodologies that are generally referred to as *nonparametric* models. In these techniques, the underlying unknown function is *driven* from the data itself and no assumption regarding its statistical distribution is made. Such an approach has been shown to work well by Ballester and Mitchell [7].

Various nonparametric ML methods inspired from statistical learning theory are examined in this work to model the unknown function that maps structural and physicochemical information of a protein-ligand complex to a corresponding BA value. Ours is the first work to perform a comprehensive assessment of the scoring and ranking accuracies of conventional and ML SFs side by side across a diverse test set of proteins using a common diverse set of features across the ML SFs. We show that the best ML SF has a scoring power of 0.807 (in terms of Pearson correlation coefficient) and a ranking power of 62.2% (in terms of success rate of correctly ranking ligands within clusters of protein-ligand complexes featuring a common protein) compared to 0.644 and 58.5%, respectively, for the best conventional SF for a benchmark test set—this is a significant improvement in predictive power. The only prior related work of this kind is the recent one by Ballester and Mitchell [7]. However, they only considered the scoring problem by constructing one ML SF (random forests) on a subset of the features we consider, although they also report very good results with their method (which has a scoring power of 0.776). Here, we consider a diverse test set and assess the impact of increasing training set size and the number of features used in different ways. Our results show the similarities and differences in behavior of conventional and ML SFs when they are used for scoring vs. when they are used for ranking ligands and shed light on whether improving scoring accuracy necessarily improves ranking accuracy and to what extent.

The remainder of the paper is organized as follows. Section 2 presents the compound database used for the comparative assessment of SFs, the features extracted to characterize the compounds, the training and test datasets employed, and the conventional and ML SFs we study. Next, in Section 3, we describe measures used to assess the performance of SFs and present results comparing the scoring and ranking accuracies of conventional and ML SFs on a diverse test set. We also analyze how ML SFs are impacted by training set size and the number of features used. Conclusions are in Section 4.

2   Materials and Methods

Compound Database: We used the same complex database that Cheng et al. used as a benchmark in their recent comparative assessment of sixteen popular conventional SFs [4]. They obtained the data from the 2007 version of PDBbind [8], which is a selective compilation of the Protein Data Bank (PDB) database [9]. Both databases are publicly accessible and continually updated. The PDB is periodically mined and only complexes that are suitable for drug discovery are filtered into the PDBbind database. In PDBbind, a number of filters are imposed to obtain high-quality protein-ligand complexes with both experimentally-known BA and three-dimensional structure from PDB. A set of 1300 protein-ligand complexes meeting the filtering criteria are compiled into a set referred to as the *refined set*. The PDBbind curators compiled another list out of the refined set. It is called the *core set* and is mainly intended to be used for benchmarking docking and scoring systems. The core set is composed of diverse protein families and diverse BAs. BLAST [10] was employed to cluster the refined set based on protein sequence similarity with a 90% cutoff. From each resultant cluster, three protein-ligand complexes were selected to be its representatives in the core set. A cluster must fulfill the following criteria to be admitted into the core set: (i) it has at least four members and (ii) the BA of the highest-affinity complex must be at least 100-fold of that of the complex with the lowest one. The representatives were then chosen based on their BA rank: the complex having the highest rank, the middle one, and the one with the lowest rank. The approach of constructing the core set guarantees unbiased, reliable, and biologically rich test set of complexes. We also take advantage of the newly-deposited complexes in the 2010 version of the database in some of our experiments.

Training and Test Sets: In order to be consistent with the comparative frame-work used to assess the sixteen conventional SFs mentioned above [4], we too consider the 2007 version of PDBbind. We extracted physicochemical and geometrical features for the complexes in the core set of PDBbind 2007 and we stored them in a dataset called the *core test set* which is denoted by Cr and includes 195 complexes. A *primary training set*, denoted by Pr, was built by removing all Cr complexes from the 1300 complexes in the refined set of the same year. As a result, Pr (for PDBbind 2007) contains 1105 complexes that are completely disjoint from Cr complexes. The refined sets of PDBbind 2007 and 2010 (with 2061 complexes), after removing Cr complexes, were the source of another set that comprises 1989 complexes. We refer to this set as the *secondary training set* or Sc for short.

Compound Characterization: For each protein-ligand complex, we extracted physicochemical features used in the empirical SFs X-Score [2] (a set of 6 features denoted by X) and AffiScore [11] (a set of 30 features denoted by A), and geometrical features used in the ML SF RF-Score [7] (a 36-feature set denoted by R). The software packages that calculate X-Score, AffiScore (from SLIDE), and RF-Score features were available to us in an open-source form from their authors and a full list of these features are provided in the appendix of [6]. By considering all seven combinations of these three types of features (i.e., X, A, R, $X \cup A$, $X \cup R$, $A \cup R$, and $X \cup A \cup R$), we generated seven versions of the Pr, Sc, and Cr datasets, which we distinguish by using appropriate subscripts identifying the features used. For instance, Pr_{XR} denotes the version of Pr comprising the set of features $X \cup R$ (referred to simply as XR) and experimentally-determined BA data for complexes in the Pr dataset.

Conventional Scoring Functions: A total of sixteen popular SFs are compared to ML-based SFs in this study. The sixteen functions are either used in mainstream commercial docking tools and/or have been developed in academia. The functions were recently compared against each other in a study conducted by Cheng et al. [4]. This set includes five SFs in the Discovery Studio software [12]: LigScore, PLP, PMF, Jain, and LUDI. Five SFs in SYBYL software [13]: D-Score, PMF-Score, G-Score, ChemScore, and F-Score. GOLD software [14] contributes three SFs: GoldScore, ChemScore, and ASP. GlideScore in the Schrödinger software [15]. Besides, two standalone SFs developed in academia are also assessed, namely, DrugScore [16] and X-Score [2]. Some of the SFs have several options or versions, these include LigScore (LigScore1 and LigScore2), PLP (PLP1 and PLP2), and LUDI (LUDI1, LUDI2, and LUDI3) in Discovery Studio; GlideScore (GlideScore-SP and GlideScore-XP) in the Schrödinger software; DrugScore (DrugScore-PDB and DrugScore-CSD); and X-Score (HP-Score, HMScore, and HSScore). For brevity, we only report the version and/or option that yields the best performance on the PDBbind benchmark that was considered by Cheng et al. [4].

Machine Learning Methods: We utilize a total of six regression techniques in our study: multiple linear regression (MLR), multivariate adaptive regression splines (MARS), k-nearest neighbors (kNN), support vector machines (SVM), random forests (RF), and boosted regression trees (BRT) [17]. We choose MLR because it resembles empirical SFs that are in fact multiple linear regression models. MARS is an extension of MLR that can model non-linearities and hence it is selected to show how scoring and ranking accuracies are affected by dropping the linearity constraint of empirical SFs. We build kNN and SVM SFs due to their solid theoretical properties and successful applications in drug-development related problems such as QSAR modeling. The state-of-the-art performance of RF and BRT is the primary reason behind our choosing them to construct SFs. In addition to their excellent predictive power, they also offer exceptional descriptive and interpretive capabilities to drug designers to gain invaluable insights regarding different complex interactions between proteins and ligands. These six ML techniques are implemented in the following R language packages that we use [18]: the package *stats* readily available in R for MLR, *earth* for MARS [19], *kknn* for kNN [20], *e1071* for SVM [21], *random forests* for RF [22], and *gbm* for BRT [23]. These methods benefit from some form of parameter tuning prior to their use in prediction. The optimal parameters we use to build our models resulted from a grid search associated with 10-fold cross validation over the training set Pr and are provided in [6]. We applied these six ML methods to all seven combinations of the X, A, and R features (viz., X, A, R, XA, XR, AR, and XAR) for the datasets Pr, Sc, and Cr. Therefore, for each dataset, we considered ($6 \times 7 =$) 42 different ML SFs. We distinguish them using the notation *ML technique::tools used to calculate features*. For instance, kNN::XA implies that the SF is a k-nearest neighbor (kNN) model that is trained and tested on datasets (say primary training set and the core set, respectively) described by *XA* features (i.e., features extracted using the X-Score and AffiScore tools). Again, as in the case of conventional SFs, for brevity, for each ML technique (unless otherwise noted), we report results only for the feature combination (out of the seven possible) that yields the best performance.

3   Results and Discussion

In this section, we compare the scoring and ranking accuracies of conventional and ML SFs. We first define the metrics we use to gauge these accuracies and then present results on a diverse test set. This is followed by an assessment of the impact of training set size and the number of features used.

3.1   Evaluation of Scoring Functions

This study focuses on comparing SFs in terms of their scoring and ranking powers. In our experiments, we measure the scoring power of SFs using Pearson correlation coefficient (denoted by R_p) between predicted and measured BAs. Values of R_p range between -1 and 1. The SF that achieves the highest correlation coefficient for some dataset is considered more accurate (in terms of scoring power) than its counterparts that realize smaller R_p values.

Calculating the ranking power of an SF on the core test set is straightforward due to its construction. First, each protein-ligand complex is scored, i.e., its BA is predicted. Then, for each protein cluster (i.e., the three protein-ligand complexes associated with a common protein), complexes are ordered according to their predicted BA scores. Any given cluster is considered correctly ranked if its predicted-affinity-based order matches its corresponding measured-affinity-based order. We denote this order by "1-2-3" which implies that the strongest-binding ligand is ranked as the first candidate drug, the second strongest binding one is ranked second by the SF, and the weakest binder is ranked last—the number in the ordering corresponds to the true measured rank and the position in the ordering corresponds to the predicted rank. The percentage of clusters with "1-2-3" ordering is referred to as the *1-2-3 ranking rate*, denoted by $R_{1\text{-}2\text{-}3}$, and is used as a measure of the *ranking power* of a given SF as in [4].

3.2   ML vs. Conventional Approaches on the Core Test Set

In Table 1, we report the scoring and ranking performances of six ML and sixteen conventional SFs on the core test Cr comprising 65 three-complex clusters corresponding to 65 diverse protein families. The ML SFs were trained on Pr and then used to predict BAs for the core test Cr, while the results for the conventional SFs were calculated in [4] based on their predicted BA values for the same test complexes in Cr. The table is divided into two column groups. In the left column group, SFs are ordered based on their scoring power in terms of Pearson correlation coefficient. For each SF, the option and/or variation that results in the highest R_p value is used as a representative for it. In the right column group, SFs are ordered based on their $R_{1\text{-}2\text{-}3}$ ranking statistic. Similar to scoring, for each SF we report results only for the option that yields the highest $R_{1\text{-}2\text{-}3}$ value. For example, among all the seven possible feature sets for RF, the feature set XR results in the highest scoring power of $R_p = 0.807$. However, the feature set XAR yields a higher ranking success rate of 62.2% for RF compared

Table 1. Comparison of the scoring and ranking powers of 6 ML and 16 conventional SFs on the core test set Cr

Scoring Power Based Order			Ranking Power Based Order		
Scoring Function	R_p	$R_{1\text{-}2\text{-}3}$ (%)	Scoring Function	$R_{1\text{-}2\text{-}3}$ (%)	R_p
RF::XR	0.807	57.4	RF::XAR	62.2	0.800
BRT::XAR	0.799	58.2	SVM:XAR	61.5	0.773
SVM::XAR	0.773	61.5	BRT::X	59.4	0.718
kNN::XA	0.740	44.6	X-Score::HSScore	58.5	0.619
MARS::XAR	0.710	44.6	DS::PLP1	54.7	0.545
MLR::XA	0.689	41.5	kNN::XR	53.8	0.736
X-Score::HMScore	0.644	52.3	DrugScore^{CSD}	52.3	0.569
DrugScore^{CSD}	0.569	52.3	MARS::AR	50.8	0.657
SYBYL::ChemScore	0.555	47.7	MLR::R	50.8	0.600
DS::PLP1	0.545	54.7	SYBYL::ChemScore	47.7	0.555
GOLD::ASP	0.534	43.1	SYBYL::G-Score	46.2	0.492
SYBYL::G-Score	0.492	46.2	SYBYL::D-Score	46.2	0.392
DS::LUDI3	0.487	43.1	GOLD::ASP	43.1	0.534
DS::LigScore2	0.464	35.4	DS::LUDI3	43.1	0.487
GlideScore-XP	0.457	33.8	DS::PMF	41.5	0.445
DS::PMF	0.445	41.5	DS::Jain	41.5	0.316
GOLD::ChemScore	0.441	36.9	SYBYL::PMF-Score	38.5	0.268
SYBYL::D-Score	0.392	46.2	GOLD::ChemScore	36.9	0.441
DS::Jain	0.316	41.5	DS::LigScore2	35.4	0.464
GOLD::GoldScore	0.295	23.1	GlideScore-XP	33.8	0.457
SYBYL::PMF-Score	0.268	38.5	SYBYL::F-Score	29.2	0.216
SYBYL::F-Score	0.216	29.2	GOLD::GoldScore	23.1	0.295

to a ranking success rate of 57.4% obtained using XR features. Hence, RF::XAR is reported in the right column group.

From the table, it is evident that ML SFs perform very well on both scoring and ranking tasks compared to conventional SFs, with RF, BRT, and SVM being the top three in both cases and the six ML SFs being the best ones at scoring and among the nine best ones out of 22 at ranking. In particular, the best ML SF and the best conventional SF on both tasks are RF and X-Score, respectively, and the former provides more than 25% better scoring accuracy and more than 6% better ranking accuracy compared to the latter. When we consider the SF method as well as its feature set or variant, there are a total of 28 different SFs in the two lists (in the left and right column groups) of Table 1. We find that better scoring accuracy, in general, leads to better ranking accuracy. For example, the Pearson correlation coefficient and the Spearman correlation coefficient (which measures correlation in terms of the position in two ranked lists) of R_p to $R_{1\text{-}2\text{-}3}$ values of the 28 different SFs in Table 1 are 0.795 and 0.788, respectively. Similarly, we find that among the top 14 (50%) of the 28 SFs based on scoring power, 11 out of 14 appear in the top 50% of SFs based on ranking power. Therefore, in relative terms, across all 28 SFs, an SF with a higher scoring power compared to that of another SF is also likely to have higher ranking power. However, upon closer inspection, we find that high scoring performance does not necessarily translate to high ranking performance in absolute terms. While the highest scoring accuracy is 0.807 for RF, its ranking accuracy, which is also the highest (although for a different feature set), is only 62.2%. Consequently, the improvement obtained by using the best ML SF compared to a conventional one in terms of scoring power is significantly higher ($> 25\%$) than in terms ranking power ($> 6\%$). Furthermore, when considering the actual feature set used, the most accurate SF in predicting absolute BA values, RF::XR, is outperformed by the less accurate XScore::HSScore in ranking ligands for 65 protein targets. It should be noted that correct ranking of ligands *for each protein* does not necessarily follow from linear correlation between predicted and measured BAs *across all proteins*. This could partially explain why some empirical SFs such as XScore::HSScore and DS::PLP1 outperformed *k*NN and MARS even though these ML SFs perform better in scoring. The particular features employed by the aforementioned empirical SFs could have also played a role in giving them an edge over some ML techniques in ranking. Focusing specifically on the 42 ML SFs that we built, the correlation between their scoring and ranking accuracies is only a modest 0.55. This underlines the challenge of obtaining highly accurate ranking even using an SF with high scoring accuracy.

3.3   Impact of Training Set Size

Experimental information about 3D structure and BA of new protein-ligand complexes is regularly determined. This contributes to the growing size of public biochemical repositories and corporate compound banks. To assess the impact that a larger training set size would have on the scoring as well as the ranking accuracies of ML SFs, we consider the annually updated database PDBbind that is based on PDB—more than 750 new protein-ligand complexes have been deposited into this database from the year 2007 to 2010. To be able to use a greater range of training set sizes, we choose the larger dataset Sc to build and test different ML scoring models. For a given number of training complexes x, $x = 1 \times 178, 2 \times 178, \ldots, 10 \times 178$, we select x complexes randomly (without replacement) from Sc to train the six ML models that were tested on the disjoint core set Cr. This process is repeated 100 times to obtain robust average scoring (R_p) and ranking ($R_{1\text{-}2\text{-}3}$) performance values, which are plotted in Figure 2.

In both scoring and ranking cases, we observe that performance of almost all models improves as the size of the training data increases. When the size of training data is small relative to the number of features (XAR corresponds to a maximum of 72 features), we notice relatively poor accuracy for some SFs such as MLR and SVM. This is perhaps a consequence of overfitting since the ratio of training data size to number of features is small. The slopes of RF and BRT based SFs indicate that they benefit the most from increasing the size of training set in scoring and ranking. On the other hand, the ranking performance of SFs

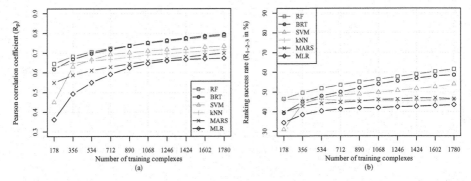

Fig. 2. Dependence of (a) scoring and (b) ranking accuracies of ML scoring models on training set size when the training complexes are selected randomly from Sc_{XAR} and then tested on the disjoint core set Cr_{XAR}

based on kNN, MARS, and MLR appear to be less responsive to including more and more training complexes.

SF designers can conduct similar experiments to estimate accuracy enhancement when their proposed functions are recalibrated on larger number of data instances. Take, for example, the RF model. We can approximately project its scoring and ranking powers on the core test set after a few years from now. If we averaged its improvement slope over the last three increments in training size, we obtain roughly 0.01 and 1.26% increase in terms of R_p and $R_{1\text{-}2\text{-}3}$, respectively, for each 178 increase in number of training records. By assuming that extra 1780 protein-ligand complexes will be deposited into PDBbind in the next few years, one can then optimistically expect (because of diminishing returns with increasing training data size) the scoring and ranking powers of RF::XAR on the test set Cr to go up to about $R_p = 0.9$ and $R_{1\text{-}2\text{-}3} = 74.8\%$ from their current values of 0.8 and 62.2%, respectively. Such an enhancement would certainly have a great impact when the goal is to only choose promising drugs from databases that contain millions of drug-like molecules.

3.4 Impact of the Number of Features

The BA of a protein-ligand complex depends on many physicochemical interaction factors that are too complex to be accurately captured by any one approach. We perform an experiment to investigate how utilizing different types of features from different scoring tools, X-Score, AffiScore, and RF-Score, with an increasing number of them affects the performance of the various ML models. A pertinent issue when considering a variety of features is how well different SF models exploit an increasing number of features. The features we consider are the X, A, and a larger set of geometrical features than the R feature set available from the RF-Score tool. RF-Score counts the number of occurrences of 36 different protein-ligand atom pairs within a distance of 12 Å. In order to have more features of this kind for this experiment, we produce 36 such counts

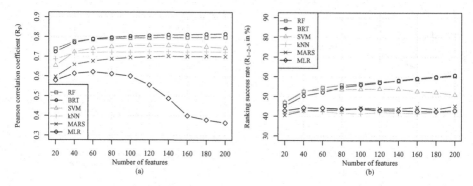

Fig. 3. Dependence of (a) scoring and (b) ranking accuracies of ML scoring models on the number of features, with the features drawn randomly (without replacement) from a pool of X, A, and R-type features and used to train the ML models on the *Sc* dataset and then tested on the disjoint core set *Cr*

for five contiguous distance intervals of 4 Å each: (0 Å, 4 Å], (4 Å, 8 Å], ..., (16 Å, 20 Å]. This provides us 6 X, 30 A, and (36 × 5 =) 180 geometrical features or a total of 216 features. We randomly select (without replacement) x features from this pool, where $x = 20, 40, 60, ..., 200$, and use them to characterize the *Sc* dataset, which we then use to train the six ML models. These models are subsequently tested on the *Cr* dataset characterized by the same sampled features. This process is repeated 100 times to obtain robust average R_p and $R_{1\text{-}2\text{-}3}$ statistics, which are plotted in Figure 3.

Clearly, the various SFs have very different response in scoring and ranking performance to increase in the number of features. For several of them, peak scoring performance is attained at 60 (*k*NN and MLR) or 120 (SVM) features and then there generally tends to be a drop or saturation in performance at larger number of features. Although the features we used are distinct, they have varying degrees of correlation between them. This combined with larger number of features may lead to overfitting problems for some of the SFs. The overfitting problem is clearly manifested in the scoring performance plot of MLR model. The ranking accuracies of MLR as well as those based on *k*NN and MARS are consistently poor over all feature sizes. SVM ranking performance peaks at 60 features, flattens, and then starts to fall at 120 features. The meta-parameters of all ML models are not tuned for every number of features chosen. This especially affects the performance of SVM which we have found to be very sensitive to its parameter values. However, tuning SVM parameters for every number of features is computationally intensive, and therefore we did not attempt to search for the optimal parameter values for every feature set size for it. In contrast to these models, the performance of RF and BRT benefits from increasing number of features especially in their ranking accuracies. Based on these results, utilizing as many relevant features as possible in conjunction with ensemble based approaches like BRT and RF is the best option. The higher accuracy of these two ensemble techniques in this experiment and the previous ones are due to their low bias and variance errors and their resilience to overfitting even with the use of large number of features [24].

4   Conclusion

In this paper, we considered the problem of scoring and ranking ligands based on how tightly they bind to their receptor proteins and comprehensively assessed sixteen conventional and six ML SFs in various test scenarios. We found that better scoring accuracy, in general, leads to better ranking accuracy. However, high scoring performance does not necessarily translate to high ranking performance in absolute terms. Our results showed that ML SFs based on ensemble prediction methods (viz. RF and BRT) surpass others in scoring and ranking accuracy. On the diverse core test set, using the best ML SF as opposed to the best conventional SF improved scoring performance in terms of Pearson correlation coefficient from 0.644 to 0.807. The improvement, however, is less pronounced in ranking accuracy in which case the best ML model achieved 62.2% ranking success rate compared to the 58.5% rate obtained by the best conventional SF. We also observed steady gains in scoring and ranking performance of some ML SFs, in particular for those based on RF and BRT, as the training set size and type and number of features were increased. SFs based on MLR, which resemble empirical models, and to some extent kNN and SVM, for a variety of reasons, were not as effective at exploiting a larger training set size and their performance degraded due to overfitting as the number of features was increased. Overall, based on our results, we expect continued gains in scoring and ranking accuracy of ML SFs, especially RF- and BRT-based, as the training set size and number of features used increase in the future.

Acknowledgments. This material is based upon work supported by US National Science Foundation under Grant No. 1117900.

References

1. Ewing, T.J.A., Makino, S., Skillman, A.G., Kuntz, I.D.: Dock 4.0: search strategies for automated molecular docking of flexible molecule databases. Journal of Computer-Aided Molecular Design 15(5), 411–428 (2001)
2. Wang, R., Lai, L., Wang, S.: Further development and validation of empirical scoring functions for structure-based binding affinity prediction. Journal of Computer-Aided Molecular Design 16, 11–26 (2002), doi:10.1023/A:1016357811882
3. Gohlke, H., Hendlich, M., Klebe, G.: Knowledge-based scoring function to predict protein-ligand interactions. Journal of Molecular Biology 295(2), 337 (2000)
4. Cheng, T., Li, X., Li, Y., Liu, Z., Wang, R.: Comparative assessment of scoring functions on a diverse test set. Journal of Chemical Information and Modeling 49(4), 1079–1093 (2009)
5. Ashtawy, H.M., Mahapatra, N.R.: A comparative assessment of conventional and machine-learning-based scoring functions in predicting binding affinities of protein-ligand complexes. In: 2011 IEEE International Conference on Bioinformatics and Biomedicine (BIBM), pp. 627–630. IEEE (2011)

6. Ashtawy, H.M., Mahapatra, N.R.: A comparative assessment of ranking accuracies of conventional and machine-learning-based scoring functions for protein-ligand binding affinity prediction. IEEE/ACM Transactions on Computational Biology and Bioinformatics (TCBB) 9(5), 1301–1313 (2012)
7. Ballester, P., Mitchell, J.: A machine learning approach to predicting protein-ligand binding affinity with applications to molecular docking. Bioinformatics 26(9), 1169 (2010)
8. Wang, R., Fang, X., Lu, Y., Wang, S.: The PDBbind database: Collection of binding affinities for protein-ligand complexes with known three-dimensional structures. Journal of Medicinal Chemistry 47(12), 2977–2980 (2004); PMID: 15163179
9. Berman, H.M., Westbrook, J., Feng, Z., Gilliland, G., Bhat, T.N., Weissig, H., Shindyalov, I.N., Bourne, P.E.: The protein data bank. Nucleic Acids Research 28(1), 235–242 (2000)
10. Madden, T.: The blast sequence analysis tool. The NCBI Handbook. National Library of Medicine (US), National Center for Biotechnology Information (2002)
11. Schnecke, V., Kuhn, L.A.: Virtual screening with solvation and ligand-induced complementarity. In: Klebe, G. (ed.) Virtual Screening: An Alternative or Complement to High Throughput Screening? pp. 171–190. Springer, Netherlands (2002)
12. Accelrys Inc., The Discovery Studio Software, San Diego, CA (2001) (version 2.0)
13. Tripos, Inc., The SYBYL Software, 1699 South Hanley Rd., St. Louis, Missouri, 63144, USA (2006) (version 7.2)
14. Jones, G., Willett, P., Glen, R., Leach, A., Taylor, R.: Development and validation of a genetic algorithm for flexible docking. Journal of Molecular Biology 267(3), 727–748 (1997)
15. Schrödinger, L.: The Schrödinger Software, New York (2005) (version 8.0)
16. Velec, H.F.G., Gohlke, H., Klebe, G.: DrugScore CSD - knowledge-based scoring function derived from small molecule crystal data with superior recognition rate of near-native ligand poses and better affinity prediction. Journal of Medicinal Chemistry 48(20), 6296–6303 (2005)
17. Hastie, T., Tibshirani, R., Friedman, J.: The elements of statistical learning (2001)
18. R Development Core Team: R: A Language and Environment for Statistical Computing. R Foundation for Statistical Computing, Vienna, Austria (2010)
19. Stephen Milborrow, T.H., Tibshirani, R.: earth: Multivariate Adaptive Regression Spline Models (2010) (R package version 2.4-5)
20. Hechenbichler, K.S.K.: Kknn: Weighted k-Nearest Neighbors (2010) (R package version 1.0-8)
21. Dimitriadou, E., Hornik, K., Leisch, F., Meyer, D., Weingessel, A.: e1071: Miscellaneous Functions of the Department of Statistics (e1071), TU Wien (2010) (R package version 1.5-24)
22. Breiman, L.: Random forests. Machine Learning 45, 5–32 (2001)
23. Ridgeway, G.: Gbm: Generalized Boosted Regression Models (2010) (R package version 1.6-3.1)
24. Breiman, L.: Bias, variance, and arcing classifiers (technical report 460). Statistics Department, University of California (1996)

Author Index